Impact of Climate and Socio-Economic on Irrigation Water Management and Agricultural Water Productivity

Impact of Climate and Socio-Economic on Irrigation Water Management and Agricultural Water Productivity

Guest Editors

Chenglong Zhang
Xiaojie Li

Basel • Beijing • Wuhan • Barcelona • Belgrade • Novi Sad • Cluj • Manchester

Guest Editors

Chenglong Zhang
China Agricultural University
Beijing
China

Xiaojie Li
Chinese Academy of Sciences
Beijing
China

Editorial Office
MDPI AG
Grosspeteranlage 5
4052 Basel, Switzerland

This is a reprint of the Special Issue, published open access by the journal *Water* (ISSN 2073-4441), freely accessible at: https://www.mdpi.com/journal/water/special_issues/VT84LV57FA.

For citation purposes, cite each article independently as indicated on the article page online and as indicated below:

Lastname, A.A.; Lastname, B.B. Article Title. *Journal Name* **Year**, *Volume Number*, Page Range.

ISBN 978-3-7258-2853-1 (Hbk)
ISBN 978-3-7258-2854-8 (PDF)
https://doi.org/10.3390/books978-3-7258-2854-8

© 2024 by the authors. Articles in this book are Open Access and distributed under the Creative Commons Attribution (CC BY) license. The book as a whole is distributed by MDPI under the terms and conditions of the Creative Commons Attribution-NonCommercial-NoDerivs (CC BY-NC-ND) license (https://creativecommons.org/licenses/by-nc-nd/4.0/).

Contents

About the Editors vii

Xiaojie Li and Chenglong Zhang
Impact of Climate and Socio-Economic on Irrigation Water Management and Agricultural Water Productivity
Reprinted from: *Water* **2024**, *16*, 3149, https://doi.org/10.3390/w16213149 1

Suneeporn Suwanmaneepong, Kulachai Kultawanich, Lampan Khurnpoon, Phatchara Eamkijkarn Sabaijai, Harry Jay Cavite, Christopher Llones, et al.
Alternate Wetting and Drying as Water-Saving Technology: An Adoption Intention in the Perspective of Good Agricultural Practices (GAP) Suburban Rice Farmers in Thailand
Reprinted from: *Water* **2023**, *15*, 402, https://doi.org/10.3390/w15030402 5

Yidi Sun, Jigan Xie, Huijing Hou, Min Li, Yitong Wang and Xuetao Wang
Effects of Zeolite on Physiological Characteristics and Grain Quality in Rice under Alternate Wetting and Drying Irrigation
Reprinted from: *Water* **2023**, *15*, 2406, https://doi.org/10.3390/w15132406 18

Juzhen Xu, Yanbo Wang, Yuanquan Chen, Wenqing He, Xiaojie Li and Jixiao Cui
Identifying the Influencing Factors of Plastic Film Mulching on Improving the Yield and Water Use Efficiency of Potato in the Northwest China
Reprinted from: *Water* **2023**, *15*, 2279, https://doi.org/10.3390/w15122279 32

Wanru Li, Mekuanent Muluneh Finsa, Kathryn Blackmond Laskey, Paul Houser, Rupert Douglas-Bate and Kryštof Verner
Optimizing Well Placement for Sustainable Irrigation: A Two-Stage Stochastic Mixed Integer Programming Approach
Reprinted from: *Water* **2024**, *16*, 2715, https://doi.org/10.3390/w16192715 44

Taeseong Kang, Yongchul Shin, Minhwan Shin, Dongjun Lee, Kyoung Jae Lim and Jonggun Kim
Evaluation of the Effect of Agricultural Return Flow on Water Quality, Water Quantity and Aquatic Ecology in Downstream Rivers
Reprinted from: *Water* **2024**, *16*, 1604, https://doi.org/10.3390/w16111604 67

Alexandre Troian, Mário Conill Gomes, Tales Tiecher, Marcos Botton Piccin, Danilo dos Santos Rheinheimer and José Miguel Reichert
Participatory Analysis of Impacts of Agricultural Production Systems in a Watershed Depicting Southern Brazilian Agriculture
Reprinted from: *Water* **2024**, *16*, 716, https://doi.org/10.3390/w16050716 88

Linlin Zhao, Rensheng Chen, Yong Yang, Guohua Liu and Xiqiang Wang
A New Tool for Mapping Water Yield in Cold Alpine Regions
Reprinted from: *Water* **2023**, *15*, 2920, https://doi.org/10.3390/w15162920 114

Morag Hunter, D. H. Nimalika Perera, Eustace P. G. Barnes, Hugo V. Lepage, Elias Escobedo-Pacheco, Noorhayati Idros, et al.
Landscape-Scale Mining and Water Management in a Hyper-Arid Catchment: The Cuajone Mine, Moquegua, Southern Peru
Reprinted from: *Water* **2024**, *16*, 769, https://doi.org/10.3390/w16050769 135

Ping Miao, Dagula, Xiaojie Li, Shahid Naeem, Amit Kumar, Hongli Ma, et al.
Runoff Decline Is Dominated by Human Activities
Reprinted from: *Water* **2023**, *15*, 4010, https://doi.org/10.3390/w15224010 . . . . **163**

Wen Liu
The Response of NDVI to Drought at Different Temporal Scales in the Yellow River Basin from 2003 to 2020
Reprinted from: *Water* **2024**, *16*, 2416, https://doi.org/10.3390/w16172416 . . . . **181**

Yitong Wang, Qiujie Shan, Chuan Wang, Shaoyuan Feng and Yan Li
Research Progress and Application Analysis of the Returning Straw Decomposition Process Based on CiteSpace
Reprinted from: *Water* **2023**, *15*, 3426, https://doi.org/10.3390/w15193426 . . . . **200**

About the Editors

Chenglong Zhang

Chenglong Zhang, Ph.D., is an Associate Professor at the College of Water Resources and Civil Engineering and State Key Laboratory of Efficient Utilization of Agricultural Water Resources at the China Agricultural University. His current research interests include modeling and methods for the agricultural water management in changing environments, mainly focusing on the intelligent control of agricultural water systems. To date, he has co-authored more than 60 scientific publications and collaborated with more than 10 scientific institutes in research and consulting. He is a member of the Chinese Society of Agricultural Engineering (CSAE) and Chinese Hydraulic Engineering Society (CHES) and a reviewer for different papers in international journals.

Xiaojie Li

Xiaojie Li, Ph.D. in Agriculture, Water, and Soil Engineering, is an Assistant Professor at the Institute of Geographic Sciences and Natural Resources Research, Chinese Academy of Sciences. Her research focuses on understanding the impact of environmental changes, including climate change, land use change, vegetation greening, etc., on agricultural and hydrological water cycles. She employs advanced hydrological models and crop models to address these challenges. Up to now, she has participated in ten scientific projects, being the principal in five projects, and two scientific rewards. The knowledge developed in these research projects has contributed to more than 20 publications, which are published in international journals, conferences, and theses.

Editorial

Impact of Climate and Socio-Economic on Irrigation Water Management and Agricultural Water Productivity

Xiaojie Li [1,*] and Chenglong Zhang [2]

[1] Institute of Geographic Sciences and Natural Resources Research, Chinese Academy of Sciences, Beijing 100101, China
[2] College of Water Resources & Civil Engineering, China Agricultural University, Beijing 100083, China; zhangcl1992@cau.edu.cn
* Correspondence: lixiaojie@igsnrr.ac.cn

Citation: Li, X.; Zhang, C. Impact of Climate and Socio-Economic on Irrigation Water Management and Agricultural Water Productivity. *Water* **2024**, *16*, 3149. https://doi.org/10.3390/w16213149

Received: 29 October 2024
Revised: 31 October 2024
Accepted: 31 October 2024
Published: 4 November 2024

Copyright: © 2024 by the authors. Licensee MDPI, Basel, Switzerland. This article is an open access article distributed under the terms and conditions of the Creative Commons Attribution (CC BY) license (https://creativecommons.org/licenses/by/4.0/).

1. Introduction to the Special Issue

Water security and food security are fundamental pillars of sustainable social and economic development [1]. As the human activity with the greatest water requirement, agricultural production consumes 70% of total water use worldwide and intensifies global pressure on water resources [2]. Therefore, improving agricultural water productivity is an important measure to ensure global water security and food security. Among all agricultural inputs, irrigation plays the most important role in ensuring stable agricultural production. However, increasing water demand leads to further strain on irrigation systems, while growing water scarcity poses a serious threat to food security and sustainable development in agricultural regions, endangering the livelihood of 3.2 billion people [3]. Without appropriate measures, freshwater use may soon reach its limit, underlining the urgent need for sustainable irrigation water management practices [4].

Climate change, human activities, and economic development pose serious threats to agricultural water productivity and irrigation water management due to changes in water supply and demand. These factors alter the spatial and temporal distribution of rainfall, impact water availability and allocation, and affect various other aspects of agricultural production [5,6]. For example, as global warming continues, extreme weather events such as severe droughts and floods are intensifying, negatively impacting global agricultural production and irrigation management [7]. Additionally, enhanced human activities, including land use and land cover changes, groundwater overexploitations, and stream flow regulations, have altered the hydrological cycle and water allocation for agricultural irrigation use [8]. Furthermore, socio-economic growth, driven by a rising global population, creates a persistent need for food production, increased urban and industrial water supply, suitable water quality, and environmental protection, leading to significant pressure on water resources and management [9]. Therefore, understanding the impact of climate change, human activity, and socio-economic development on agricultural water productivity and irrigation water management, along with pursuing targeted improvements, is a key challenge for ensuring water and food security and enhancing agriculture resilience against future challenges.

To this end, in this Special Issue, we attempted to publish related research papers on the following topics: (1) the impacts of climate change, human activities, and socio-economic development—including increasing temperature, extreme weather events, flooding, irrigation management, cropping patterns, water conservation measures, economic growth, infrastructure development, mining, governance and policy support, etc.—on the agricultural system; (2) management strategies and assessment approaches to help improve the sustainability of the agricultural and hydrological system under the challenging environment. We have collected eleven high-quality papers and summarize them in the following section.

2. A Summary of the Special Issue

Alternate wetting and drying (AWD) technology is an innovative and effective way to save water and improve agricultural water productivity. The study by Suwanmaneepong et al. [Contribution 1] explores the adoption of AWD among suburban rice farmers in Thailand. The findings indicate that variable cost is positively associated with higher AWD adoption intention in the short-run, while a higher fixed cost lowers the probability of AWD adoption. It is essential for farmers to apply the AWD method safely and correctly, together with the assistance of crop insurance, to encourage efficient water use in agriculture.

While the AWD method in rice farming may cause mild soil moisture stress, the use of zeolite could improve crop growth and yields due to its strong ion exchange capacity and high affinity for water and fertilizer. The study by Sun et al. [Contribution 2] examines the role of zeolite in improving rice cultivation, particularly under AWD conditions. The results indicate that zeolite application increased rice dry matter and grain yield, while both AWD and zeolite improved water productivity. Combining AWD and zeolite could offer several benefits, especially under water stress conditions.

In addition to AWD technology, plastic film mulching is another crucial technique to prevent water evaporation and improve agricultural water productivity. Xu et al. [Contribution 3] conducted a meta-analysis to quantify the effect of mulching properties, different levels of natural conditions, fertilizer application, and cultivation measures on potato yield and its water productivity. The results show that plastic film mulching significantly increased yield and water productivity, especially under less favorable natural and fertilizer conditions, which provides valuable reference for improving agricultural water management.

Sustainable agriculture goes beyond improving agricultural water productivity; it is also highly related to irrigation water management. In areas with limited surface water availability, well drilling and placement are essential to maximize groundwater resources. Li et al. [Contribution 4] developed two-stage stochastic mixed-integer programming models to optimize groundwater well placement for agricultural irrigation in Ethiopia. Their study highlights that well layouts vary slightly in different scenarios and that the model achieves lower costs in out-of-sample tests—11% and 4% lower than in deterministic cases—making it a robust strategy for sustainable groundwater management in irrigation.

As a vital resource for crop production, agricultural water is both affected by the environment and has significant environmental impacts. It plays a crucial role in maintaining ecosystem balance, affecting river flows, water quality, and aquatic life Kang et al. [Contribution 5] evaluated agricultural return flow effects on downstream rivers using reservoir data, the SWAT model, and the PHABSIM model. Their findings show that agricultural return flow significantly increases river flow during non-rainy seasons, but fails to meet optimal ecological flow rates. Additionally, the impact on river water quality is minimal, except during rice paddy drainage. Overall, agricultural water contributes positively to downstream aquatic ecosystems.

In addition to agricultural water, agricultural production systems also exert pressure on aquatic ecosystems. The study of Troian et al. [Contribution 6] used a participatory approach, utilizing a multidimensional model to evaluate land use, soil management, and agricultural waste disposal based on 7 criteria and 25 sub-criteria, applied across 14 farms in a large watershed in southern Brazil. The results highlight that key factors influencing environmental risks include the fragility of cultivated areas and the lack of conservation practices, contributing to 73% of the risks. Their study suggests incorporating a cost-effectiveness analysis to better manage environmental conflicts in future research.

Water management is important in cold alpine regions. However, traditional models often neglect cryosphere elements such as frozen ground and snow cover, which significantly impact hydrological processes and lead to inaccurate estimates. Zhao et al [Contribution 7] incorporated these elements into the Seasonal Water Yield model to better assess water yield in the Three-River Headwaters Region of the Qinghai–Tibetan Plateau. Frozen ground reduces water infiltration by acting as a low-permeability layer, while snow cover influences water yield through melting and sublimation. The improved model helps in making more reliable water resource management decisions in cold alpine regions, especially in the context of climate change.

Mining provides significant benefits to society but can also pose environmental risks. Hunter et al. [Contribution 8] investigated the impact of the Cuajone copper mine on water resources in the hyper-arid region of Southern Peru, focusing on the Torata river catchment. Based on water chemistry analyses from 16 sites over three seasons, they found that the mine does not significantly affect water quality. Instead, urban effluents and agricultural runoff are identified as the primary contributors of water contamination, especially in the lower catchment. However, elevated levels of total dissolved solids still pose risks for agricultural use and domestic consumption.

Runoff is one of the key water resources for irrigation, and irrigation practices, as a major human activity, can significantly affect the volume and quality of runoff. Miao et al. [Contribution 9] investigated the factors of declining runoff in the Xiliugou basin, a semi-arid region in northern China. By using six methods, including statistical and hydrological models, they found that human activities, particularly water conservation and land use changes, are responsible for 76% of the decline, while climate change contributes to only 24%. Their study highlights the significant role of soil and water conservation projects in altering runoff patterns, offering valuable insights for water resource management and sustainable development in the region.

Drought has become a major climatic concern affecting ecology and agriculture. Liu [Contribution 10] examined the sensitivity of vegetation growth to drought in the Yellow River Basin from 2003 to 2020, using the Standardized Precipitation Evapotranspiration Index and the Normalized Difference Vegetation Index. The results show that vegetation in arid and semi-arid areas is positively correlated with drought, while humid regions exhibit negative correlations, especially in high-altitude areas where vegetation is more sensitive to heat than water. Vegetation responds strongly to short-term (1 to 3 months) droughts, with sensitivity peaking in summer. The study provides insights into vegetation's response to drought and informs strategies for ecological protection in the Yellow River Basin under climate change conditions.

With the rapid growth of agricultural production, the amount of crop straw has significantly increased, making its decomposition a key challenge in agricultural production management. Wang et al. [Contribution 11] reviewed the literature on straw returning and decomposition processes, focusing on improving soil fertility through the use of straw decomposition agents. Utilizing bibliometric analysis via the CiteSpace software, their study visualized research trends and progress from 2002 to 2022, highlighting the importance of microorganisms, such as Pseudomonas, in enhancing decomposition rates. Their study also explored how returning straw to the soil can improve crop yields and mitigate environmental issues, such as heavy metal pollution. Their findings emphasize the growing interest in improving nutrient release from straw using bacterial agents to accelerate the decomposition process for sustainable agricultural development.

Author Contributions: Conceptualization, X.L. and C.Z.; validation, C.Z.; formal analysis, X.L. and C.Z.; writing—original draft preparation, X.L.; writing—review and editing, C.Z.; supervision, X.L.; funding acquisition, X.L. and C.Z. All authors have read and agreed to the published version of the manuscript.

Funding: This Special Issue was funded by the Beijing Natural Science Foundation (No. 8222032) and the National Natural Science Foundation of China (No. 52279050).

Acknowledgments: As the Guest Editors of this Special Issue, we would like to extend our gratitude to all of the authors and reviewers whose contributions have greatly enhanced its quality.

Conflicts of Interest: The authors declare no conflicts of interest.

List of Contributions:

1. Suwanmaneepong, S.; Kultawanich, K.; Khurnpoon, L.; Sabaijai, P.E.; Cavite, H.J.; Llones, C.; Lepcha, N.; Kerdsriserm, C. Alternate Wetting and Drying as Water-Saving Technology: An Adoption Intention in the Perspective of Good Agricultural Practices (GAP) Suburban Rice Farmers in Thailand. *Water* **2023**, *15*, 402. https://doi.org/10.3390/w15030402.
2. Sun, Y.; Xie, J.; Hou, H.; Li, M.; Wang, Y.; Wang, X. Effects of Zeolite on Physiological Characteristics and Grain Quality in Rice under Alternate Wetting and Drying Irrigation. *Water* **2023**, *15*, 2406. https://doi.org/10.3390/w15132406.
3. Xu, J.; Wang, Y.; Chen, Y.; He, W.; Li, X.; Cui, J. Identifying the Influencing Factors of Plastic Film Mulching on Improving the Yield and Water Use Efficiency of Potato in the Northwest China. *Water* **2023**, *15*, 2279. https://doi.org/10.3390/w15122279.
4. Li, W.; Finsa, M.M.; Laskey, K.B.; Houser, P.; Douglas-Bate, R.; Verner, K. Optimizing Well Placement for Sustainable Irrigation: A Two-Stage Stochastic Mixed Integer Programming Approach. *Water* **2024**, *16*, 2715. https://doi.org/10.3390/w16192715.
5. Kang, T.; Shin, Y.; Shin, M.; Lee, D.; Lim, K.J.; Kim, J. Evaluation of the Effect of Agricultural Return Flow on Water Quality, Water Quantity and Aquatic Ecology in Downstream Rivers. *Water* **2024**, *16*, 1604. https://doi.org/10.3390/w16111604.
6. Troian, A.; Gomes, M.C.; Tiecher, T.; Piccin, M.B.; Rheinheimer, D.d.S.; Reichert, J.M. Participatory Analysis of Impacts of Agricultural Production Systems in a Watershed Depicting Southern Brazilian Agriculture. *Water* **2024**, *16*, 716. https://doi.org/10.3390/w16050716.
7. Zhao, L.; Chen, R.; Yang, Y.; Liu, G.; Wang, X. A New Tool for Mapping Water Yield in Cold Alpine Regions. *Water* **2023**, *15*, 2920. https://doi.org/10.3390/w15162920.
8. Hunter, M.; Perera, D.H.N.; Barnes, E.P.G.; Lepage, H.V.; Escobedo-Pacheco, E.; Idros, N.; Arvidsson-Shukur, D.; Newton, P.J.; de los Santos Valladares, L.; Byrne, P.A.; et al. Landscape-Scale Mining and Water Management in a Hyper-Arid Catchment: The Cuajone Mine, Moquegua, Southern Peru. *Water* **2024**, *16*, 769. https://doi.org/10.3390/w16050769.
9. Miao, P.; Dagula; Li, X.; Naeem, S.; Kumar, A.; Ma, H.; Ding, Y.; Wang, R.; Luan, J. Runoff Decline Is Dominated by Human Activities. *Water* **2023**, *15*, 4010. https://doi.org/10.3390/w15224010.
10. Liu, W. The Response of NDVI to Drought at Different Temporal Scales in the Yellow River Basin from 2003 to 2020. *Water* **2024**, *16*, 2416. https://doi.org/10.3390/w16172416
11. Wang, Y.; Shan, Q.; Wang, C.; Feng, S.; Li, Y. Research Progress and Application Analysis of the Returning Straw Decomposition Process Based on CiteSpace. *Water* **2023**, *15*, 3426. https://doi.org/10.3390/w15193426.

References

1. Huntington, H.P.; Schmidt, J.I.; Loring, P.A.; Whitney, E.; Aggarwal, S.; Byrd, A.G.; Dev, S.; Dotson, A.D.; Huang, D.; Johnson, B.; et al. Applying the food–energy–water nexus concept at the local scale. *Nat Sustain.* **2021**, *4*, 672–679. [CrossRef]
2. D'Odorico, P.; Chiarelli, D.D.; Rosa, L.; Bini, A.; Zilberman, D.; Rulli, M.C. The global value of water in agriculture. *Proc. Natl. Acad. Sci. USA* **2020**, *117*, 21985–21993. [CrossRef] [PubMed]
3. FAO. *The State of Food and Agriculture 2020: Overcoming Water Challenges in Agriculture*; FAO: Rome, Italy, 2020. [CrossRef]
4. Steffen, W.; Richardson, K.; Rockström, J.; Cornell, S.E.; Fetzer, I.; Bennett, E.M.; Biggs, R.; Carpenter, S.R.; De Vries, W.; De Wit, C.A.; et al. Planetary boundaries: Guiding human development on a changing planet. *Science* **2015**, *347*, 1259855 [CrossRef] [PubMed]
5. Alcamo, J.; Dronin, N.; Endejan, M.; Golubev, G.; Kirilenko, A. A new assessment of climate change impacts on food production shortfalls and water availability in Russia. *Glob. Environ. Chang.* **2007**, *17*, 429–444. [CrossRef]
6. Hanjra, M.A.; Qureshi, M.E. Global water crisis and future food security in an era of climate change. *Food Policy* **2010**, *35*, 365–377 [CrossRef]
7. El-Nashar, W.; Elyamany, A. Adapting Irrigation Strategies to Mitigate Climate Change Impacts: A Value Engineering Approach *Water Resour. Manag.* **2023**, *37*, 2369–2386. [CrossRef]
8. Wang, R.; Xiong, L.; Xu, X.; Liu, S.; Feng, Z.; Wang, S.; Huang, Q.; Huang, G. Long-term responses of the water cycle to climate variability and human activities in a large arid irrigation district with shallow groundwater: Insights from agro-hydrological modeling. *J. Hydrol.* **2023**, *626*, 130264. [CrossRef]
9. Cetinkaya, C.P.; Gunacti, M.C. Multi-criteria analysis of water allocation scenarios in a water scarce basin. *Water Resour. Manag.* **2018**, *32*, 2867–2884. [CrossRef]

Disclaimer/Publisher's Note: The statements, opinions and data contained in all publications are solely those of the individual author(s) and contributor(s) and not of MDPI and/or the editor(s). MDPI and/or the editor(s) disclaim responsibility for any injury to people or property resulting from any ideas, methods, instructions or products referred to in the content.

Article

Alternate Wetting and Drying as Water-Saving Technology: An Adoption Intention in the Perspective of Good Agricultural Practices (GAP) Suburban Rice Farmers in Thailand

Suneeporn Suwanmaneepong [1], Kulachai Kultawanich [1], Lampan Khurnpoon [1], Phatchara Eamkijkarn Sabaijai [1], Harry Jay Cavite [2], Christopher Llones [1,*], Norden Lepcha [1] and Chanhathai Kerdsriserm [1]

[1] School of Agricultural Technology, King Mongkut's Institute of Technology Ladkrabang, Bangkok 10520, Thailand
[2] Sasin School of Management, Chulalongkorn University, Bangkok 10330, Thailand
* Correspondence: christopher.allones@gmail.com

Abstract: The alternate wetting and drying (AWD) as water-saving technology aligns with the good agricultural practices (GAP) principles, particularly in the environmental management of water conservation. Thus, GAP adopters as farmer groups are seen as viable AWD adopters in the initial stages of scaling out the adoption in Thailand. However, the understanding of integrating AWD as water-saving management among GAP adopters remains scant. Using the case of rice GAP farmers in Thailand, the study found a higher probability of adoption intention among GAP compared to non-GAP. AWD perceived advantage, knowledge, and the suitability of rice farms for AWD adoption trials are positively associated with higher adoption intention. While higher fixed cost lowers the probability of adoption, variable cost is positively associated with higher adoption intention in the short-run production decision. In order to scale out the adoption of AWD, farmers' understanding of the safe and proper application of AWD, together with assistance for crop insurance in the case of crop failure, will be crucial. Risks connected with the adoption decision continue to be the biggest barrier to adoption, especially among small-scale farmers.

Keywords: AWD; GAP; rice; adoption; water-saving

1. Introduction

Consumers' growing environmental and health concerns have driven national and international policymakers to take several actions to improve farm management to increase food safety and sustainability, which has been challenging for the agricultural sector [1]. Globally, the concept of sustainable and safe food production is expanding rapidly, involving more complex technologies, trade-offs (e.g., productivity vs. sustainability), stringent standards, and sending differing messages among farmers, particularly in developing countries [1–4].

Across Asian countries, good agricultural practices (GAP) have been widely promoted to meet the shifting customer demand for safe and sustainable food crops [5]. In addition, GAP responds to international trade requirements where Asian countries export different food crops (such as rice, jackfruits, and mangosteen) to other continents [2,5,6]. For instance, several Asian countries have local versions of GAP, such as Philippine-GAP, Malaysian SALM, Singapore GAP-VF, Indon-GAP, and Q-GAP in Thailand [2].

In Thailand, through the national GAP program, which focuses on food safety, the country has been proactively addressing the issues of meeting the market's demand for food crops [4,7]. The national GAP (Q-GAP) development is driven mainly by the national government under the Ministry of Agriculture and Cooperatives (MOAC). Farmers that can meet the requirements may use the "Q-GAP" logo, where "Q" stands for quality mark, to label their produce [7]. Promoting Q-GAP among Thai farmers seeks to increase Thai customers' trust in food sold in domestic markets and raise Thai products' competitiveness in global markets [2,6]. Aside from food safety, the GAP requirements also focus on three other areas: Quality produce, growers' health and safety, and environmental management [4,7].

Target crops for GAP certification are for exports and domestic consumption, like durian, mangosteen, mango, pineapple, coconut, and rice [7]. Among the crops produced in the country, rice production continued to be a priority for domestic and export markets [8–10]. However, under the GAP principle on environmental management concerns, rice production contributes to the national greenhouse gas emissions (GHG) by 27.19Mt CO_2-eq and accounts for 51.38% of GHG emissions since 2011 [11,12]. In addition, continuous flooding is a common method on rice farms in Thailand, which is said to be ideal for CH_4 and nitrous oxide (N_2O) [1,13]. Sriphirom et al. [12] and Srisopaporn et al. [1] pointed out that reducing GHG emissions and water usage are the two critical components of rice farming for sustainable production following GAP principles.

In 2016, Thailand introduced the adoption of alternate wetting and drying (AWD) as water-saving irrigation technology developed by the International Rice Research Institute (IRRI). The AWD is considered an alternative to continuous flooding that aims to mitigate the impact of drought, reduce GHG emissions and be cost-saving without compromising rice yields [14–16]. Several studies concluded the potential of AWD as cost-effective technology in reducing water input by as much as 25–30% without adversely impacting rice yields [12,15,17,18]. At the same time, Linquist et al. [19] found a 48% reduction in methane emissions from rice cultivation with AWD treatment [18]. Furthermore, lower CH_4 emissions were observed under alternate wetting and drying than continuous flooding [20,21]. Islam et al. [20] conclude that GHG emission-reducing technology (e.g., urea deep placement or UDP) combined with a water-saving management strategy like AWD compared to rice production under continuous flooding conditions is effective.

Given the standards stipulated under GAP, particularly on environmental management and AWD's potential advantages, the Thai government sought farmers' cooperation in AWD adoption for water-saving. GAP adopters as farmer groups are seen as potential AWD adopters in the initial stage of scaling out AWD technology as it aligns with the GAP requirements for adopting water-saving in rice cultivation. While knowledge of the potential effects of adopting AWD has been growing, the understanding of integrating AWD as water-saving management among GAP adopters remains poor globally, particularly in Thailand. Thus, the study aims to answer the following research questions:

1. What factors influences the adoption intention on alternate wetting and drying?
2. What are farmers' perceptions on the promotion of AWD as alternative water-saving technology?
3. What are the policy implications of integrating AWD adoption with good agricultural practices under GAP's principle of sustainable water management in production?

The remainder of the paper proceeds as follows. First, we provide more background on Thailand's GAP and AWD adoption. Then we present the information on the focus study sites and the methodology used. Afterward, we discuss the results and implications of the study. In the last section, we provide the conclusions and policy implications of the study.

2. Thailand's GAP and AWD Adoption

The increasing concerns about food safety and demand for high-quality food and agricultural products globally drive GAP development in farm production. In 2003, Thailand started to develop its local GAP standard, Q-GAP (Q stands for "quality"), which focuses on food safety, quality produce, environmental management, and safety for growers. Good agricultural practices vary depending on the crop type, production system, and customer needs [5]. Driven by food safety requirements, GAP consists of simple instructions and a checklist for growers to comply, as outlined in Table 1 for GAP requirements in Thailand set by the Thai Agricultural Standard. Certified GAP farm compliant are recognized by the national GAP certifying institution that farms are operating environmentally friendly, sustainable, care for workers' welfare, and safe and high-quality produce [5]. For Thailand, the nationwide promotion of good agricultural practices aims to develop farmers with the changing demand for food agriculture in the domestic and international markets to promote the inclusion of Thai farmers in the mainstream market.

Table 1. Q-GAP requirements for food crops for each production aspect and its relation to the four modules under Thai Agricultural Standard (TAS 9001-2013) on good agricultural practices.

Item	Total Requirements	Count of Related Requirements			
		Food Safety (FS)	Produce Quality (PQ)	Environmental Management (EN)	Worker Health, Safety, Welfare (WHSW)
1. Water	14	7	2	5	0
2. Planting area	11	6	1	6	1
3. Pesticides	21	14	0	4	9
4. Pre-harvest quality management	19	12	5	6	5
5. Harvest and postharvest handlings	14	10	4	0	0
6. Holding, moving produce in planting plots, and storage	9	4	0	0	0
7. Personal hygiene	14	4	0	0	6
8. Record keeping and traceability	20	17	7	4	4
9. Total	122	74	19	25	25

Notes: Table adapted from TAS 9001-213 under Appendix A of [7]. Details of each requirement can be accessed at https://www.acfs.go.th/standard/download/eng/GAP_Food_Crop.pdf accessed on 6 December 2022.

Later in 2013, amendments were made to Thailand's GAP requirements for food crops to align with the ASEAN Economic Community Blueprint (AEC Blueprint). The AEC blueprint aims to strengthen further the ASEAN members' economic integration by establishing a single market and production base to increase the region's competitiveness [22]. As a result, the ASEAN developed the ASEAN GAP, where member states aligned their local GAP standards. For example, Philippine-GAP, Malaysian SALM, Singapore GAP-VF, Indon-GAP, and Q-GAP in Thailand [2]. In addition, two more GAP standards certifying institutions are available in Thailand—ThaiGAP and GLOBAL GAP [6].

Under the Q-GAP, food safety is a priority of the Department of Agriculture and the National Bureau of Agricultural Commodity and Food Standards (ACFS). For instance, 60 percent of the total 122 requirements are related to food safety, followed by environmental management (20 percent) and growers' safety (20 percent), summarized in Table 1. In addition, growers who wish to get certified undergo assessments by an official GAP inspector on the eight production aspects checklist provided under Thai agricultural standards on GAP (e.g., water, planting area, record keeping) outlined in Table 1. At the same time, recertification is done every two years.

Promoting GAP standards in Thailand offers incentives and disincentives for adoption among Thai farmers. Hobbs [3] discussed the possible incentives of GAP adoption (e.g., economic, human capital, regulatory, and legal incentives. On the other hand, several studies have been concerned that stringent requirements under GAPs could marginalize smallholder farmers due to the needed investment in adopting good practices [1–3]. Therefore, to help Thai farmers with the Q-GAP certification, the Department of Agricultural Extension (DAE) provides GAP training services to improve farmers' understanding of Q-GAPs processes and achieve higher potential benefits that could outweigh the cost of certification.

In the current Q-GAP, the focus food crops include major exporting agricultural commodities such as asparagus, mangoes, corn, and rice [1,4]. For rice, Q-GAP farmers must register their rice plots and follow a set of practices under the Q-GAP guidelines. Participation in the program is voluntary and registered farmers will undergo training for good agricultural practices on a specific crop which the Ministry of Agriculture will conduct. During the implementation of GAP, the government provides the inspection and certification to aid farmers in complying with the national GAP standard. Q-GAP for rice has been widely promoted in the country (71 out of 76 provinces in Thailand), and more than 40 thousand farmers have registered already [1]. Most are irrigated rice farms relying on irrigation for cultivation, especially during the dry season.

Under the Q-GAP standards, water conservation and water-saving related strategies are stipulated under 1.6 and 1.8–1.10 list of GAPs requirements. However, most Thai farmers continually practice continuous flooding in rice cultivation, which uses much water. As part of water-saving irrigation, alternate wetting and drying (AWD) are integrated with GAPs, mainly irrigated rice farms. For rainfed rice farms, direct seeding and selection of water-saving and drought resistance rice varieties are the primary recommended water-saving techniques in aligning with good agricultural practices on water conservation for more sustainable rice production. Direct seeding ensures crop establishment and does not require pre-saturation irrigation, thus reducing water input use [23]. At the same time, irrigated rice field practices direct seeding or alternate wetting and drying, or a combination of both practices. Under an AWD, rice fields are not flooded continuously but allow the soil to dry for several days and be flooded again [15]. The AWD implementation follows the standard guidelines for safe alternate wetting and drying recommended by IRRI [24]. In addition, designated local agricultural extension officers provide training to farmers. Moreover, the training targets farmers who are members and officers of water user groups with more control and participation in irrigation management. As a result, an estimated 38% less water is used under AWD while maintaining the same crop management.

Several irrigated rice farms often experience water shortages during the dry season, especially in downstream areas [25,26]. Integrating AWD under Q-GAP or into farmers water management practices could allow better water allocation. For instance, safe AWD implementation among upstream irrigated fields could increase available water for downstream areas [15,17,26]. Despite the increasing evidence of safe AWD showing an increase in yield and water production, AWD adoption in Thailand is still scant due to Thai farmers being risk-averse, contradicting the traditional practices of continuous flooding. Moreover, AWD adoption requires collective actions among farmers, especially water users under the same irrigation infrastructure. Controlled irrigation in one area of the irrigation canal will affect the entire field. The situation adds to existing barriers among irrigation officers and extension workers in promoting AWD adoption in Thailand. As Enriquez et al. [14] emphasized, AWD adoption is not straightforward as it involves not only monitoring of the irrigation surface but a collective action among farmers involved in the irrigation system. Hence, GAP farmers as a farmer group could exhibit higher acceptance towards AWD as water-saving technology in rice farming, especially in the early stages of scaling out AWD adoption in the nation.

3. GAP Study Sites and Data Analysis

3.1. GAP Study Sites

Throughout the first half of 2022, a total of 26 GAP-certified farms and 30 non-certified farms were closely interviewed in the suburban areas of Khlong Sam Wa district in Bangkok, Thailand. The information included in the survey is the input and output of rice production and farmers' perception regarding the adoption of alternate and wetting as a water-saving mechanism being promoted in the study area. The area is in upper central Thailand and is bordered by Lam Luk Ka, Nong Chok, Min Buri, Khan Na Yao, Bang Khen, and Sai Mai. The study sites often face water shortages, and irrigation remains challenging for most farms. The introduction of AWD is seen to be a potential water-saving technology among rice farms in the study area. Figure 1 demonstrates the rice cropping calendar integrated with AWD introduced in the study area.

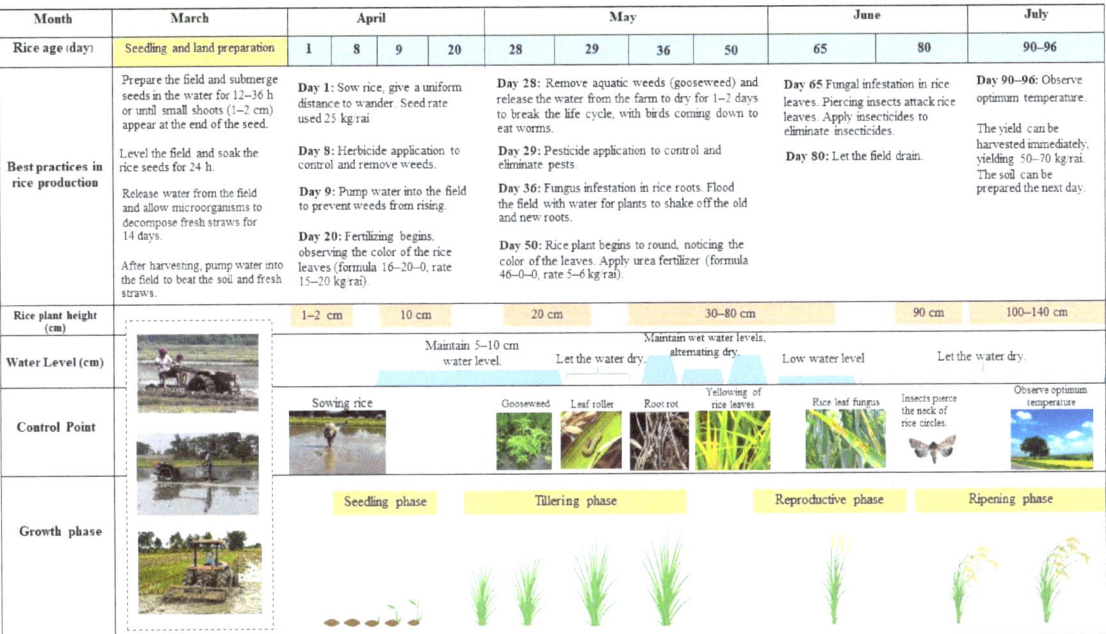

Figure 1. GAP study site's cropping calendar with alternate wetting and drying (AWD) for GAP-certified rice production.

The first part of the production process is seedling and land preparations. During land preparation, farmers drain the field to allow microorganisms to decompose fresh straws. The farmer used a seeding rate of 25 kg/rai. Herbicide and pesticide applications are made during the seedling phase (8th day) and tillering phase (29th day). The tillering phase is a critical phase where aquatic weeds should be removed, and fungal infestations should be controlled. During tillering and reproductive stages, alternate wetting and drying of the field are observed by maintaining a water level of 5–10 cm. A critical control point during the ripening phase is to monitor the optimum field temperature where the height of the plants is around 100–140 cm. Rice yields are harvested immediately within days 90–96, with yields around 50–70 kg/rai.

Based on mapping the cropping calendar practiced among rice farms in the study sites, Table 2 shows the difference in the cost and return of rice production between the certified and non-certified GAP farmers. The total fixed cost incurred by non-GAP farmers (702.67 THB/rai) was significantly higher than GAP farmers (528.51 THB/rai). The highest among the two groups was land rent, followed by depreciation cost, which was highly significant in non-GAP farmers. As for variable cost, there was no significant difference in total variable cost incurred between the two farmer groups. However, looking at the individual cost component items, significant differences exist except for organic fertilizer and bio-fermented water costs. Generally, non-GAP farmers incurred significantly higher variable costs, particularly on seed, chemical fertilizer, herbicides, and fuel. Meanwhile, GAP farmers incurred significantly higher labor, pesticides, and fuel expenses.

Overall, non-GAP farmers have significantly higher total costs than GAP farmers. Moreover, although both farmer groups did not significantly differ in yield per rai, GAP farmers had significantly higher total income and net profit owing to lower total cost and higher selling price. This finding indicates that production costs and product prices affect the profitability of GAP and non-GAP farmers in the study area.

Table 2. Cost and return of rice production by GAP and non-GAP farmers.

Cost Items	GAP	Non-GAP	t-Value
Fixed cost			
Land rent	408.43	425.79	−0.28
Tax	0.30	0.17	0.60
Opportunity cost of land use	59.59	16.50	1.33
Depreciation	60.18	260.21	−10.09 ***
Total fixed cost (TFC)	528.51	702.67	−3.24 ***
Variable cost			
Labor	1085.11	992.91	2.20 **
Seed	347.88	536.79	−4.10 ***
Organic fertilizer	12.51	0.00	-
Bio-fermented water	1.83	0.00	-
Chemical fertilizer	414.49	587.83	−2.73 ***
Herbicides	114.49	172.20	−2.69 ***
Pesticides	80.85	11.24	4.32 ***
Fuel	285.61	421.93	−2.99 ***
Other expenses	182.12	0.00	4.31 ***
Total variable cost (TVC)	2524.90	2722.90	−1.00
Total cost (TFC+TVC)	3053.41	3425.56	−1.91 *
Yield (kg/rai)	755.84	752.21	1.03
Selling price (THB/rai)	7.38	6.77	4.30 ***
Total income (THB/rai)	5710.55	5088.27	3.35 ***
Net profit (THB/rai)	2388.99	1662.71	3.89 ***

Notes: 1 rai = 0.16 hectare; 1 THB = 0.030 USD (6 months average exchange rate from July–December 2019); *** $p < 0.01$; ** $p < 0.05$; * $p < 0.10$.

3.2. Data Analysis

To answer the study's first research question, the logit model following the model specification in the study of Aluddin et al. [17] and Rejesus et al. [27] was employed in estimating the probability of AWD adoption intention based on selected predictors. The logit model specified in Equation (1) allows the estimation of the probability of logit change due to a unit change of the predictors. The left-hand side of Equation (1) represents the ratio of the AWD adoption probability $P_r(y = 1)$ and the probability of non-adoption $1 - P_r(y = 1)$. On the right-hand side, the parameters β are the coefficients to be estimated, and X_j is a vector of the selected predictors related to the sociodemographic characteristics of farmers, such as age, education, marital status, household size, and GAP adoption. In addition, farmers' perceptions of the advantages, knowledge, and trial adoptability of AWD technology were also included as predictors in the logit model specified in Equation (1).

$$\ln\left[\frac{P_r(y=1)}{1 - P_r(y=1)}\right] = \beta_0 + \beta_j X_j \quad (1)$$

Whereas Equation (2) specifies the marginal effects estimated from the logit model in Equation (1) that allows the interpretation of a logit of the odds ratio in terms of marginal change of the selected predictors to the outcome variable on AWD adoption intention.

$$\text{Marginal effects} = \frac{\partial P_r(y=1)}{\partial x_j} \quad (2)$$

The selected predictors for the logit model are summarized in Table 3. The predictors used are sociodemographic characteristics of the focus farmers, derived cost and return variables (presented in Table 2), and perceptions of farmers on AWD technology. After a series of AWD information drives and demonstrations, farmers were asked whether they intended to adopt or not the alternate wetting and drying as a water-saving strategy on their farm. The adoption intention variable was measured as a binary variable taking the value of 1 for farmers who intend to implement AWD and 0 if otherwise. Likewise, a binary variable is coded as 1 if a farmer is GAP certified and 0 if otherwise. Sociodemographic variables considered in the study include age, education, marital status, and household size. Farmers' age was hypothesized to have a negative relationship with the probability of adoption. This implies that younger farmers have a higher probability of adoption intention than older farmers. Given that AWD is a knowledge-based technology, higher education attainment will influence the probability of adoption among sampled farmers. As most Thai farmers employ family labor in production, the study considered the household size to affect farmers' decision to adopt AWD. A larger household size allows more household members to be available for farm labor.

Table 3. Definitions of variables.

Variables	Description	Type
AWD adoption	=1 if adopted AWD, 0 otherwise	Binary
Age	Age of household head	Continuous
Education	Years of education	Continuous
Marital status	=1 if married, 0 otherwise	Binary
Household size	Number of household members	Continuous
GAP adoption	=1 if adopted GAP, 0 otherwise	Binary
Yield	Yield per rai	Continuous
Variable cost	Total variable cost per rai	Continuous
Fixed cost	Total fixed cost per rai	Continuous
Net profit	Net profit per rai	Continuous
AWD advantages	Farmers' perception of AWD advantages	Composite score
AWD knowledge	Knowledge on AWD	Composite score
AWD trial adoptability	Perceived adoptability on AWD trials	Composite score

Notes: A reliability test was conducted for the items used in generating the composite scores. Alpha values are 0.74, 0.84, and 0.85 for perceived advantage, perceived knowledge, and trial adoptability, respectively.

For farmers' perceptions, multiple items were used to reflect the intended factors (i.e., farmers' perceptions) using exploratory factor analysis. Items under the perceived advantages, knowledge, and trial adoptability factors of AWD technology were measured using a five-point Likert scale from 1 as strongly disagree and 5 as strongly agree. Using an exploratory factor analysis, we derived the composite factor score in measuring farmers' perception of AWD advantages, knowledge, and trial adoptability. Figure 2 in the result and discussion section presents the items that underwent the reliability test in the factor analysis.

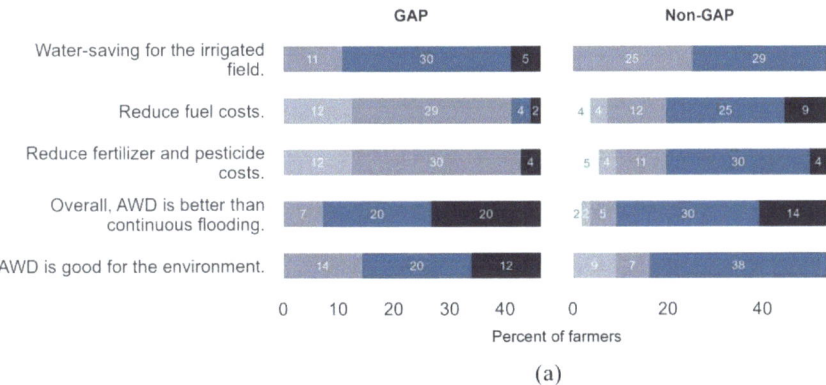

(a)

Figure 2. *Cont.*

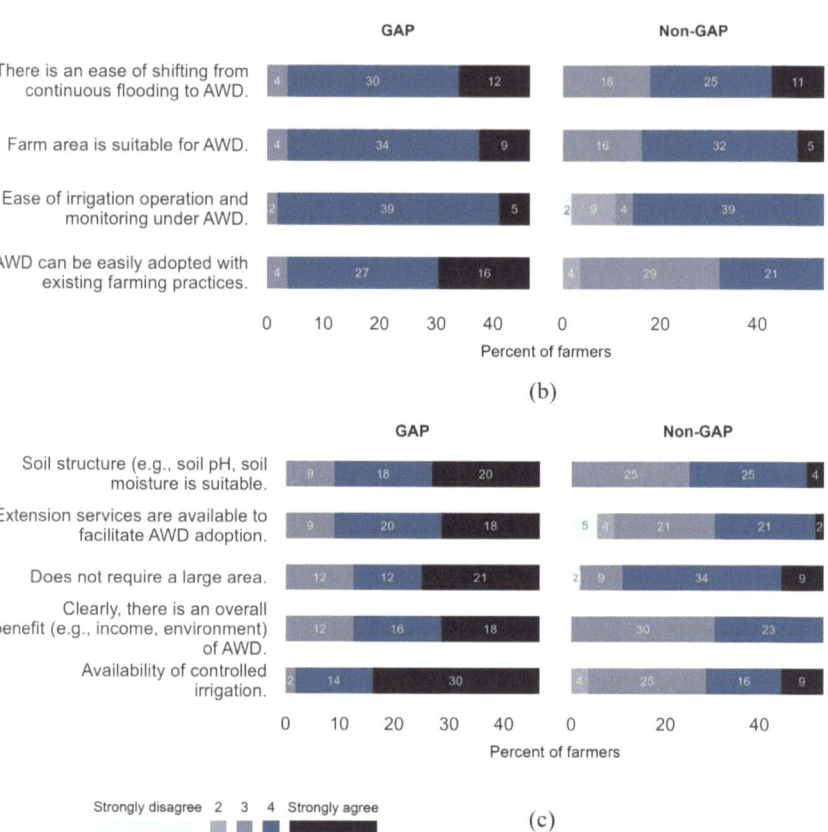

Figure 2. GAP and non-GAP farmers' perceptions of AWD adoption intention regarding the (**a**) innovation advantages, (**b**) ease of adoption, and (**c**) trial capability.

4. Results and Discussion

4.1. Innovation Advantages, Ease of Adoption, and Trial Capability Perceptions on AWD

Introducing change, such as adopting new farming technology, requires an awareness of farmers' behavioral traits [14]. Hence, through a series of interviews, the GAP and non-GAP farmers' perceptions of AWD adoption intention regarding the innovation advantages, ease of adoption, and trial capability are summarized in Figure 2. Before the farmer interviews, AWD-related rice farming demonstrations were conducted to facilitate leveling understanding, particularly for farmers who were less familiar with the technology. A total of 26 GAP and 30 non-GAP farm owners participated in the alternate wetting and drying information drive and farm demonstration.

The promotion of AWD to Thai farmers is due to the technology's potential to reduce water input. Studies on AWD water-saving potential capacity vary from 15–30%, which translates to less irrigation and a reduction in irrigation cost [14,15]. Between groups, GAP adopters consider AWD a better water-saving technology than its counterpart. However, AWD adoption involves additional investment, such as additional time required to monitor the field and the irrigation surface. Managing and monitoring irrigation can be complex, given the high cases of deflection in a communal resource [28,29]. Enriquez [14] views AWD adoption as viable to a collective organization such as farmer groups (e.g., cooperatives, GAP farmers) to minimize transaction costs. Several studies on water user groups along irrigation systems with high interrelationships among group members exhibit a high level of participation in collective action in irrigation management [25,30,31].

Regarding ease of AWD adoption, Thai GAP farmers show higher perceived readiness than non-GAP. However, a high proportion of farmers still have low awareness about safety, environmental and social impact, or implications of agricultural practices under GAP [4]. In addition, the perceived capacity for AWD trials is more prevalent among GAP than non-GAP farmers. The observed behavior among GAP is due to farmers being accustomed to stringent processes stipulated under the GAP requirements. Moreover, water management under the environmental management module of the TAS requires farmers to innovate in saving water in production. This implies that the promotion of AWD as a water-saving technology will be more suitable and feasible for GAP farmers. Hence, the overall benefits of AWD adoption are perceived to be higher among GAP than non-GAP. In addition, both groups agree that AWD could reduce fuel and fertilizer costs. The application of AWD reduces the irrigation hours and lowers fuel needs in operating the water pumps. In a study by Djaman et al. [32], they found an increase in rice yield and a 30 percent increase in the efficiency of nitrogen use under AWD compared to continuous flooding. The increased efficient use of the inputs improves the nitrogen application rate, which could partially contribute to input cost savings. Similar results were also found by Song et al. [33], where AWD combined with reduced fertilizer application promotes phosphorous use efficiency without yield loss in rice plants.

4.2. Logit Results on the Determinants of AWD Adoption

Table 4 presents the result of the logit model with the associated marginal effects on AWD adoption. Results show that age is significant and negatively associated with the probability of AWD adoption intention. This implies the potential of young Thai farmers to adopt water-saving technology like AWD in rice farming. The negative association of age on AWD adoption intention in the study area was also observed in Bangladesh [17] and the Philippines [27]. Alauddin et al. [17] found that older household heads in Bangladesh show a lower probability of adopting AWD technology. At the same time, marital status and household size are positively associated, but not significantly, with a higher probability of adoption intention in the sampled farmers. The non-significant effect of marital status and household size in adoption studies was found in Northwestern China [34] and Tarlac province in the Philippines [27].

Table 4. Determinants of alternate wetting and drying (AWD).

Variables	Logit	p-Value	Marginal Effects	p-Value
Intercept	−46.5372 ***	0.0037		
	(16.0507)			
Age	−0.1295 *	0.0926	−0.0315 *	0.0874
	(0.0770)		(0.0184)	
Education	0.014	0.9733	0.0034	0.9733
	(0.4184)		(0.1017)	
Marital status	0.3703	0.8206	0.0914	0.8216
	(1.6330)		(0.4054)	
Household size	0.3059	0.2509	0.0744	0.2516
	(0.2664)		(0.0649)	
GAP	3.2204 **	0.0304	0.6485 ***	0.0011
	(1.4878)		(0.1979)	
AWD advantages	0.4405 *	0.0755	0.1071 *	0.0616
	(0.2478)		(0.0573)	
AWD knowledge	3.5292 **	0.0145	0.8582 **	0.0107
	(1.4438)		(0.3364)	
AWD trial adoptability	1.7661 ***	0.0018	0.4295 ***	0.0017
	(0.5658)		(0.1371)	
AIC	44.1798			
BIC	62.408			
Pseudo R-square	0.8			

Notes: *** $p < 0.01$; ** $p < 0.05$; * $p < 0.10$.

On the other hand, the study did not find enough statistical evidence on the significance of years spent on education, however, the observed positive effect of education on the probability of AWD adoption corresponds with other related studies (e.g., [17,27,34,35]). At the same time, the observed non-significance of formal years in education among focus groups can be compensated with farmers' tacit knowledge accumulated from years of farming experience. Moreover, perceived AWD knowledge is found to be significant among sampled farmers and positively associated with the probability of adopting AWD. As most farmers show low years of formal education, tacit or experiential knowledge among farmers will be necessary for AWD adoption. The timing of implementing AWD involves years of experience in crop management, and improper implementation of AWD can lead to a reduction in rice yield [12,17].

An increase in AWD adoption's perceived advantages improves the probability of AWD adoption intention. Moreover, the probability of adoption intention is higher for farmers who perceive the current farming area is suitable for AWD trial. This implies that more evidence of the positive impact of AWD adoption and wider information dissemination is beneficial in promoting AWD. As demonstrated by the focus group discussion, farmers' reluctance is rooted in how different the AWD approach is compared to the traditional practices of continuous flooding.

On the other hand, adoption intention between GAP and non-GAP shows a higher probability of AWD adoption among GAP farmers. The observed higher adoption intention among GAP farmers can be attributed to the efforts to comply with the stringent requirement of the Ministry of Agriculture under Thai GAP standards. The results provide evidence to support Thailand's agricultural sector in targeting GAP farmers by integrating AWD in the GAP standards for water-saving technology, particularly among rice GAP.

4.3. Cost and Return Effects on GAP Farmer' AWD Adoption Intention

Table 5 presents the estimated effects of the cost and return variables on the probability of AWD adoption. Results reveal that farmers with relatively high yields have a lower probability of AWD adoption. The hesitance of farmers is a natural reaction when confronted with change, especially when farmers have high yields before the proposed intervention. Thus, Thai farmers with relatively high yields may exhibit a lower probability of AWD adoption due to high opportunity costs and risks from the uncertainty of adoption outcomes, while several studies support the positive effect of AWD on yield (e.g., [12,17,18]). Moreover, there are studies reporting the potential reduction in yield with improper AWD application (e.g., [19,36]).

Table 5. Estimates of the effect of cost and return on AWD adoption intention.

Variables	Logit	p-Value	Marginal Effects	p-Value
Intercept	3.5931	0.4276		
	(4.5296)			
Yield	−0.0257 **	0.0275	−0.0064 **	0.0262
	(0.0116)		(0.0029)	
Total variable cost	0.0029 **	0.0319	0.0007 **	0.0302
	(0.0014)		(0.0003)	
Total fixed cost	−0.0014	0.6612	−0.0003	0.6617
	(0.0031)		(0.0008)	
Net profit	0.0043 ***	0.0047	0.0011 ***	0.0042
	(0.0015)		(0.0004)	
AIC	55.4857			
BIC	65.6124			
Pseudo R-square	0.5795			

Notes: *** $p < 0.01$; ** $p < 0.05$.

For the cost components, higher fixed cost is associated with a lower probability of adoption, while variable cost is positively associated with AWD adoption intention

Changes in variable inputs occur in the short-run production decision and are positively associated with potential yield. Thus, changes in the short-run farming practices (e.g., AWD adoption) that will increase the potentially higher yield could viably increase the adoption intention. On the other hand, farmers with higher fixed inputs may face a higher risk of adopting new technology, such as the AWD, given that higher sunk costs are incurred when faced with crop failure.

5. Conclusions and Policy Implications

Promoting water conservation through AWD can be challenging given the low literacy rate among farmers in developing countries, as measured by the years spent in formal education. The study assumes that the technology can be integrated with GAP standards under the environmental management module of TAS 9001-2013 for irrigated rice farms in the early stages of scaling out AWD usage in Thailand. In the current state of implementing good agricultural practices in the country, chemical contamination has been the focus of the national GAP. The focus should be gradually extended to other components, such as environmental management, particularly the adoption of water-saving management for rice GAP. The potential of the alternate wetting and drying to meet the criteria related to the water requirement of GAP TAS 9001-2003 will be a great opportunity in scaling out the promotion of AWD adoption. This notion was supported by the study's findings, which showed that GAP farmers were more likely to intend to implement AWDs than non-GAP farmers. This can be attributed to GAP adopters accustomed to the demanding standards they must conform to receive certification. At the same time, the adoption of AWD aligns with the priority of GAP principles for water management. Moreover, we found that perceived advantages, knowledge, and suitability of rice farm areas for AWD trials are positively associated with a higher intention for AWD adoption. In light of these findings, the following discusses the study's recommendations and policy implications.

Since AWD adoption can only be put into practice with a controlled water source like the national irrigation projects, the choice to irrigate will often be taken collectively among water users, and it is anticipated that the implementation will be costly. This suggests that, especially in the early phases of adoption, AWD should be introduced and implemented in an imposed manner through the irrigation authority as a water-saving irrigation method. However, as rice farmers get accustomed to the processes involved in irrigation under AWD, participatory irrigation management integrated with AWD will more likely be viable among farmers.

Moreover, the perceived risks involved with changing irrigation practices are another impediment for the Thai government in scaling out the usage of AWD. Although the adoption of AWD was determined to have a favorable benefit, AWD's adverse effects on rice yield were also reported. Therefore, to mitigate the possible impact of the perceived risks, such as crop failure, it is imperative to develop further farmers' understanding of the safe and proper implementation of AWD (also referred to as "safe AWD" in most studies) and to provide crop insurance.

Poor knowledge, lack of information, and awareness regarding AWD technology among farmers will be additional barriers to promoting AWD adoption. Since AWD is a knowledge-based technology, the first recommendation will depend heavily on the availability of local agricultural extension agencies that can provide accurate information and close coaching toward a safe and effective implementation of AWD. Second, enrolling farmers in crop insurance will be an essential safety net against perceived risks, particularly in situations where crop loss will be unavoidable over the length of AWD adoption trials. Finally, more research is required to understand AWD better as rice yields vary not only on irrigation management but also on soil type, fertility, climate, and combinations of these factors.

Author Contributions: Conceptualization, S.S., H.J.C. and C.L.; Data curation, C.L.; Formal analysis, C.L. and S.S.; Funding acquisition, S.S.; Investigation, S.S.; Methodology, C.L.; Project administration, S.S.; Software, C.L.; Supervision, S.S.; Validation, S.S., K.K., P.E.S., L.K., C.K., H.J.C., C.L. and N.L.; Visualization, H.J.C. and C.L.; Writing—original draft, C.L., H.J.C. and S.S.; Writing—review & editing, C.L., H.J.C. and S.S. All authors have read and agreed to the published version of the manuscript.

Funding: This work was financially supported by King Mongkut's Institute of Technology Ladkrabang Research Fund [grant number 2563-02-04-001]

Institutional Review Board Statement: The study was conducted in accordance with the Declaration of Helsinki, and approved by the Research Ethics Committee of King Mongkut's Institute of Technology Ladkrabang (EC-KMITL-65-70 and approved on 26 May 2022).

Informed Consent Statement: Informed consent was obtained from all subjects involved in the study

Data Availability Statement: The datasets and code used in this paper are available from the author(s) upon reasonable request.

Acknowledgments: The authors would like to acknowledge King Mongkut's Institute of Technology Ladkrabang (KMITL) for supporting this study. The authors would also like to thank the editors and the anonymous reviewers for their comments, which helped improve the paper's quality.

Conflicts of Interest: The authors declare no conflict of interest.

References

1. Srisopaporn, S.; Jourdain, D.; Perret, S.R.; Shivakoti, G. Adoption and continued participation in a public Good Agricultural Practices program: The case of rice farmers in the Central Plains of Thailand. *Technol. Forecast. Soc. Chang.* **2015**, *96*, 242–253. [CrossRef]
2. Amekawa, Y. Can a public GAP approach ensure safety and fairness? A comparative study of Q-GAP in Thailand. *J. Peasant Stud.* **2013**, *40*, 189–217. [CrossRef]
3. Hobbs, J. *Incentives for the Adoption of Good Agricultural Practices*; FAO GAP Working Paper Series; Food and Agriculture Organization: Rome, Italy, 2013.
4. Sardsud, V. National experiences: Thailand. In *Challenges and Opportunities Arising from Private Standards on Food Safety and Environment for Exporters of Fresh Fruit and Vegetables in Asia: Experience of Malaysia, Thailand, and Vietnam*; United Nations: New York, NY, USA, 2007; pp. 53–69.
5. Premier, R.; Ledger, S. Good Agricultural Practices in Australia and Southeast Asia. *Horttech* **2006**, *16*, 552–555. [CrossRef]
6. Wongprawmas, R.; Canavari, M.; Waisarayutt, C. A multi-stakeholder perspective on the adoption of good agricultural practices in the Thai fresh produce industry. *Br. Food J.* **2015**, *117*, 2234–2249. [CrossRef]
7. TAS 9001–2013; Good Agricultural Practices for Food Crop 2013. Ministry of Agriculture and Cooperatives Thai Agricultural Standard: Bangkok, Thailand, 2013.
8. Cavite, H.J.; Kerdsriserm, C.; Suwanmaneepong, S. Strategic guidelines for community enterprise development: A case in rural Thailand. *J. Enterprising Communities People Places Glob. Econ.* **2021**, 1–21. [CrossRef]
9. Ebers, A.; Nguyen, T.T.; Grote, U. Production efficiency of rice farms in Thailand and Cambodia: A comparative analysis of Ubon Ratchathani and Stung Treng provinces. *Paddy Water Environ.* **2017**, *15*, 79–92. [CrossRef]
10. Suwanmaneepong, S.; Kerdsriserm, C.; Lepcha, N.; Cavite, H.J.; Llones, C.A. Cost and return analysis of organic and conventional rice production in Chachoengsao Province, Thailand. *Org. Agric.* **2020**, *10*, 369–378. [CrossRef]
11. ONEP. *Second Biennial Update Report*; ONEP (Office of the Natural Resources and Environmental Policy and Planning): Bangkok, Thailand, 2018; pp. 1–108.
12. Sriphirom, P.; Chidthaisong, A.; Towprayoon, S. Effect of alternate wetting and drying water management on rice cultivation with low emissions and low water used during wet and dry season. *J. Clean. Prod.* **2019**, *223*, 980–988. [CrossRef]
13. Baggs, E.M. Soil microbial sources of nitrous oxide: Recent advances in knowledge, emerging challenges and future direction. *Curr. Opin. Environ. Sustain.* **2011**, *3*, 321–327. [CrossRef]
14. Enriquez, Y.; Yadav, S.; Evangelista, G.K.; Villanueva, D.; Burac, M.A.; Pede, V. Disentangling Challenges to Scaling Alternate Wetting and Drying Technology for Rice Cultivation: Distilling Lessons From 20 Years of Experience in the Philippines. *Front Sustain. Food Syst.* **2021**, *5*, 675818. [CrossRef]
15. Lampayan, R.M.; Rejesus, R.M.; Singleton, G.R.; Bouman, B.A.M. Adoption and economics of alternate wetting and drying water management for irrigated lowland rice. *Field Crops Res.* **2015**, *170*, 95–108. [CrossRef]
16. Malumpong, C.; Ruensuk, N.; Rossopa, B.; Channu, C.; Intarasathit, W.; Wongboon, W.; Poathong, K.; Kunket, K. Alternate Wetting and Drying (AWD) in Broadcast rice (*Oryza sativa* L.) Management to Maintain Yield, Conserve Water, and Reduce Gas Emissions in Thailand. *Agric. Res.* **2021**, *10*, 116–130. [CrossRef]
17. Alauddin, M.; Rashid Sarker, M.A.; Islam, Z.; Tisdell, C. Adoption of alternate wetting and drying (AWD) irrigation as a water-saving technology in Bangladesh: Economic and environmental considerations. *Land Use Policy* **2020**, *91*, 104430. [CrossRef]

18. Pearson, K.A.; Millar, G.M.; Norton, G.J.; Price, A.H. Alternate wetting and drying in Bangladesh: Water-saving farming practice and the socioeconomic barriers to its adoption. *Food Energy Secur.* **2018**, *7*, e00149. [CrossRef]
19. Linquist, B.A.; Anders, M.M.; Adviento-Borbe, M.A.A.; Chaney, R.L.; Nalley, L.L.; da Rosa, E.F.F.; Kessel, C. Reducing greenhouse gas emissions, water use, and grain arsenic levels in rice systems. *Glob. Chang. Biol.* **2015**, *21*, 407–417. [CrossRef]
20. Islam, S.M.M.; Gaihre, Y.K.; Islam, M.R.; Ahmed, M.N.; Akter, M.; Singh, U.; Sander, B.O. Mitigating greenhouse gas emissions from irrigated rice cultivation through improved fertilizer and water management. *J. Environ. Manag.* **2022**, *307*, 114520. [CrossRef]
21. Cheng, H.; Shu, K.; Zhu, T.; Wang, L.; Liu, X.; Cai, W.; Qi, Z.; Feng, S. Effects of alternate wetting and drying irrigation on yield, water and nitrogen use, and greenhouse gas emissions in rice paddy fields. *J. Clean. Prod.* **2022**, *349*, 131487. [CrossRef]
22. ASEAN. *Economic Community Blueprint*; ASEAN Secretariat: Jakarta, Indonesia, 2008; ISBN 978-979-3496-77-1.
23. Cabangon, R.J.; Tuong, T.P.; Abdullah, N.B. Comparing water input and water productivity of transplanted and direct-seeded rice production systems. *Agric. Water Manag.* **2002**, *57*, 11–31. [CrossRef]
24. Siopongco, J.; Wassmann, R.; Sander, B.O. *Alternate Wetting and Drying in Philippine Rice Production: Feasibility Study for a Clean Development Mechanism*; IRRI Bulletin: Manila, Philippines, 2013; p. 14.
25. Llones, C.; Mankeb, P.; Wongtragoon, U.; Suwanmaneepong, S. Production efficiency and the role of collective actions among irrigated rice farms in Northern Thailand. *Int. J. Agric. Sustain.* **2022**, *20*, 1047–1057. [CrossRef]
26. Wongtragoon, U.; Kubo, N.; Tanji, H. Performance diagnosis of Mae Lao Irrigation Scheme in Thailand (I) Development of Unsteady Irrigation Water Distribution and Consumption model. *Paddy Water Environ.* **2010**, *8*, 1–13. [CrossRef]
27. Rejesus, R.M.; Palis, F.G.; Rodriguez, D.G.P.; Lampayan, R.M.; Bouman, B.A.M. Impact of the alternate wetting and drying (AWD) water-saving irrigation technique: Evidence from rice producers in the Philippines. *Food Policy* **2011**, *36*, 280–288. [CrossRef]
28. Llones, C.; Mankeb, P.; Wongtragoon, U.; Suwanmaneepong, S. Bonding and bridging social capital towards collective action in participatory irrigation management. Evidence in Chiang Rai Province, Northern Thailand. *Int. J. Soc. Econ.* **2021**, *49*, 296–311. [CrossRef]
29. Ricks, J.I. Pockets of Participation: Bureaucratic Incentives and Participatory Irrigation Management in Thailand. *Water Altern.* **2015**, *8*, 193–214.
30. Araral, E. What Explains Collective Action in the Commons? Theory and Evidence from the Philippines. *World Dev.* **2009**, *37*, 687–697. [CrossRef]
31. Chaudhry, A.M. Improving on-farm water use efficiency: Role of collective action in irrigation management. *Water Resour. Econ.* **2018**, *22*, 4–18. [CrossRef]
32. Djaman, K.; Mel, V.; Diop, L.; Sow, A.; El-Namaky, R.; Manneh, B.; Saito, K.; Futakuchi, K.; Irmak, S. Effects of Alternate Wetting and Drying Irrigation Regime and Nitrogen Fertilizer on Yield and Nitrogen Use Efficiency of Irrigated Rice in the Sahel. *Water* **2018**, *10*, 711. [CrossRef]
33. Song, T.; Xu, F.; Yuan, W.; Chen, M.; Hu, Q.; Tian, Y.; Zhang, J.; Xu, W. Combining alternate wetting and drying irrigation with reduced phosphorus fertilizer application reduces water use and promotes phosphorus use efficiency without yield loss in rice plants. *Agric. Water Manag.* **2019**, *223*, 105686. [CrossRef]
34. Feike, T.; Khor, L.Y.; Mamitimin, Y.; Ha, N.; Li, L.; Abdusalih, N.; Xiao, H.; Doluschitz, R. Determinants of cotton farmers' irrigation water management in arid Northwestern China. *Agric. Water Manag.* **2017**, *187*, 1–10. [CrossRef]
35. Rehman, H.U.; Kamran, M.; Basra, S.M.A.; Afzal, I.; Farooq, M. Influence of Seed Priming on Performance and Water Productivity of Direct Seeded Rice in Alternating Wetting and Drying. *Rice Sci.* **2015**, *22*, 189–196. [CrossRef]
36. Chidthaisong, A.; Cha-un, N.; Rossopa, B.; Buddaboon, C.; Kunuthai, C.; Sriphirom, P.; Towprayoon, S.; Tokida, T.; Padre, A.T.; Minamikawa, K. Evaluating the effects of alternate wetting and drying (AWD) on methane and nitrous oxide emissions from a paddy field in Thailand. *Soil Sci. Plant Nutr.* **2018**, *64*, 31–38. [CrossRef]

Disclaimer/Publisher's Note: The statements, opinions and data contained in all publications are solely those of the individual author(s) and contributor(s) and not of MDPI and/or the editor(s). MDPI and/or the editor(s) disclaim responsibility for any injury to people or property resulting from any ideas, methods, instructions or products referred to in the content.

Article

Effects of Zeolite on Physiological Characteristics and Grain Quality in Rice under Alternate Wetting and Drying Irrigation

Yidi Sun *, Jigan Xie, Huijing Hou, Min Li, Yitong Wang and Xuetao Wang

College of Hydraulic Science and Engineering, Yangzhou University, Yangzhou 225009, China; hjhou@yzu.edu.cn (H.H.)
* Correspondence: yidisun0626@outlook.com; Tel.: +86-15698824626

Abstract: Background: Zeolite (Z) is gradually used in rice production due to its holding ability for water and nutrients, but limited information is available on how its physiological function affects rice grain yield and quality under water stress. Methods: This study aimed to investigate the effect of Z application on rice physiological characteristics, dry matter and nitrogen accumulation, grain yield and quality under continuous flooding (CF) and alternate wetting and drying irrigation (AWD). Results: The results showed that, compared with CF, AWD reduced leaf SPAD, root bleeding intensity, aboveground dry matter and nitrogen accumulation, resulted in lower grain yield without Z application, but improved root–shoot ratio and root N accumulation. Z application increased dry matter and N accumulation, and subsequent grain yield by improving leaf SPAD and root bleeding intensity. Both AWD and Z application improved water use efficiency. AWD reduced head rice rate, chalky rice rate and chalkiness, but improved the taste value by increasing the breakdown and reducing the setback. Z application improved protein content, reduced breakdown and setback, but increased chalky rice rate and chalkiness. Conclusions: These results indicated that AWD and Z application could achieve several benefits including improved grain yield, grain quality and water use efficiency.

Keywords: zeolite; alternate wetting and drying irrigation; rice; physiological characteristics; grain quality

1. Introduction

Rice has an important strategic position in global food production. China is a major rice producer and rice consumer; it is necessary to improve the rice grain yield to meet the future demand for rice as the planting area is decreasing year by year [1]. Rice is a water-intensive crop and its traditional flooded irrigation regime leads to high water consumption and low water use efficiency, which is contrary to the increasingly short supply of water resources in China [2]. Therefore, it is necessary to develop a water-saving irrigation regime in paddy fields in order to ensure water security and food security with minimum water consumption in China.

Alternate wetting and drying irrigation is the most widely used as an effective water-saving irrigation regime, which reduces the water consumption by 15~30%, relative to traditional flooding irrigation [3], but its effect on the physiological characteristics and grain yield are still controversial [4–6]. Some studies suggested the soil was in a state of water saturation under flooding irrigation, which might reduce the oxidative activity of rice roots, and then inhibit root growth and development, thus it was difficult to ensure high grain yield and excellent rice grain quality [5,6]. Paddy soil moisture was in an alternating saturated and unsaturated environment, and the soil aeration was improved under AWD. The improved soil aeration was beneficial in maintaining strong root activity, and then the ability of roots to absorb nutrients was enhanced, which improved the rice physiological function and grain yield under AWD [7,8]. Additionally, both NH_4^+-N and NO_3^--N were present in soil under AWD due to the enhanced nitrification reaction caused by the increase in dissolved oxygen in the paddy field, which enhanced root growth and improved nitrogen accumulation and yield of rice [9]. However, some studies indicated AWD caused nitrogen loss in the form of N_2 or N_2O due to the enhanced soil nitrification–denitrification process, which reduced nitrogen accumulation in plants and thus reduced rice grain yield. There are also some reports that showed AWD had no significant effect on rice grain yield. Rice quality is the second most important factor in determining the economic benefits of rice after yield [10]. The rice grain quality was not only affected by its own genetic characteristics, but also related to external environmental conditions such as cultivation patterns. The previous study found moderate AWD significantly improved head rice rate while it decreased chalkiness, and thus the rice quality was better [11]. Other studies found that moderate alternate wetting and drying irrigation at the grain filling stage improved appearance quality, increased breakdown and decreased setback of rice, but severe alternate wetting and drying irrigation showed the opposite effects [12]. Although most studies have shown that alternate wetting and moderate drying irrigation was beneficial to the improvement in rice grain quality, previous research results showed that it reduced the appearance and nutritional quality of rice [13]. At present, the effects of AWD irrigation on quality are contradictory [14,15]. Thus, it is necessary to clarify the impact of AWD irrigation on rice quality, and solve the possible negative effects on rice grain yield.

Zeolite, as a soil conditioner, is a type of aqueous porous aluminosilicate crystal mineral, with a lattice-like structure composed of (Si, Al, O) tetrahedra; thus, it has an intensive affinity for water and fertilizer due to its strong ion exchange capacity, high adsorption and hydration–dehydration properties, so it had significant advantages in improving crop yield [16]. Therefore, zeolite is widely used in the agricultural and environmental fields [16–21]. Many scholars pointed out that zeolite had remarkable results on improving the growth and physiological traits of crops and yields such as soybean [17], corn [18] and wheat [19] under different soil conditions. In addition, zeolite could continuously provide water molecules during drought stress periods, and promoted the diffusion of water molecules to the root zone during the irrigation period, so as to alleviate the negative effects of drought stress on crop growth. Some field experiments showed that zeolite had a more significant yield increase effect under drought stress conditions [20–23]. We anticipate that the high ion exchange and adsorption properties of zeolite for ammonium ion would have an impact on soil nitrogen supply in paddy fields, which, in turn, might affect rice grain quality. However, there were few studies on the effect of zeolite on rice grain quality. Whether an integrated AWD and zeolite management pattern could make full use of the incentivizing effect of water and fertilizer in improving rice yield, water use efficiency and grain quality needed to be further verified. Therefore, the objectives of this study were to further investigate the effects of Z on the physiological characteristics, grain yield and quality of rice in the AWD paddy field through a field experiment.

2. Materials and Methods

2.1. Experimental Site

The experiment was carried out at the Donggang Irrigation Experiment Station in Dandong, Liaoning Province (113°34′43″ E, 39°52′38″ N). Donggang has a continental monsoon climate in the north temperate humid region, with an annual average temperature of 8.4 °C, an annual precipitation of 900–1000 mm and a sunlight duration of 2484.3 h. The soil texture is silt loam, and the initial physical and chemical properties of soil were shown in Table 1 [24]. The daily average temperature and reference crop evapotranspiration during the rice growing period were shown in Figure 1. Meteorological data were observed by an automatic weather station located on the west side of the experimental station. ET_0 is estimated by the Penman–Monteith method recommended by the Food and Agriculture Organization of the United Nations.

Table 1. Physicochemical properties of the soil before the field experiment and of the clinoptilolite zeolite.

Soil Properties		Chemical Composition of Zeolite (%)			
Bulk density	1.39 g m^{-3}	SiO$_2$	67.09	FeO	0.07
pH	6.76	Al$_2$O$_3$	12.44	MnO	0.03
Total N,	2833 kg ha^{-1}	CaO	6.51	P$_2$O$_5$	0.03
NH$_4^+$-N,	8.0 kg ha^{-1}	MgO	1.22	Loss on ignition	10.1%
NO$_3^-$-N	9.2 kg ha^{-1}	K$_2$O	1.2	CEC	142 cmol$_c$ kg^{-1}
Available P	53 mg kg^{-1}	Fe$_2$O$_3$	0.78	SSA	670 m^2 g^{-1}.
Exchangeable K	165 mg kg^{-1}	Na$_2$O	0.26		
CEC	15 cmol$_c$ kg^{-1}	TiO$_2$	0.09		

Notes: CEC, cation exchange capacity; SSA, specific surface area.

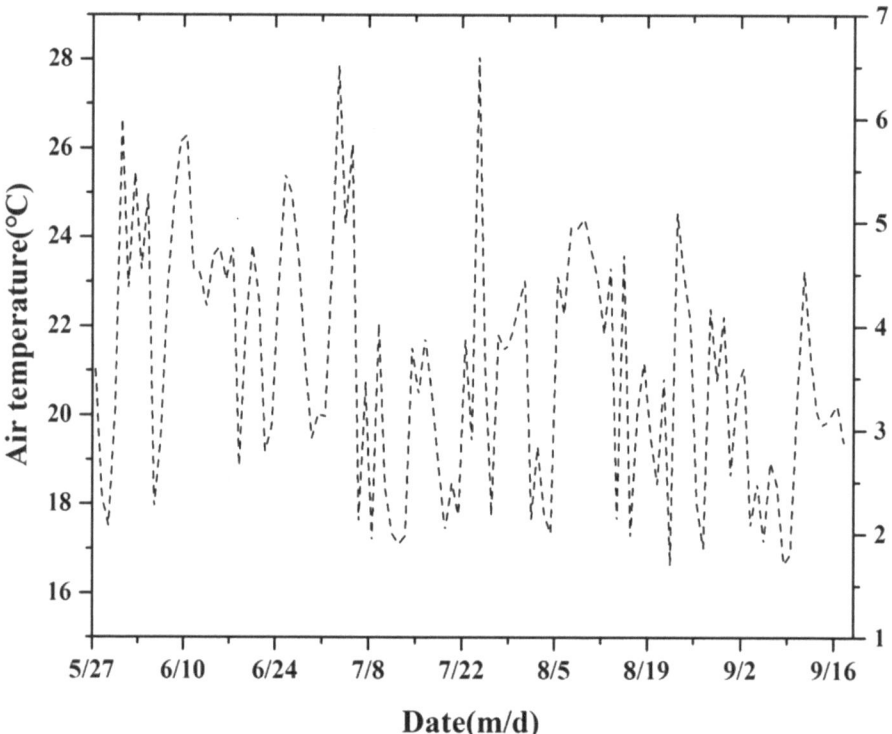

Figure 1. Daily average air temperature during rice-growing period. Air temperature was measured at a weather station close to the experimental site.

2.2. Experimental Design and Management

The experiment was adopted in a split-plot design with two factors of irrigation regime(I) and zeolite application rate (Z), with three replicates (lysimeters) in 2017 and 18 treatments in total, I$_{CF}$Z$_0$, I$_{CF}$Z$_5$, I$_{CF}$Z$_{10}$, I$_{AWD}$Z$_0$, I$_{AWD}$Z$_5$, I$_{AWD}$Z$_{10}$, respectively. The main-plots were irrigation regimes with continuous flooding irrigation (CF) and alternate wetting and drying irrigation (AWD); the sub-plots were rates of Z addition with 0 t·ha^{-1} (Z$_0$), 5 t·ha^{-1} (Z$_5$) and 10 t·ha^{-1} (Z$_{10}$). The experiment was carried out in a non-weighing lysimeter with automatic canopy, and the specifications of each lysimeter were 2.55 m (length) × 2.62 m (width) × 2.55 m (height) = 6.67 m^2. The lysimeter was cast from reinforced concrete to prevent lateral cross flow of moisture and nutrients. The experimental design and layout are shown in Figure 2.

	I_{CF}	I_{AWD}
Z_0	$I_{CF}Z_0$	$I_{AWD}Z_0$
Z_5	$I_{CF}Z_5$	$I_{AWD}Z_5$
Z_{10}	$I_{CF}Z_{10}$	$I_{AWD}Z_{10}$
Z_{10}	$I_{CF}Z_{10}$	$I_{AWD}Z_{10}$
Z_5	$I_{CF}Z_5$	$I_{AWD}Z_5$
Z_0	$I_{CF}Z_0$	$I_{AWD}Z_0$
Z_0	$I_{CF}Z_0$	$I_{AWD}Z_0$
Z_5	$I_{CF}Z_5$	$I_{AWD}Z_5$
Z_{10}	$I_{CF}Z_{10}$	$I_{AWD}Z_{10}$

Figure 2. The experimental design (**A**) and layout (**B**) of split-plot experiment.

For the CF regime, a 1–5 cm water layer was maintained in the field during the whole rice-growing period, except for natural drying about 15 days before harvest. For the AWD regime, a 2–3 cm water layer was maintained in the field during the regreening stage, and then an alternating cycle of drying and wetting was conducted, the plots were irrigated to a water depth of 30–40 mm when the soil moisture was dry to $-10\sim15$ kPa, dried naturally about 15 days before harvest. The depth of water layer was monitored by water gauge at 8:00 am every day. The soil water potential at 15 cm depth was monitored by soil moisture tensiometers (Institute of Soil Science of Chinese Academy of Sciences, Nanjing, China) at 8:00 am and 2:00 pm every day. The amount of irrigation water was measured by a volumetric water meter installed on the irrigation pipe. In order to simulate deep leaching in the field, 1.5 mm d^{-1} of water was drained by a PVC pipe at soil depth of about 200 cm. Local fertilization standard was adopted. Nitrogen fertilizer (150 kg ha^{-1}) was applied in a ratio of 5:3:2 as base fertilizer (28 May), the first top dressing (10 June) and the second top dressing (21 July). Phosphate fertilizer (75 kg·ha^{-1}) was applied as base fertilizer at one time; potassium fertilizer (60 kg·ha^{-1}) was applied in a ratio of 5:5 as base fertilizer and the second top dressing. Zeolite was applied via surface application after soaking, and was incorporated into near-surface (0–5 cm) soil along with the base fertilizer at one time, which were mixed by turning over. Regular manual weeding was conducted to prevent weeds. Other cultivation techniques and field management followed local traditional methods. The row spacing of rice transplanting was 30 cm, the plant spacing was 14 cm and 3 plants were planted in each hole. The rice was transplanted on 28 May and harvested on 18 September.

2.3. Sampling and Measurements

2.3.1. Leaf SPAD and Root Bleeding Intensity

Leaf SPAD was measured by a chlorophyll meter (SPAD-502, Minolta Camera Co, Osaka, Japan) at tillering, joint-booting, panicle-initiation and grain-filling stages. A total of 3 representative plants with 3 uppermost fully expanded leaves were selected for measurement per treatment. Root bleeding intensity at the base was collected at a distance of 10 cm from the ground, absorbent cotton was used to absorb the bleeding fluid. Three holes of representative plants were selected for measurement per treatment at tillering, joint-booting, panicle-initiation and grain-filling stages; the measurement time was from 18:00 pm to 7:00 am the next day and root bleeding intensity was calculated from the weight difference of absorbent cotton before and after the measurement.

2.3.2. Dry Matter and Nitrogen Accumulation

At maturity, three holes of representative plants were selected for measuring dry matter and nitrogen accumulation per treatment. The samples were washed with distilled water and the root, stem, leaf and panicle were separated. The samples were placed in an oven at 105 °C for 30 min, and then adjusted to 80 °C for drying to constant weight. Dry matter accumulation was manually weighed by electronic scale. Root shoot ratio was calculated by the ratio of root dry weight to shoot dry weight.

2.3.3. Grain Yield and Water Use Efficiency

At maturity stage, each plot was harvested separately, then rice grain was naturally dried to a moisture content of about 14% to determine the yield. Water usage for each plot was calculated according to the principle of water balance. Water use efficiency was calculated using the following equation:

$$\text{WUE} = \frac{Y}{W} \tag{1}$$

where Y is the grain yield (kg·ha^{-1}) and W is the total seasonal water use (m^3·ha^{-1}).

2.3.4. Grain Quality

After harvesting, about 300 g of rice grains were taken from each treatment and air-dried for about 60 days for rice grain quality determination. The rice grain was hulled with an FC-2K Husker (Yamamoto, Japan) to obtain brown rice and brown rice rate was calculated by dividing that brown rice weight by rice grain weight. Brown rice was reprocessed by VP-32T Whitener (Yamamoto, Japan) to obtain milled rice and milled rice rate was calculated by dividing that milled rice weight by rice grain weight. Rice appearance quality (head rice rate, chalky rice rate and chalkiness degree) were measured by a ES-1000 Rice Inspector (Shizuoka, Japan). Protein concentration, amylose concentration and eating score were determined using a Infratec^{TM} 1241 Grain Analyzer (Foss, Tecator, Japan). Rice flour was obtained using a JFS-13A Tornado Crush Mill and a 3 g sample was mixed with 25 mL distilled water to determine starch viscosity properties. Starch viscosity properties (peak viscosity, through viscosity, breakdown, final viscosity, setback, peak time and pasting temp) were measured using a RVA-4 Rapid Visco Analyzer (Newport Scientific, Australia). The eating score was determined by a 1241 near infrared rapid quality analyzer (Foss, Tecator, Japan).

3. Result

3.1. Leaf SPAD and Root Bleeding Intensity

The changes in the leaf SPAD value and root bleeding intensity in different growth stages are shown in Table 2. The leaf SPAD value was the highest at the tillering stage due to the application of tillering fertilizer, and then there was a decreasing trend with the consumption of nitrogen in soil. The leaf SPAD value then increased gradually from the joint-booting to panicle-initiation stage, because the nitrogen in the soil was replenished again after the topdressing of panicle fertilizer, then tended to stabilize at the grain-filling stage. The irrigation regime only had a significant effect on the SPAD value at the grain-filling stage, AWD decreased the SPAD value by 5.0% at this stage compared with CF. Z addition significantly affected the SPAD value at the joint-booting and panicle-initiation stages. Compared with no Z addition, the application of 5 and 10 t·ha^{-1} Z increased the SPAD value by 4.7% and 7.2% at the joint-booting stage, respectively. At the panicle-initiation stage, the application of 10 t·ha^{-1} Z improved the SPAD value by 4.0%, but no significant difference was found between the Z_0 and Z_5.

Table 2. Effects of irrigation regimes and Z application rates on leaf SPAD and root bleeding intensity at different rice-growing stages.

Main Effect	Leaf SPAD Value				Root Bleeding Intensity (g·h^{-1})			
	Tillering Stage	Joint-Booting Stage	Panicle-Initiation Stage	Grain-Filling Stage	Tillering Stage	Joint-Booting Stage	Panicle-Initiation Stage	Grain-Filling Stage
I_{CF}	45.3 a	42.1 a	43.9 a	44.0 a	6.24 a	7.01 a	6.13 a	5.90 a
I_{AWD}	45.5 a	41.3 a	43.1 a	41.8 b	5.11 a	6.68 a	4.73 b	4.52 b
Z_0	44.8 a	40.1 c	42.6 b	42.5 a	5.49 a	6.37 b	5.04 b	4.74 b
Z_5	45.6 a	42.0 b	43.6 ab	42.9 a	5.70 a	6.88 ab	5.53 a	5.29 a
Z_{10}	45.9 a	43.0 a	44.3 a	43.2 a	5.85 a	7.31 a	5.71 a	5.59 a
ANOVA								
I	ns	ns	ns	*	ns	ns	**	**
Z	ns	**	**	ns	ns	**	**	**
I*Z	ns	ns	ns	ns	ns	ns	ns	ns

Notes: Means followed by the same letter within the same column are not significantly different at $p < 0.05$ by Tukey's HSD test. *, ** and ns represent significance at $p < 0.05$, 0.01 and not significant. I: irrigation regime; Z: clinoptilolite zeolite application.

The root bleeding intensity increased first and then decreased during the whole rice-growing stage, and reached a peak at the joint-booting stage. The root bleeding intensity was not affected by the interaction effect of the irrigation regime and Z addition at all the observation stages. The irrigation regime had a significant effect on the root bleeding intensity at the panicle-initiation and grain-filling stages. Compared with CF, AWD reduced root bleeding intensity by 22.8% and 23.4% at the panicle-initiation and grain-filling stages, respectively. Z addition had a significant effect on the root bleeding intensity from the joint-booting to grain-filling stages. Compared with no Z addition, 10 t·ha^{-1} Z amendment increased the root bleeding intensity by 14.8% at the joint-booting stage. The application of 5 and 10 t·ha^{-1} Z improved the root bleeding intensity by 9.72% and 13.3% at the panicle-initiation stage, improved the root bleeding intensity by 11.6% and 17.9% at the grain-filling stage, respectively, relative to no Z addition, but no significant difference occurred between Z_5 and Z_{10}.

3.2. Dry Matter Accumulation and Distribution at Maturity Stage

The effect of irrigation regime and Z amendment on the dry matter accumulation and distribution at the maturity stage are shown in Table 3. The interaction between the irrigation regime and Z amendment had no significant effect on the dry matter accumulation and distribution proportion at the maturity stage. The irrigation regime had a significant effect on the dry matter accumulation in the stem-leaf and the dry matter distribution proportion in root and root–shoot ratio. Compared with CF, AWD reduced the dry matter accumulation in the stem-leaf by 6.9%, but had no significant effect in other tissues. AWD improved the proportion of dry matter distribution in root and root–shoot ratio by 11.0% and 11.2%, respectively, relative to CF.

Z amendment had a significant effect on the dry matter accumulation in different parts of the plant and the proportion of dry matter distribution in the stem-leaf and panicle. Compared with Z_0, Z_5 and Z_{10} improved the dry matter accumulation in the stem-leaf and panicle by 17.6% and 25.9%, and 8.9% and 13.5%; Z_5 and Z_{10} increased the aboveground dry matter accumulation by 13.1% and 18.7%, increased the total dry matter accumulation by 12.7% and 18.4%, respectively, and there was no significant difference between Z_5 and Z_{10}. Compared with Z_0, Z_5 and Z_{10} improved the dry matter distribution proportion in the stem-leaf by 4.2% and 6.5%, reduced the dry matter distribution proportion in the panicle by 2.7% and 4.4%, respectively, and there was no significant difference between Z_5 and Z_{10}.

Table 3. Effects of irrigation regimes and Z application rates on dry matter accumulation and distribution at maturity stage.

Main Effect	Dry Matter Accumulation (kg·ha^{-1})					Dry Matter Distribution (%)			
	Root	Stem-Leaf	Panicle	Aboveground	Whole Plant	Root	Stem-Leaf	Panicle	Root–Shoot Ratio
I_{CF}	6.10 a	30.3 a	38.2 a	68.5 a	74.6 a	8.17 b	40.5 a	51.3 a	0.089 b
I_{AWD}	6.48 a	28.2 b	36.9 a	65.1 a	71.6 a	9.07 a	39.2 a	51.7 a	0.099 a
Z_0	5.79 b	25.5 b	34.9 b	60.4 b	66.2 b	8.77 a	38.5 b	52.7 a	0.096 a
Z_5	6.33 ab	30.0 a	38.3 a	68.3 a	74.6 a	8.49 a	40.2 a	51.3 b	0.093 a
Z_{10}	6.74 a	32.1 a	39.6 a	71.7 a	78.4 a	8.60 a	41.0 a	50.4 b	0.094 a
ANOVA									
I	ns	*	ns	ns	ns	**	ns	ns	**
Z	**	**	**	**	**	ns	**	**	ns
I*Z	ns	ns	ns	ns	ns	ns	ns	ns	ns

Notes: Means followed by the same letter within the same column are not significantly different at $p < 0.05$ by Tukey's HSD test. *, ** and ns represent significance at $p < 0.05, 0.01$ and not significant.

3.3. Nitrogen Accumulation and Distribution at Maturity Stage

The effect of the irrigation regime and Z amendment on nitrogen accumulation and distribution at maturity stage are shown in Table 4. Nitrogen (N) accumulation and distribution was not significantly affected by the interaction between the irrigation regime and Z amendment. The irrigation regime had a significant effect on N accumulation in the root, stem-leaf and aboveground parts, and the proportion of N distribution in the root and stem-leaf.

Table 4. Effects of irrigation regimes and Z application rates on nitrogen accumulation and distribution at maturity stage.

Main Effect	N Accumulation (kg·ha^{-1})					N Distribution (%)		
	Root	Stem-Leaf	Panicle	Aboveground	Whole Plant	Root	Stem-Leaf	Panicle
I_{CF}	8.84 b	28.0 a	78.4 a	106.4 a	115.3 a	7.66 b	24.2 a	68.1 a
I_{AWD}	9.69 a	24.7 b	76.1 a	100.6 b	110.3 a	8.63 a	22.3 b	69.1 a
Z_0	8.36 c	22.7 c	71.5 b	94.1 b	102.5 b	8.07 a	22.1 b	69.8 a
Z_5	9.39 b	27.0 b	78.7 a	105.6 a	115.0 a	8.06 a	23.5 a	68.5 a
Z_{10}	10.05 a	29.3 a	81.6 a	110.9 a	121.0 a	8.31 a	24.2 a	67.5 a
ANOVA								
I	*	*	ns	*	ns	**	*	ns
Z	**	**	**	**	**	ns	**	ns
I*Z	ns	ns	ns	ns	ns	ns	ns	ns

Notes: Means followed by the same letter within the same column are not significantly different at $p < 0.05$ by Tukey's HSD test. *, ** and ns represent significance at $p < 0.05, 0.01$ and not significant.

Compared with CF, AWD significantly increased the root N accumulation by 9.6%, reduced the stem-leaf and aboveground N accumulation by 11.8% and 5.5%, improved the proportion of N distribution in the root by 12.7% and reduced the proportion of N distribution in the stem-leaf by 7.9%, respectively.

The N accumulation in different parts of the plants, and the N distribution ratio in the stem-leaf, were significantly affected by Z amendment. Compared with no Z addition, the addition of 5 and 10 t·Z ha^{-1} improved N accumulation in the root, stem-leaf, panicle, aboveground parts and whole pant by 12.3% and 20.2%, 18.9% and 29.1%, 10.1% and 14.1%, 10.9% and 15.1% and 12.2% and 18.0%, respectively, but no significant difference between Z_5 and Z_{10} occurred for N accumulation in the panicle, aboveground parts and whole plant. The application of 5 and 10 t·Z ha^{-1} improved the propoertion of N distribution in the stem-leaf by 6.3% and 9.5%, respectively, but no significant difference between Z_5 and Z_{10} occurred.

3.4. Grain Yield and Water Use Efficiency

Rice grain yield and water use efficiency were significantly affected by Z amendment and I*Z interaction (Figure 3). The water use efficiency was significantly affected by the irrigation regime's main effect, but this did not occur in the rice grain yield. In the condition of 0 and 5 t·Z ha^{-1} application, the rice grain yield was significantly reduced under AWD, relative to CF, but there was no significant difference between CF and AWD when 10 t·Z ha^{-1} was applied. Under CF, the addition of 5 and 10 t·Z ha^{-1} improved the grain yield by 6.4% and 6.5% and, under AWD, the addition of 5 and 10 t·Z ha^{-1} improved the grain yield by 7.8% and 12.8%, relative to no Z addition. The highest grain yields were obtained in the $I_{CF}Z_5$, $I_{CF}Z_{10}$ and $I_{AWD}Z_{10}$ treatments, and no significant difference among them occurred.

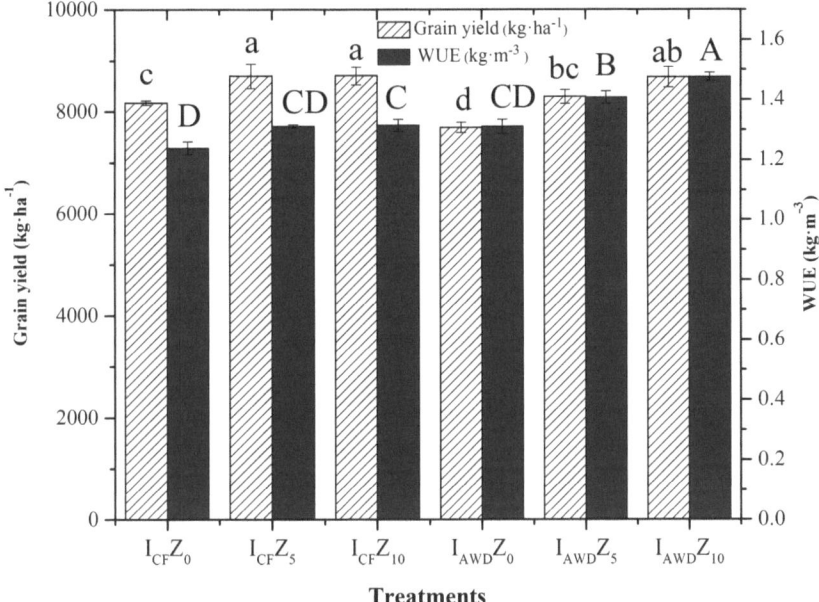

Figure 3. Rice grain yield and water use efficiency under different treatments. Different letters on the top of bars indicate significant difference at $p < 0.05$ between variables within the same group.

Compared with CF, AWD significantly improved the water use efficiency by 8.5%. Under CF, the addition of 5 and 10 t·Z ha^{-1} improved the water use efficiency by 5.6% and 6.5%. Under AWD, the addition of 5 and 10 t·Z ha^{-1} improved the water use efficiency by 7.6% and 12.2%, relative to no Z addition. The highest water use efficiency was obtained in the $I_{AWD}Z_{10}$ treatment; it improved the water use efficiency by 19.4% relative to $I_{CF}Z_0$ (the traditional cultivation mode).

The rice grain yield had a significant positive correlation with the aboveground dry matter accumulation, nitrogen accumulation, SPAD and root bleeding intensity at the joint-booting, panicle initiation and grain filling stages, especially with the aboveground dry matter accumulation, nitrogen accumulation, and SPAD at the joint-booting stage, but had a significant negative correlation with root–shoot ratio (Table 5). Nitrogen accumulation had a very significant correlation with dry matter accumulation; both of them were positively correlated with SPAD and root bleeding intensity, especially at the joint-booting stage. SPAD was positively correlated with root bleeding intensity.

Table 5. Correlation analysis among grain yield, root–shoot ratio, aboveground dry matter accumulation, aboveground nitrogen accumulation, SPAD and root bleeding intensity at joint-booting, panicle initiation and grain filling stages.

	Yield	RST	ADM	AN	JBS	PIS	GFS	JBR	PIR	GFR
Yied	1									
RST	−0.56 *	1								
ADM	0.88 **	−0.39	1							
AN	0.89 **	−0.46 *	0.99 **	1						
JBS	0.90 **	−0.44	0.86 **	0.86 **	1					
PIS	0.65 **	−0.4	0.68 **	0.71 **	0.77 **	1				
GFS	0.53 *	−0.83 **	0.52 *	0.55 *	0.53 *	0.53 *	1			
JBR	0.69 **	−0.35	0.81 **	0.82 **	0.63 **	0.53 *	0.48 *	1		
PIR	0.65 **	−0.80 **	0.57 *	0.61 **	0.57 *	0.53 *	0.88 **	0.59 **	1	
GFR	0.66 **	−0.76 **	0.59 **	0.63 **	0.63 **	0.67 **	0.85 **	0.53 *	0.95 **	1

Notes: RST, ADM, AN, JBS, PIS, GFS, JBR, PIR, GFR represent root–shoot ratio, aboveground dry matter accumulation, aboveground nitrogen accumulation, SPAD and root bleeding intensity at joint-booting, panicle initiation and grain filling stages, respectively. * and ** represent significance at $p < 0.05, 0.01$.

3.5. Grain Quality

The rice grain quality was not significantly affected by the interaction of the irrigation regime and Z amendment. The effects of the irrigation regime and Z amendment on the milling, appearance, nutritional and eating quality of rice grain are shown in Table 6. The irrigation regime had no significant effect on the brown rice rate, milled rice rate and chalkiness, but a significant difference in the head rice rate and chalky rice rate were noted between CF and AWD. AWD reduced the head rice rate by 4.7% and the chalky rice rate by 5.0%, relative to CF. Z application had no significant effect on the brown rice rate, milled rice rate and head rice rate, but a significant difference in the chalky rice rate and chalkiness were noted when Z was applied. Compared with no Z application, the addition of 10 t·Z ha^{-1} improved the chalky rice rate by 14.3% and the addition of 5 and 10 t·Z ha^{-1} increased the chalkiness by 15.0% and 35.2%, respectively.

Table 6. Effect of irrigation regimes and zeolite application on milling, appearance, nutritional and eating quality of rice grain.

Main Factor	Brown Rice Rate (%)	Milled Rice Rate (%)	Head Rice Rate (%)	Chalky Rice Rate (%)	Chalkiness Degree (%)	Protein Concentration (%)	Amylose Concentration (%)	Eating Score
ICF	84.3 a	78.3 a	73.1 a	6.41 a	1.43 a	6.62 a	25.2 a	69.4 b
IAWD	84.0 a	76.4 a	69.7 b	6.09 b	1.34 a	6.95 a	26.3 a	70.8 a
Z_0	84.2 a	76.8 a	70.5 a	5.82 b	1.20 c	6.52 b	25.8 a	70.4 a
Z_5	84.1 a	76.6 a	71.8 a	6.28 ab	1.38 b	6.77 ab	25.8 a	70.3 a
Z_{10}	84.2 a	78.8 a	71.8 a	6.65 a	1.59 a	7.00 a	25.6 a	69.7 a
ANOVA								
I	ns	ns	ns	**	**	*	ns	ns
Z	ns	ns	ns	**	**	*	ns	ns
I*Z	ns	ns	ns	ns	ns	ns	ns	ns

Notes: Means followed by the same letter within the same column are not significantly different at $p < 0.05$ by Tukey's HSD test. *, ** and ns represent significance at $p < 0.05, 0.01$ and not significant.

The irrigation regime had no significant effect on the protein concentration and amylose concentration, but a significant difference in the eating score was found between CF and AWD. AWD significantly improved the eating score relative to CF. Z application had no significant effect on the amylose concentration and eating score, but a significant difference in the protein concentration was noted when Z was applied. The addition of 10 t·ha^{-1} Z increased the protein content by 7.4%, relative to no Z addition.

The effect of the irrigation regime and Z amendment on the starch RVA profile characteristics of the rice grain are shown in Table 7. The irrigation regime and Z amendment both had a significant effect on the breakdown and setback, but such a significant effect did not occur in the peak viscosity, through viscosity, final viscosity, peak time and pasting temp. AWD increased the breakdown by 5.9% and decreased the setback by 6.0%, relative to CF. Compared with no Z amendment, 5 and 10 t·ha^{-1} Z application both reduced the breakdown by 4.3% and 10 t·ha^{-1} Z application decreased the setback by 4.3%.

Table 7. Effect of irrigation regimes and zeolite application on starch RVA profile characteristics of rice grain.

Main Factos	Peak Viscosity (cP)	Through Viscosity (cP)	Breakdown (cP)	Final Viscosity (cP)	Setback (cP)	Peak Time (min)	Pasting Temp (°C)
I_{CF}	3159 a	2390 a	748 b	3518 a	1173 a	6.63 a	71.2 a
I_{AWD}	3218 a	2454 a	792 a	3557 a	1103 b	6.63 a	71.2 a
Z_0	3208 a	2416 a	793 a	3537 a	1164 a	6.62 a	71.3 a
Z_5	3160 a	2407 a	759 b	3520 a	1136 ab	6.65 a	71.1 a
Z_{10}	3208 a	2444 a	759 b	3555 a	1114 b	6.62 a	71.2 a
ANOVA							
I	ns	ns	*	ns	*	ns	ns
Z	ns	ns	**	ns	*	ns	ns
I*Z	ns	ns	ns	ns	ns	ns	ns

Notes: Means followed by the same letter within the same column are not significantly different at $p < 0.05$ by Tukey's HSD test. *, ** and ns represent significance at $p < 0.05, 0.01$ and not significant.

4. Discussion

4.1. Effects of AWD on Rice Physiological Characteristics, Yield and Water Use Efficiency

Although the water-saving effect of AWD is significant, there is still no consistent conclusion regarding its effects on rice physiological characteristics and grain yield. Crop roots are the main organ for absorbing nutrients and water; the present study also found that the rice root bleeding intensity was positively correlated with the grain yield (Table 5). Moderate drought stress significantly improved the rice root vigor, and then improved the grain yield, while severe drought stress had opposing effects [25]. In our study, AWD significantly reduced the root bleeding intensity at the grain filling stage (Table 2); although the drought stress was not severe, the main reason for this result was that the sustained low temperature at the panicle-initiation stage prolonged the drought stress duration, which inhibited the growth of the rice roots at a later stage. The leaves played two important roles in the rice-growing stage; on the one hand, they were the "source" organs for grain filling and, on the other hand, they were essential for dry matter production and the subsequent grain yield [26]. In the present study, AWD significantly reduced the SPAD at the grain filling stage, relative to CF; the main reason for this lower SPAD was that nutrient supply was inhibited due to the lower root bleeding intensity under AWD. A correlation analysis also indicated that the leaf SPAD was positively correlated with the root bleeding intensity (Table 5). Therefore, the lower root bleeding intensity and leaf SPAD would inhibit rice growth and development, which resulted in a significantly lower grain yield under AWD than CF without zeolite application in the present study.

As shown in a previous study [27–29], the rice grain yield was positively correlated with the aboveground dry matter accumulation (Table 5). The previous study found that the irrigation regime had no significant effect on the aboveground dry matter accumulation and yield [30]. Another study found that the aboveground dry matter accumulation under AWD treatment was significantly lower than that of CF before the grain filling stage, but that AWD significantly improved the dry matter accumulation and the subsequent grain yield at maturity stage [31]. In our study, the difference in the dry matter accumulation at maturity stage was not significant between CF and AWD (Table 3), which is the main reason that the grain yield was not significantly affected by the irrigation regime's main effect. One study examined 528 side-by-side comparisons of AWD with CF by meta-analysis; they found that mild AWD significantly improved the water productivity by an average of 24.2%, relative to CF [32]. The present study showed that AWD improved the water productivity by 8.5% (Figure 3), which was below average; this is ascribed to the fact that the yield of the AWD treatment was not higher than that of the CF and that AWD did not reduce water leakage due to the bottomed lysimeter used in this experiment.

4.2. Effects of AWD on Rice Grain Quality

Irrigation regime had a significant effect on the rice grain quality. Controlled irrigation, which is similar to the AWD used in our study, significantly improved the head rice rate [33]. Other studies also demonstrated the fact that water-saving irrigation can improve the head rice rate [34,35]. However, mild drought stress had no significant effect on the head rice rate, and severe drought stress significantly reduced the head rice rate [36]. In the present study, AWD significantly reduced the head rice rate of rice (Table 6). The main reason for this was that the grain filling was affected by prolonged drought stress in the AWD paddy due to the cold and humid climate during the panicle-initiation and grain-filling stages (Figure 1). In addition, our study showed that AWD significantly reduced the chalky rice rate and chalkiness degree, which concurred with previous published studies [33,34]; however, contrary to the research results of another study [37], these contradictory findings were mostly related to the different thresholds of water stress and rice varieties, etc.

Generally, rice with a high breakdown and a low setback is soft and sticky, and thus has a high taste quality [38]. The previous study indicated that the rice grain quality was improved when the soil-water potential was −15 kPa during the grain-filling stage [39]. In our study, the soil water potential of the AWD was −10 kPa in the present experiment, which significantly improved the eating score of the rice due to the higher breakdown and lower setback under AWD than CF. In conclusion, the effects of water-saving irrigation on rice quality are still unclear. Therefore, studies on the effects of water-saving irrigation on rice quality should be carried out simultaneously around multi-point experiments, multi-variety experiments and multiple water stress regimes.

4.3. Effects of Zeolite on Rice Physiological Characteristics, Yield and Water Use Efficiency

The application of zeolite to paddies could improve rice yield to varying degrees [20,40–42]. The present study also demonstrated that the addition of zeolite significantly improved the rice grain yield under both CF and AWD. Furthermore, the large specific surface area and high hydration–dehydration properties of zeolite enhanced the water-holding capacity for soil, especially under drought stress conditions, which was beneficial in offsetting the negative effects of drought stress on rice growth and subsequent yield [43]. Our study showed that the grain yield increased significantly with the increase in Z addition rate under AWD; similar conclusions were found in previously published studies [23,44,45]. In addition, the application of Z enhanced root bleeding intensity, mainly because of the enhanced the availability of nutrients in the soil by Z amendment, which, in turn, increased the leaf SPAD in our study (Table 2), delayed leaf senescence and promoted nitrogen accumulation in plants. The higher root vigor at the grain-filling stage is closely related to the synthesis of phytohormones, especially cytokinins, which help to regulate photosynthesis and then promote dry matter accumulation [46]. Rice physiological characteristics, dry matter and nitrogen accumulation are closely related to rice yield (Table 5). Therefore, the mechanism of increasing the yield via Z application could be explained by the fact that Z promoted the dry matter and nitrogen accumulation by improving the physiological function of rice. Additionally, the increase in nitrogen accumulation by zeolite might be due to its large specific surface area and high nitrogen exchange ability.

Zeolite application could significantly improve crop water use efficiency [20]. In the present study, the water use efficiency improved significantly with the increase in Z addition rate; this could be attributed to the fact that Z application significantly increased the rice yield without increasing the water consumption. Additionally, the water saving effect of zeolite was more significant under AWD. Therefore, the application of 10 t·ha^{-1} Z could offset the adverse effects of yield reduction, further improve the water use efficiency and achieve the win–win goal of increasing yield and saving water in the AWD paddy field, which is of great significance for the promotion of the AWD cultivation regime.

4.4. Effects of Zeolite on Rice Grain Quality

Rice grain quality is closely related to cultivation measures. For example, increasing nitrogen fertilizer application could reduce the chalky rice rate and chalkiness degree, and thus improve the appearance quality of rice [10], which is consistent with another study [47]. However, some studies obtained the opposite results and found that an increasing rate of nitrogen fertilizer addition significantly increased the chalky rice rate and chalkiness degree [35,48]. The above studies showed that nitrogen fertilizer management is one of the main factors affecting rice quality. Therefore, zeolite could affect rice grain quality by improving the nitrogen and water availability in soil. A study found that Z had no significant effect on the appearance quality of rice [49], but another study showed that Z application reduced the chalky rice rate and chalkiness [50]. In the present study, Z application significantly increased the chalky rice rate and chalkiness degree (Table 6). The formation of chalkiness is closely related to the level of "source-sink" during the grain filling stage, so the main reason for increasing chalkiness with Z amendment might be that an increased "sink" and insufficient "source" of rice resulted in more chalkiness during the grain-filling stage due to the higher nitrogen proportion in the second topdressing. In line with our finding, a high addition rate of nitrogen fertilizer during the panicle-initiation stage increased the chalkiness degree of rice [35]. Our study found that zeolite application improved the protein concentration in rice grain; this finding is mostly related to promoting the absorption and utilization of nitrogen in plants due to the fertilizer retention properties by zeolite amendment. In the present study, although zeolite application improved the protein concentration, it did not affect the amylose concentration, nor did it reduce the eating score. Zeolite application reduced both the breakdown and setback (Table 7), which may be one of the reasons why it did not reduce eating score. To sum up, the effect of zeolite on rice grain quality should be further explored, in combination with different nitrogen dosages and application methods. These results indicated that the integration of AWD with 10 t·Z ha^{-1} addition could be an effective approach for enhancing rice grain yield and quality with high water use efficiency.

5. Conclusions

AWD significantly reduced the root bleeding intensity and leaf SPAD, increased the root–shoot ratio, reduced the aboveground N accumulation and resulted in a lower rice grain yield without zeolite application, relative to CF. Zeolite amendment significantly improved the rice grain yield by improving the root bleeding intensity, leaf SPAD, dry matter and subsequent N accumulation. Both AWD and zeolite amendment significantly improved the water use efficiency. Compared with CF, although AWD reduced the head rice rate, it improved the appearance quality by reducing the chalkiness, and improved eating score by increasing the breakdown and reducing the setback. Zeolite amendment improved the chalkiness and protein concentration. Zeolite amendment reduced both the breakdown and setback, resulting in no significant effect on the rice amylose concentration and eating score. In general, the integration of 10 t·Z ha^{-1} with AWD improved the rice grain yield and grain quality with an increased water use efficiency. Furthermore, the impact of modified zeolite on crop yield and quality deserves attention in future research.

Author Contributions: Y.S., H.H. conceived and designed the research plans. Y.S. and J.X. wrote the manuscript. X.W., M.L. and Y.W. designed the figures. All authors have read and agreed to the published version of the manuscript.

Funding: The study was jointly funded by the Natural Science Foundation of Jiangsu province, China (No. BK20220594 and No. BK20210824).

Data Availability Statement: The data used to support the findings of this study are available from the corresponding author upon request.

Conflicts of Interest: The authors declare that no conflict of interest exists in the submission of this manuscript, and that the manuscript is approved by all authors for submission.

References

1. FAO. 2022. Available online: http://faostat3.fao.org/browse/FB/CC/E (accessed on 19 December 2022).
2. Du, T.S.; Kang, S.Z.; Sun, J.S.; Zhang, X.Y.; Zhang, J.H. An improved water use efficiency of cereals under temporal and spatial deficit irrigation in north China. *Agric. Water Manag.* **2010**, *97*, 66–74. [CrossRef]
3. Lampayan, R.M.; Rejesus, R.M.; Singleton, G.R.; Bouman, B.A.M. Adoption and economics of alternate wetting and drying water management for irrigated lowland rice. *Field Crops Res.* **2015**, *170*, 95–108. [CrossRef]
4. Nie, T.Z.; Huang, J.Y.; Zhang, Z.X.; Chen, P.; Li, T.C.; Dai, C.L. The inhibitory effect of a water-saving irrigation regime on CH_4 emission in Mollisols under straw incorporation for 5 consecutive years. *Agric. Water Manag.* **2023**, *278*, 108163. [CrossRef]
5. Jayakumar, B.; Subathra, C.; Velu, V.; Ramanathan, S. Effect of integrated crop management practices on rice (*Oryza sativa* L.) volume and rhizosphere redox potential. *J. Agron.* **2005**, *40*, 311–314. [CrossRef]
6. Zhao, X.; Zhou, Y.; Wang, S.; Xing, G.; Shi, W.; Xu, R.; Zhu, Z. Nitrogen balance in a highly fertilized rice–wheat double-cropping system in southern China. *Soil Sci. Soc. Am. J.* **2012**, *76*, 1068–1078. [CrossRef]
7. Chapagain, T.; Yamaji, E. The effects of irrigation method, age of seedling and spacing on crop performance, productivity and water-wise rice production in Japan. *Paddy Water Environ.* **2010**, *8*, 81–90. [CrossRef]
8. Islam, S.M.M.; Gaihre, Y.K.; Biswas, J.C.; Jahan, M.S.; Singh, U.; Adhikary, S.K. Different nitrogen rates and methods of application for dry season rice cultivation with alternate wetting and drying irrigation: Fate of nitrogen and grain yield. *Agric. Water Manag.* **2018**, *196*, 144–153. [CrossRef]
9. Li, Z.; Letuma, P.; Zhao, H.; Zhang, Z.; Lin, W.; Chen, H.; Lin, W. A positive response of rice rhizosphere to alternate moderate wetting and drying irrigation at grain filling stage. *Agric. Water Manag.* **2018**, *207*, 26–36. [CrossRef]
10. Chen, S.Q.; Xue, J.F.; Pan, G.J.; Wang, Q.Y. Relationship between RVA profile characteristics and other quality traits in grain positions of japonica rice. *J. Nucl. Agric. Sci.* **2015**, *29*, 244–251.
11. Zhang, J.S.; Cheng, S.Z.; Liu, D.H.; You, C.; Zhou, G.S.; Jin, D.M. Effects of drought stress in mid- late stage on grain yield and quality of cultivated rice (*Oryza sativa* L). *Hubei Agric. Sci.* **2007**, *46*, 3.
12. Huang, D.F.; Xi, L.L.; Wang, Z.Q.; Liu, L.J.; Yang, J.C. Effects of Irrigation Patterns during Grain Filling on Grain Quality and Concentration and Distribution of Cadmium in Different Organs of Rice. *Acta Agron. Sin.* **2008**, *34*, 456–464. [CrossRef]
13. Lv, Y.D.; Zheng, G.P.; Guo, X.H.; Yin, D.W.; Ma, D.R.; Xu, Z.J.; Chen, W.F. Effects of lower limit of soil water potential on grain quality of rice in cold region. *Chin. J. Rice Sci.* **2011**, *25*, 515–522.
14. Wang, Y.; Cui, J.; Wang, X.B.; Zhao, M.; Zhu, C.J.; Shi, L.L.; Zhang, X. Effect of fertilization method on soil available nutrients and taste of Japanese and Chinese rice. *Chin. J. Eco-Agric.* **2010**, *18*, 286–289. [CrossRef]
15. Chen, M.Y.; Li, X.F.; Cheng, J.Q.; Ren, H.R.; Liang, J.; Zhang, H.C.; Huo, Z.Y. Effects of total straw returning and nitrogen application regime on grain yield and quality in mechanical transplanting *Japonica* rice with good taste quality. *Acta Agron. Sin.* **2017**, *43*, 1802–1816. [CrossRef]
16. Noori, M.; Zendehdel, M.; Ahmadi, A. Using natural zeolite for improvement of soil salinity and crop yield. *Toxicol. Environ. Chem. Rev.* **2006**, *88*, 77–84. [CrossRef]
17. Khan, A.Z.; Nigar, S.; Khalil, S.K.; Wahab, S.; Rab, A.; Khattak, M.K.; Henmi, T. Influence of synthetic zeolite application on seed development profile of soybean grown on allophanic soil. *Pak. J. Bot.* **2013**, *45*, 1063–1068.
18. Malekian, R.; Abedi-Koupai, J.; Eslamian, S.S. Influences of clinoptilolite and surfactant-modified clinoptilolite zeolite on nitrate leaching and plant growth. *J. Hazard. Mater.* **2011**, *185*, 970–976. [CrossRef]
19. Khodaei Joghan, A.; Ghalavand, A.; Aghaalikhani, M.; Gholamhoseini, M.; Dolatabadian, A. How organic and chemical nitrogen fertilizers, zeolite, and combinations influence wheat yield and grain mineral content. *J. Crop Improv.* **2012**, *26*, 116–129. [CrossRef]
20. Chen, T.T.; Wilson, L.T.; Liang, Q.; Xia, G.M.; Chen, W.; Chi, D.C. Influences of irrigation, nitrogen and zeolite management on the physicochemical properties of rice. *Arch. Agron. Soil Sci.* **2017**, *63*, 1210–1226. [CrossRef]
21. Hazrati, S.; Tahmasebi-Sarvestani, Z.; Mokhtassi-Bidgoli, A.; Modarres-Sanavy, S.A.M.; Mohammadi, H.; Nicola, S. Effects of zeolite and water stress on growth, yield and chemical compositions of *Aloe vera* L. *Agric. Water Manag.* **2017**, *181*, 66–72. [CrossRef]
22. Najafinezhad, H.; Tahmasebi Sarvestani, Z.; Modarres Sanavy, S.A.M.; Naghavi, H. Evaluation of yield and some physiological changes in corn and sorghum under irrigation regimes and application of barley residue, zeolite and superabsorbent polymer. *Arch. Agron. Soil Sci.* **2015**, *61*, 891–906. [CrossRef]
23. Ozbahce, A.; Tari, A.F.; Gönülal, E.; Simsekli, N.; Padem, H. The effect of zeolite applications on yield components and nutrient uptake of common bean under water stress. *Arch. Agron. Soil Sci.* **2015**, *61*, 615–626. [CrossRef]
24. Sun, Y.D.; He, Z.L.; Wu, Q.; Zheng, J.L.; Li, Y.H.; Wang, Y.Z.; Chen, T.T.; Chi, D.C. Zeolite amendment enhances rice production, nitrogen accumulation and translocation in wetting and drying irrigation paddy field. *Agric. Water Manag.* **2020**, *235*, 106–126. [CrossRef]
25. Xu, G.W.; Wang, H.Z.; Zhai, Z.H.; Sun, M.; Li, Y.J. Effect of water and nitrogen coupling on root morphology and physiology, yield and nutrition utilization for rice. *Trans. Chin. Soc. Agric. Eng.* **2015**, *31*, 132–141.

26. Wang, L.; Xue, C.; Pan, X.; Chen, F.; Liu, Y. Application of controlled-release urea enhances grain yield and nitrogen use efficiency in irrigated rice in the Yangtze River Basin, China. *Front. Plant Sci.* **2018**, *9*, 999. [CrossRef] [PubMed]
27. Mahajan, G.; Chauhan, B.S. Performance of dry directseeded rice in response to genotype and seeding rate. *Agron. J.* **2016**, *108*, 257. [CrossRef]
28. Pal, R.; Mahajan, G.; Sardana, V.; Chauhan, B.S. Impact of sowing date on yield, dry matter and nitrogen accumulation, and nitrogen translocation in dry-seeded rice in North-West India. *Field Crops Res.* **2017**, *206*, 138–148. [CrossRef]
29. Wu, W.; Nie, L.X.; Liao, Y.C.; Shah, F.; Cui, K.H.; Wang, Q.; Lian, Y.; Huang, J.L. Toward yield improvement of early-season rice, Other options under double rice-cropping system in central China. *Eur. J. Agron.* **2013**, *45*, 75–86. [CrossRef]
30. Dong, N.M.; Brandt, K.K.; Sorensen, J.; Hung, N.N.; Chu, V.H.; Tan, P.S.; Dalsgaard, T. Effects of alternating wetting and drying versus continuous flooding on fertilizer nitrogen fate in rice fields in the mekong delta, vietnam. *Soil Biol. Biochem.* **2012**, *47*, 166–174. [CrossRef]
31. Ye, Y.S.; Liang, X.Q.; Chen, Y.X.; Liu, J.; Gu, J.T.; Guo, R.; Li, L. Alternate wetting and drying irrigation and controlled-release nitrogen fertilizer in late-season rice. Effects on dry matter accumulation, yield, water and nitrogen use. *Field Crops Res.* **2013**, *144*, 212–224. [CrossRef]
32. Carrijo, D.R.; Lundy, M.E.; Linquist, B.A. Rice yields and water use under alternate wetting and drying irrigation: A meta-analysis. *Field Crops Res.* **2017**, *203*, 173–180. [CrossRef]
33. Peng, S.Z.; Hao, S.R.; Liu, Q.; Liu, Y.; Xu, N.H. Study on the mechanisms of yield-raising and quality-improving for paddy rice under water-saving irrigation. *J. Irrig. Drain.* **2000**, *3*, 3–7.
34. Ke, C.Y. Effect of Different Water Treatment on Rice Growth, Yield and Quality. Master's Thesis, Huazhong Agricultural University, Wuhan, China, 2010.
35. Liu, L.J.; Li, H.W.; Zhao, B.H.; Wang, Z.Q.; Yang, J.C. Effects of alternate drying-wetting irrigation during grain filling on grain quality and its physiological mechanisms in rice. *Chin. J. Rice Sci.* **2012**, *26*, 77–84.
36. Hu, Y.F. Study on the effect of water-saving irrigation on the yield and quality of rice in cold regions. *J. Agric. Sci.* **2017**, *14*, 38–39.
37. Wang, C.A.; Wang, B.L.; Zhang, W.X.; Zh, L.; Zhao, X.Z.; Gao, L.W. Effects of water stress of soil on rice yield and quality. *Acta Agron. Sin.* **2006**, *32*, 131–137.
38. Champagne, E.T.; Bett-Garber, K.L.; Grimm, C.C.; McClung, A.M. Effects of organic fertility management on physicochemical properties and sensory quality of diverse rice cultivars. *Cereal Chem.* **2007**, *84*, 320–327. [CrossRef]
39. Cai, Y.X.; Zhu, Q.S.; Wang, Z.Q.; Yang, J.C.; Zhen, L.; Qian, W.C. Effects of soil moisture on rice quality during grain-filling period. *Acta Agron. Sin.* **2002**, *17*, 1201–1206.
40. Gholamhoseini, M.; Ghalavand, A.; Khodaei-Joghan, A.; Dolatabadian, A.; Zakikhani, H.; Farmanbar, E. Zeolite-amended cattle manure effects on sunflower yield, seed quality, water use efficiency and nutrient leaching. *Soil Tillage Res.* **2013**, *126*, 193–202. [CrossRef]
41. Sepaskhah, A.R.; Barzegar, M. Yield, water and nitrogen-use response of rice to zeolite and nitrogen fertilization in a semi-arid environment. *Agric. Water Manag.* **2010**, *98*, 38–44. [CrossRef]
42. Wu, Q.; Xia, G.M.; Chen, T.T.; Chi, D.C.; Jin, Y.; Sun, D.H. Impacts of nitrogen and zeolite managements on yield and physico-chemical properties of rice grain. *Int. J. Agric. Biol. Eng.* **2016**, *9*, 93–100.
43. Ippolito, J.A.; Tarkalson, D.D.; Lehrsch, G.A. Zeolite soil application method affects inorganic nitrogen, moisture, and corn growth. *Soil Sci.* **2011**, *176*, 136–142. [CrossRef]
44. Ghanbari, M.; Ariafar, S. The effects of water deficit and zeolite application on growth traits and oil yield of medicinal peppermint (*Mentha piperita* L.). *Int. J. Med. Aromat. Plants* **2013**, *3*, 32–39.
45. Ghanbari, M.; Ariafar, S. The study of different levels of zeolite application on quantitative and qualitative parameters in basil (*Ocimum basilicum* L.) under drought conditions. *Int. J. Agric. Res. Rev.* **2013**, *3*, 844–853.
46. Guo, E.J.; Yang, X.G.; Wang, X.Y.; Zhang, T.Y.; Huang, W.H.; Liu, Z.Q. Spatial-temporal distribution of double cropping rice's yield gap in Hunan province. *Sci. Agric. Sin.* **2017**, *50*, 399–412.
47. Cong, X.H.; Shi, F.Z.; Ruan, X.M.; Luo, Y.X.; Ma, T.C.; Luo, Z.X. Effects of nitrogen fertilizer application rate on nitrogen use efficiency and grain yield and quality of different rice varietie. *Chin. J. Appl. Ecol.* **2017**, *28*, 1219–1226.
48. Jiang, H.F.; Guo, X.H.; Hu, Y.; Li, M.; Lv, Y.D.; Xu, S.L.; Xu, L.Q.; Wang, J.Y. Effects of nitrogen fertilization managements on rice quality under Soda-Saline-Alkali soil. *Southwest China J. Agric. Sci.* **2019**, *32*, 1223–1229.
49. Chen, T.T.; Xia, G.M.; Wu, Q.; Zheng, J.L.; Jin, Y.; Sun, D.H.; Wang, S.C.; Chi, D.C. The Influence of Zeolite Amendment on Yield Performance, Quality Characteristics, and Nitrogen Use Efficiency of Paddy Rice. *Crop Sci.* **2018**, *57*, 2777. [CrossRef]
50. Zheng, J.L.; Chen, T.T.; Wu, Q.; Yu, J.M.; Chen, W.; Chen, Y.L.; Siddique, K.H.M.; Meng, W.A.; Chi, D.C.; Xia, G.M. Effect of zeolite application on phenology, grain yield and grain quality in rice under water stress. *Agric. Water Manag.* **2018**, *206*, 241–251. [CrossRef]

Disclaimer/Publisher's Note: The statements, opinions and data contained in all publications are solely those of the individual author(s) and contributor(s) and not of MDPI and/or the editor(s). MDPI and/or the editor(s) disclaim responsibility for any injury to people or property resulting from any ideas, methods, instructions or products referred to in the content.

Article

Identifying the Influencing Factors of Plastic Film Mulching on Improving the Yield and Water Use Efficiency of Potato in the Northwest China

Juzhen Xu [1,2], Yanbo Wang [2], Yuanquan Chen [2], Wenqing He [1], Xiaojie Li [1,*] and Jixiao Cui [1,*]

[1] Institute of Environment and Sustainable Development in Agriculture, Chinese Academy of Agricultural Sciences, Beijing 100081, China; xjz0068@cau.edu.cn

[2] College of Agronomy and Biotechnology, China Agricultural University, Beijing 100193, China; s20223010027@cau.edu.cn (Y.W.); chenyq@cau.edu.cn (Y.C.); hewenqing@caas.cn (W.H.)

* Correspondence: lixiaojie@igsnrr.ac.cn (X.L.); cuijixiao@caas.cn (J.C.)

Abstract: Potato is an important crop in the Northwest China, however, its production is constrained by water scarcity. Plastic mulching film is an efficient technical measure to alleviate potato production restrictions. Therefore, studying the response of potato yield and water use efficiency to plastic mulching film is of great significance. The study conducted a meta-analysis to quantify the effect of plastic film on potato yield and water use efficiency in the Northwest. The study then quantified the effects of different levels of natural conditions (mean annual precipitation, mean annual accumulated temperature ≥ 10 °C), fertilizer application (nitrogen fertilizer, phosphate fertilizer, potassium fertilizer), cultivation measures (planting density, cultivation method, mulching method), and mulching properties (mulching color, mulching thickness) through subgroups analysis. Finally, the random forest model was used to quantify the importance of factors. Plastic film mulching increased yield by 27.17% and water use efficiency by 27.16%, which had a better performance under relatively lower mean annual precipitation, low mean annual accumulated temperature ≥ 10 °C, relatively lower fertilizer application, planting density of 15,000–45,000 plants·ha^{-1}, ridge, and black mulching. Natural conditions, fertilization measures were vital to improve productivity. The research results can provide reference for agricultural management strategies of potato cultivation using plastic film in the Northwest China and other potato-producing areas.

Keywords: meta-analysis; random forest; importance; natural conditions; fertilization

1. Introduction

With the increase of global population, the demand for food demand is gradually increasing, and it is that the total demand for global food will increase by 56% by 2050 [1]. Potato (*Solanum tuberosum* L.) is an annual herb of Solanaceae, which is the world's fourth largest staple crop after rice, maize, and wheat. Nowadays, China has developed into the largest potato planting country in the world [2], with a yield of 94.36 million tons and a planting area of 5.78 million ha [3]. Potato production is concentrated in Northwest China [4]. In 2020, the potato planting area in the Northwest was 1.35 million ha, with a yield of 5.19 million tons, accounting for 28.97% and 28.88% of the national total, respectively [5]. However, the Northwest China is located in the inland and has a dry climate [6]. In the Northwest, evaporation exceeds precipitation, and precipitation is irregular, posing a challenge to water resources [7]. Potato is a shallow rooting crop, which is sensitive to water, while the water deficit seriously restricts the potato in the Northwest China [8,9]. Therefore, improving water use efficiency (WUE) is crucial for potato cultivation in the Northwest China.

Plastic film mulching (PM) is an important agricultural technology extensively used in China to prevent water evaporation, improve WUE, and crop yield. At present, China is the

world's largest consumer of PM, with an annual amount of 1.36 million tons and covering an area of 17.39 million ha [10]. It is reported that PM increased the average of 51 crop yields and WUE by 45.5% and 58.0%, respectively [11]. PM increased yield and WUE by 24.3% and 28.7%, respectively, for potato cultivated in the Northwest [6]. Studies have shown that climate, soil properties, and field management measures have a significant impact on the effectiveness of PM [12]. Currently, numerous publications have used meta-analysis to comprehensively quantify the effect of PM on potato yield and WUE, in addition to the other factors. Ma et al. [13] found that PM was beneficial for the yield, WUE, and economic benefits of potato in the rainfed areas of the Northwest China. Gao et al. [14] compared the effects of PM on yield and WUE in different crops and regions, and the results showed that PM had the most significant effect on improving yield and WUE of potato, compared to other crops, and the effect of PM was the best in the Northwest among several regions. Gu et al. [15] observed that, compared with PM, degradable film insignificantly improved the yield and WUE of potato. Zhang et al. [16] noticed that ridge-furrow with PM had a better effect than only using PM; in addition, the yield and WUE increase of potato were the highest under the combination of ridge-furrow and PM. He et al. [17] compared the effects of white film and black film, and their study suggested that potato planting obtained higher economic benefits from black film, mostly because black mulching had the stronger effect of controlling weeds, which was beneficial in reducing the cost of herbicides. Overall, PM is of great significance for potato production. According to the major planting areas of potato, Li et al. [6] conducted a meta-analysis that divided China potato cultivation areas into seven major regions, and the results showed that PM performed the best on yield in the Northwest, which was consistent with Li et al. [18], and only increased WUE in the Northwest. Mo et al. [19] and Wang et al. [20] further divided the Northwest into irrigation areas and arid areas, and found that yield increase and WUE increase were both the highest in arid areas. Furthermore, the researchers introduced meta method to quantify the effects of different factors and levels of factors on PM, and the results found that climate, soil, and agricultural practices had significant impact on PM. However, although many publications studied on the response of potato yield and WUE to PM, few studies quantified the importance of influencing factors, which are an important guidance for the rational use of PM.

Meta-analysis is a method for combining the results of multiple independent studies on the same research topic for scientific statistical analysis, which can synthesize independent trial results for quantitative assessment at regional and global scales [21]. Meta-analysis is widely used in the medical field and was introduced into the ecological field at the end of the last century [22]. Researchers have also used this method to quantify the effect of PM on yield [14], WUE [23], greenhouse gas emissions [24], soil nutrients [13], root traits [25], and many other factors. In this context, random forest is a precise and powerful machine learning method for classification, which is able to quantify the importance of variables [26,27]. For example, Tseng et al. [28] adopted the method to study the factors influencing the yield gap in high-yield rice fields in Uruguay, which found that the sowing date and nitrogen application rate were the most main factors. Smidt et al. [29] reported that the elevation was the prominent factor for soybean yield. Philibert et al. [30] noticed nitrogen fertilizer and crops were the key factors in N_2O emission. Based on the advantages of both meta and random forest, the two methods can be combined. Currently, many studies introduced this method to deepen the analysis. Dang et al. [31] analyzed the importance of influence factors for soil inorganic carbon content. Liu et al. [32] determined that climatic factors were important drivers for the soil organic carbon content. Therefore, it is advantageous to adopt the combination of these two methods.

At present, there are numerous publications including field experiments and comprehensive studies exploring the impact of PM on potato production. Potato is important for food security in the Northwest China, however, there is no comprehensive analysis of potato cultivation with PM practice in the Northwest China, and the importance of influencing factors. Therefore, in this study, we utilized published literature to (a) quantify the impact of PM on potato yield and WUE in the Northwest China; (b) analyze the response of film effect to fertilization, cultivation measures, and film properties; and (c) adopt random forest model to quantify the relative importance of factors to the role of PM on potato yield and WUE. The results of this study will provide better guidance for the use of plastic film technology in potato cultivation in the Northwest China.

2. Materials and Methods

2.1. Data Collection

This study used meta-analysis to study the effect of PM on yield and WUE of potato. Using the theme of "plastic film Mulching/film Mulching & Potato", a literature search was conducted to collect field experimental research papers published domestically and internationally from 1981 to 2023 on the impact of PM on potato in China. Chinese literature mainly came from China National Knowledge Infrastructure (CNKI), while English literature mainly came from Web of Science (WOS). Excluding conference, review, macro, and model simulation literature, the first screening of the retrieved literature resulted in 189 publications.

In order to further eliminate literature that did not meet the standards and obtain more accurate data, the data screening criteria included the following: (1) The experimental site must be an outdoor field in China, and the location or longitude and latitude of the experimental site must be specified in detail in the article; (2) During the experiment, both covering and no covering treatments must be included, and other management measures, sampling and measurement time and methods must be consistent except for whether to cover the film; (3) The number of duplicates processed were known. For data presented in the form of images, we used the Get Data Graph Digitizer software to read it.

In order to explore the factors that affected the yield and WUE of potato under PM, the data were divided into different subgroups based on natural conditions, fertilizer applications, cultivation measures, and mulching properties. According to the study, the growth of crops heavily depends on climate; in the study, annual average precipitation (AP) (range of <200, 200–400, 400–600, and >600 mm) and ≥10 °C annual average accumulated temperature (AT) (range of 1500–3000, 3000–4500, and >4500 °C) are used as a part of subgroups, as shown in Table 1. Fertilizer is an indispensable agricultural resource for ensuring yield and quality in crop production. Nitrogen, phosphorus, and potassium fertilizers are important inorganic fertilizers, and the fertilizer gradients were divided into: 0–100, 100–200, and >200 kg·ha^{-1}. Planting density is another important influencing factor [20], so in this study it included 15,000–45,000, 45,000–75,000, and >75,000 plants·ha^{-1}. Agriculture in semi-arid areas heavily relies on precipitation, hence ridge and furrow cultivation is very beneficial for rainwater collection [33]. Cultivation method included ridge and flat. The mulching method of the PM included two common ones: full and partial (except full). The color and thickness of PM are important properties of PM, and important factors for the effect of PM, so white and black were chosen for color, and 0.008 and 0.01 mm were chosen for thickness.

Table 1. Gradient division of different subgroups.

Subgroup Classification	Classification Factors	Subgroups			
Natural condition	Mean annual precipitation (AP): mm	<200	200–400	400–600	>600
	≥10 °C mean annual accumulated temperature (AT): °C	1500–3000	3000–4500	>4500	
Fertilizer application	nitrogen fertilizer: kg·ha^{-1} phosphate fertilizer: kg·ha^{-1} potash fertilizer: kg·ha^{-1}	0–100	100–200	>200	
Cultivation measures	Planting density: plant ha^{-1}	15,000–45,000	45,000–75,000	>75,000	
	cultivation method	ridge	flat		
	mulching method	full	partial		
Mulching properties	mulching color	white	black		
	mulching thickness: mm	≤0.008	≥0.01		

2.2. Data Analysis

There is a hierarchical dependency between multiple observations in the study, and obtaining several effect quantities from the same publication violates the assumption that the effect quantities are independent [34]. Hierarchical meta-analysis model can be used to control the independence of data [35]. This study used "metafactor (rma.mv)" to control for the independence of different cases from a unified literature. The random effect model is used to analyze the effect of potato film mulching, and the natural logarithm of response ratio (R) is selected as the effect size [21]:

$$R = X_e / X_c \tag{1}$$

$$Ln(X_e/X_c) = LnX_e - LnX_c \tag{2}$$

where, X_e represents the average yield or WUE under PM; and X_c represents the average yield or WUE under no PM. This study uses a 95% confidence interval to determine the significance of differences in yield effects. If the confidence interval includes a 0 value, the difference is not significant; if the confidence intervals are all to the right of 0, it is a significant positive effect; if the confidence intervals are all to the left of the 0 value, it is a significant negative effect. Due to the fact that many publications did not report standard deviations and include more studies, $weight = \frac{ne \times nc}{ne + nc}$ was chosen for this study, where ne and nc represent the numbers of replicates of the treatment and control experiments, respectively [36].

In the study, random forest was adopted to analyze the relative importance of AP, AT, nitrogen fertilizer, phosphate fertilizer, potash fertilizer, planting density, cultivation method, mulching method, mulching color and mulching thickness for potato yield and WUE increase under PM. Metafor package and metaforest package in R language were used for meta and random forest analysis [37], and GraphPad Prism was used for visual mapping.

3. Results

3.1. Overview of Meta Dataset and the Effects of PM on Potato Yield and WUE

The dataset for the study came from 189 references (168 in Chinese and 21 in English), with a total of 1028 pairs of data, including 748 pairs for yield and 280 pairs for WUE. The effect size of yield and WUE followed the Gaussian normal distribution (Figure 1a,b); the R square of yield was 0.93 ($p < 0.001$), and the R square of WUE was 0.71 ($p < 0.01$). Compared with no mulching, PM significantly improved potato yield and WUE in the Northwest (Figure 1c). PM improved potato yield by 27.17% and WUE by 27.16% on average, respectively.

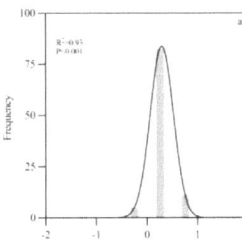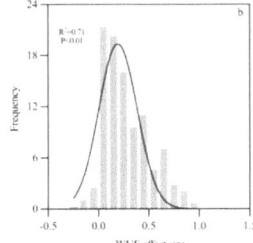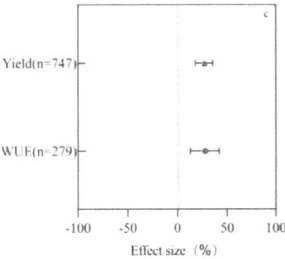

Figure 1. Frequency distribution of potato yield and water use efficiency; The curve represents a Gaussian fitting curve (**a**,**b**); Overall effect size in potato yield and WUE comparing plastic film mulching (PM) to no mulching in Northwest China (**c**). Error bars are the 95% confidence intervals (CIs). Number of samples indicated in parentheses.

3.2. Response of Potato Yield and WUE under PM to Natural Conditions

Natural conditions were the factors that affect the effectiveness of PM. As shown in Figure 2a,b, PM significantly increased yield when the AP was between 200–400 mm, 400–600 mm, but reached the highest at 200–400 mm, with yield of 30.41% and 24.8%, respectively. PM only significantly increased WUE between 200 mm and 400 mm, reaching 34.52%. PM significantly increases yield at 1500–3000 °C, 3000–4500 °C AT, but the best effect is at 1500–3000 °C, with 28.57% and 22.13%, respectively. For WUE, only the accumulated temperature between 1500 and 3000 °C significantly affected WUE, accounting for 32.53%. In general, PM performed greatly under the 200–400 mm AP and 1500–3000 °C AT.

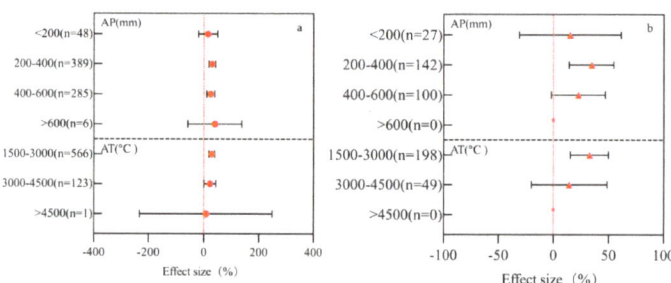

Figure 2. Effect of PM on potato yield (**a**) and WUE (**b**) comparing with no mulching in Northwest China at different AP (annual average precipitation) and AT ($\geq$10 °C annual average accumulated temperature). Error bars are the 95% confidence intervals (CIs). Number of samples are indicated in parentheses.

3.3. Response of Potato Yield and WUE under PM to Fertilizer Application

Different fertilizer applications had different effects on improving potato yield and WUE under PM (Figure 3). Under nitrogen fertilizer, yield performed well under all three gradients of nitrogen application. As the application rate increased, the effect of yield increase became more effective. PM increased yield by 32.71% at >200 kg·ha^{-1} nitrogenous application. For phosphate and potassium, PM increased yield at 0–100, 100–200 kg·ha^{-1} by 30.92% and 29.94%, respectively. For WUE, the effect of different fertilizers and their application rates on PM varied. PM only increased WUE by 32.43% at 100–200 kg·ha^{-1} nitrogenous application. For phosphate, PM was most effective at 0–100, 100–200 kg·ha^{-1} and was the highest at 100–200 kg·ha^{-1} (30.88%). For potassium, PM only improved WUE at 100–200 kg·ha^{-1} by 24.54%.

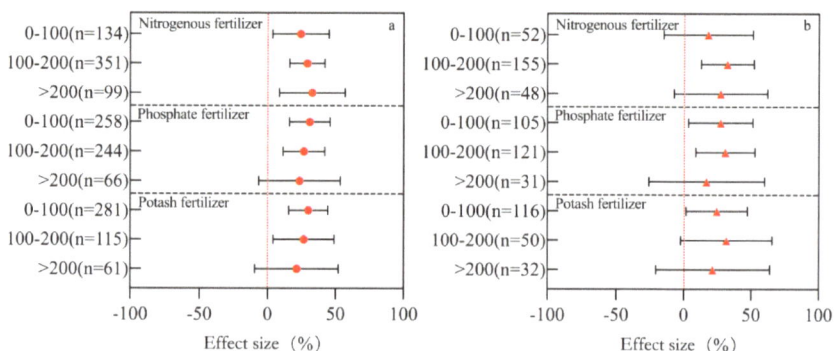

Figure 3. Effect of PM on potato yield (**a**) and WUE (**b**) comparing with no mulching in Northwest China at different nitrogenous fertilizer, phosphate fertilizer, and potash fertilizer applications. Error bars are the 95% confidence intervals (CIs). Number of samples are indicated in parentheses.

3.4. Response of Potato Yield and WUE under PM to Agriculture Practices

PM only increased yield and WUE at 45,000–75,000 plants ha^{-1}, by 30.34% and 32.42% respectively (Figure 4). The results showed that PM only increased yield and WUE at 29.57% and 32.17% respectively under ridge, a common cultivation practice in the Northwest China. The effect of PM on yield and WUE responded differently at different levels of cover, with PM increasing yield by 31.89% and 24.57% under both full and half mulching, respectively. Full mulching performed more effectively. In contrast, PM only increased WUE under full mulching by 34.44%.

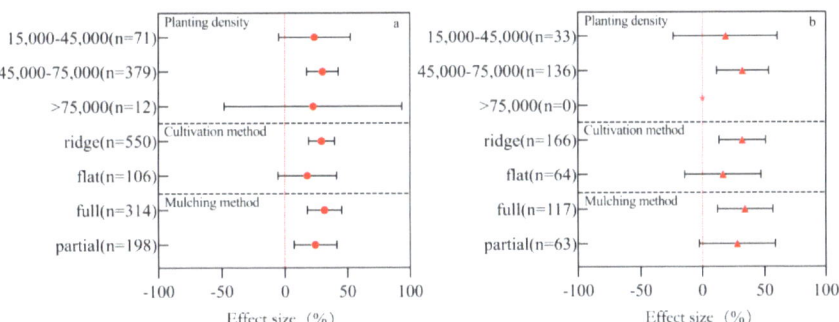

Figure 4. Effect of PM on potato yield (**a**) and WUE (**b**) comparing with no mulching in Northwest China at different planting density, cultivation method, and mulching method. Error bars are the 95% confidence intervals (CIs). Number of samples are indicated in parentheses.

3.5. Response of Potato Yield and WUE under PM to Mulching Properties

Both white and black mulching improved yield and the effect was not very different, 26.3% and 28.6% respectively, with black mulching slightly more effective (Figure 5a). Different mulching thicknesses improved yield, but ≥0.01 mm mulching increased yield by 30.44% and ≤0.008 mm mulching increased yield by 24.78%. The response of mulching color and mulching thickness to WUE differed from yield, improving WUE only with black mulching and ≥0.01 thickness, by 26.67% and 37.7%, respectively (Figure 5b).

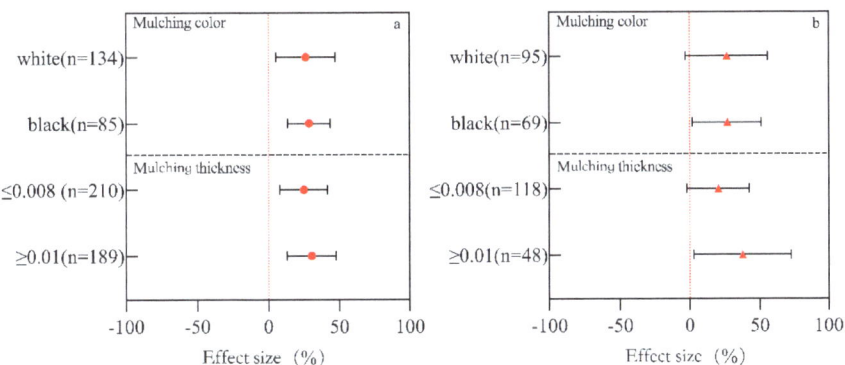

Figure 5. Effect of PM on potato yield (**a**) and WUE (**b**) comparing with no mulching in Northwest China at different mulching color, and mulching thickness (mm). Error bars are the 95% confidence intervals (CIs). Number of samples are indicated in parentheses.

3.6. The Importance of Natural Condition, Fertilizer Application, Agriculture Practice and Mulching Properties

As shown in the Figure 6a, AP was the most major influencing factor for yield increase, and the importance of mulching color was the least. The results of the random forest model showed that, for yield, the order of each factor was AP (100%), AT (56.4%), phosphate fertilizer (47.7%), potash fertilizer (47.4%), planting density (40.0%), nitrogen fertilizer (33.4%), mulching thickness (15.6%), mulching method (13.6%), cultivation method (7.3%), and mulching color (0%). For WUE (Figure 6b), AT was the most important factor, and, similar to the results for yield, the importance of mulching color was the least. The results showed that the order of each factor was AT (100%), AP (99.8%), planting density (76.9%), nitrogen fertilizer (76.3%), phosphate fertilizer (72.7%), mulching method (57.1%), mulching thickness (39.3%), potash fertilizer (34.5%), cultivation method (8.5%), and mulching color (0%). Overall, natural conditions had the most important influencing factor for the effect of PM, followed by fertilizer application and planting density. Except for the same non-importance of mulching color on yield and WUE increase, the importance of fertilizer application, cultivation measures, and mulching thickness on WUE improvement was higher than that on yield improvement.

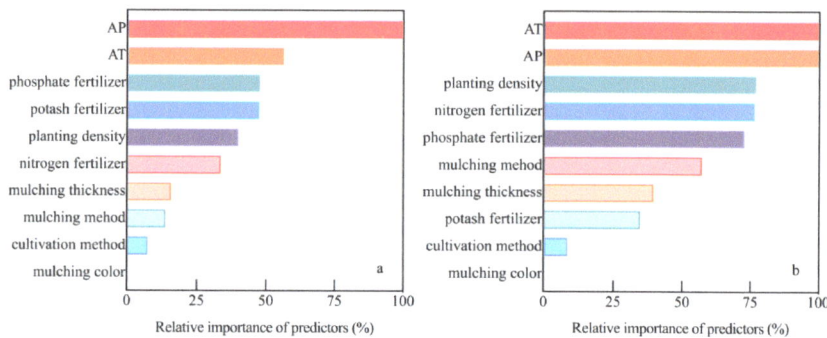

Figure 6. Relative importance of factors to the effect of PM on yield (**a**) and WUE (**b**) predicted by the random forest model. AP: annual average precipitation; and AT: ≥10 °C annual average accumulated temperature.

4. Discussion

4.1. The Effects of PM on Potato Yield and WUE

The study adopted meta-analysis to comprehensively quantify the effect of PM on potato yield and WUE in Northwest China, which showed that PM was beneficial for increasing yield by an overall average of 27.17% and increasing WUE by an overall average of 27.16%. The results were close to the findings of Li et al. [6] (Figure 1), who reported an increase of potato yield and WUE by 24.3% and 28.7%, respectively. Natural conditions were the primary factors of the effect of PM on the yield and WUE improvement (Figure 6). The results showed that PM had a greater potential to improve yield and WUE under 200–400 mm AP and 1500–3000 °C AT. The reasons mainly are that PM turns into a physical barrier on the soil surface, thereby increasing soil water and temperature, reducing soil evaporation [38], and accelerating the speed of soil drying and wetting alternation [39]. PM significantly raised soil water storage at surface by 8.4% in rainfed semi-arid areas, moreover, the effect was most significant when the AP was lower [40]. PM can reduce fluctuations in soil surface temperature and regulate temperature for crop growth [41]. Wang et al. [42] conducted a long-term positioning experiment on the Loess Plateau, which found that PM increased the average effective accumulated temperature of the 0–50 cm soil profile. Climate conditions in the Northwest are very suitable for cultivation, and the application of PM has great potential in the area [14]. Compared to the Northwest, PM performed relatively poorer in the South [16,20]; in addition, PM had inconspicuously promoted yield and WUE in the Southwest. In addition to increased WUE, PM is favorable to nitrogen use efficiency [43], and thermal use efficiency [44], both of which contribute to yield and quality promotion. However, PM increases agricultural inputs and labor costs, but ultimately significantly increases profits due to increased production [17,45]. A study in India discovered that straw mulching combined with irrigation significantly increased water productivity [46]. According to Xing et al. [47], organic mulching increased soil organic matter and potato tuber yield. This study concentrated solely on the Northwest China, and those areas with similar climatic and geographical characteristics can serve as a point of comparison.

4.2. The Influencing Factors of PM on Improving Potato Yield and WUE

The effects of PM on potato yield and WUE were significantly influenced by the application rate of synthetic fertilizer. The results showed that the yield and WUE of potato performed better under the application rate of 0–100 and 100–200 kg·ha^{-1}, which was consistent with Li et al. [6]. Wang et al. [20] studied the impact of synthetic fertilizers on the effectiveness of PM in the presence and absence of organic fertilizer application, and the results emphasized that organic and potassium fertilizers were not suitable for simultaneous use, as potassium fertilizer reduced yield at the same time, while potassium is the most needed nutrient element for potato growth. Zhang et al. [48] applied a meta-analysis to explore the effect of agricultural measures on WUE of potato cultivation in the Loess Plateau, which observed that increasing nitrogen fertilizer measures had a better effect on improving WUE than PM, possibly due to the significant effect of nitrogen fertilizer on increasing yield. Nitrogen fertilizer is an important source of greenhouse gases, although a study suggested that high nitrogen application rate combined with PM and irrigation reduced the carbon footprint [49]. However, when the amount of nitrogen fertilizer increased, the effect is weakened due to the consumption of more water to promote crop growth. Furthermore, PM increased the effectiveness of soil nutrients and improved the fertilizer use efficiency [12], which possibly occurred because that PM increased soil microbial activity, thereby increasing the decomposition and turnover of soil nutrients by microorganisms [50]. Ding et al. [51] reported that PM reduced nitrogen leaching and loss in a 28-year study. Therefore, the rational application of synthetic fertilizers is an important way to promote greater interaction between fertilizers and PM.

Because of higher WUE, the use of PM provides the possibility of increasing potato yield by higher planting density. Hou et al. conducted a field trial about the impact of planting density on yield and WUE in the Loess Plateau of China, and the results showed yield and WUE performed best in 52,500 plants ha^{-1} when considering profits [52]. Wang et al. divided the Northwest into irrigated and arid areas, and the best fertilization gradients for improving yield were 40,000–55,000 and 55,000–70,000 plants ha^{-1}, respectively [20]. In this study, the most effective planting density for yield and WUE was 45,000–75,000 plants ha^{-1}, which approached to the results. There was no linear relationship between yield and planting density, indicating that higher density did not necessarily mean higher yield, which is mainly due to the fact that, as planting density increases, the competition for resources between inter species increases, leading to a decrease in yield; in addition, the impact of planting density on WUE was closely related to rainfall [52].

Research has shown that the increase of crop yield and WUE indicates that PM can improve resource utilization efficiency through ridges and furrows [53]. The results of the present study agreed with this, which showed that ridges significantly improved yield and WUE, however, flat conditions had insignificant effects on yield and WUE. Ridge and furrow with mulching improves rainwater collection efficiency and ensures the water supply required for the entire growth period of potato [54], which is the best farming method in the Northwest China [16]. Full mulching had a more significant improvement in yield and WUE than partial mulching, which was consistent with Gao et al. [55], a study that analyzed the effect of PM on maize production. In another study, full mulching enhanced emergence rate and more stems, and at the same time increased cost investment, but finally it increased economic benefits due to increased production [45].

Studies have shown that different mulching colors of PM also affect its use in field production. Black and white mulching are currently the most widely used mulching color in agricultural production. Zhang et al. [56] found that white film had a better water-saving effect than black mulching, while black mulching was more suitable for the production of large stems like potato, which was mostly because black mulching covered the plant canopy, receiving more longwave radiation from the surface, causing the plant to grow vigorously, but at the same time increasing the water consumption. White film had a higher transmittance than black mulching, which increases soil temperature more effectively [2] A possible factor in in this type of case is that potato tubers are sensitive to soil temperature in soil. Another advantage of application of the black mulching was that it had better weed control effects than white mulching, which meant that it reduced overall herbicide and labor investment cost [57]. Li et al. [18] compared the effects of film thickness less than 0.008 mm, equal to 0.008 mm, and greater than 0.008 mm on the increase of potato yield with PM, and found that the thicker the mulching, the higher the yield increase. In this study, the results also showed that increasing the mulching thickness was favorable for increasing yield and WUE. However, using a thicker mulching would increase costs, so farmers often choose to use thinner mulching in production, resulting in serious pollution problems.

4.3. Uncertainty and Limitations

This study used the method of meta and random forest to study the effect of PM on potato yield and WUE in Northwest China, including natural conditions, fertilization application, cultivation measures, and mulching properties as influencing factors. Using random forest, the importance of factors was further quantified. However, there are still limitations to the study. Firstly, factors affecting the effect of PM include altitude, soil properties, irrigation, degradable film, variety, and so on [15,18,58]. However, due to factors such as data acquisition, they were not included in this study. Secondly, there is interaction between factors, and only the importance has been evaluated. Moreover, quantifying the degree of interaction between factors is also an important issue [6,32] Finally, while considering the impact of PM on yield and WUE, due to PM being an important source of plastic pollution in farmland [59], and the residue of PM increases the potential pathogens [60,61], the ecological effects of PM should also be considered. Moreover, it has been reported that PM increases the risk of increasing phthalate esters in the soil, which can be absorbed by plant roots and are toxic to humans and crops [62] In order to achieve sustainable development, it is necessary to balance the improvement of human life and the protection of the ecological environment by PM. Therefore, future research can incorporate more factors and consider issues such as film pollution, which is beneficial to produce a more comprehensive understanding of PM and provide guidance for other agricultural measures.

5. Conclusions

Based on the current impact of PM on potato yield and WUE in Northwest China, the results showed that PM increased yield by 27.17% and WUE by 27.16%, respectively. AP, AT, nitrogen fertilizer, phosphate fertilizer, potassium fertilizer, planting density, cultivation method, mulching method, mulching color, and mulching thickness were the influencing factors that affect the effectiveness of PM. PM had the best yield improvement and WUE effect between 200–400 mm and 1500–3000 °C. Under nitrogen fertilizer application >200 kg·ha^{-1}, phosphate fertilizer and potato fertilizer 0–100 kg·ha^{-1}, PM had the best performance on yield increase. Nitrogen, phosphate, and potassium fertilizers application had the best effect on improving WUE under 100–200 kg·ha^{-1}. With 15,000–45,000 plants·ha^{-1}, ridge, full mulching, black mulching, and ≥0.01 mm, PM had the best performance in promoting yield and WUE. The results of random forest model showed that AP and AT had high scores among the above factors, in addition, fertilizer application and planting density were also of importance. Through comprehensive agricultural management, PM can better play its role and achieve sustainable development.

Author Contributions: Writing—original draft, J.X.; Writing—review & editing, Y.W. and J.C. Supervision, Y.C., W.H. and X.L. All authors have read and agreed to the published version of the manuscript.

Funding: This work was supported by the National Key Research and Development Program of China (Grant number: 2021YFD1700700).

Data Availability Statement: Data can be obtained at the request of the corresponding authors.

Conflicts of Interest: The authors declare no conflict of interest.

References

1. Van Dijk, M.; Morley, T.; Rau, M.L.; Saghai, Y. A meta-analysis of projected global food demand and population at risk of hunger for the period 2010–2050. *Nat. Food* **2021**, *2*, 494–501. [CrossRef] [PubMed]
2. Zhang, Y.L.; Feng, R.; Nie, W.; Wang, F.X.; Feng, S.Y. Plastic Film Mulch Performed Better in Improving Heat Conditions and Drip Irrigated Potato Growth in Northwest China than in Eastern China. *Water* **2020**, *12*, 2906. [CrossRef]
3. FAO. *Crops and Livestock Products*; FAO: Rome, Italy, 2021.
4. Chang, L.; Han, F.X.; Chai, S.X.; Cheng, H.B.; Yang, D.L.; Chen, Y.Z. Straw strip mulching affects soil moisture and temperature for potato yield in semiarid regions. *Agron. J.* **2020**, *112*, 1126–1139. [CrossRef]
5. *China Rural Statistical Yearbook*; China Statistics Press: Beijing, China, 2021. (In Chinese)
6. Li, Q.; Li, H.B.; Zhang, L.; Zhang, S.Q.; Chen, Y.L. Mulching improves yield and water-use efficiency of potato cropping in China: A meta-analysis. *Field Crop. Res.* **2018**, *221*, 50–60. [CrossRef]
7. Wu, Y.; Huang, F.Y.; Jia, Z.K.; Ren, X.L.; Cai, T. Response of soil water, temperature, and maize (*Zea may* L.) production to different plastic film mulching patterns in semi-arid areas of northwest China. *Soil Till. Res.* **2017**, *166*, 113–121. [CrossRef]
8. Liang, S.M.; Ren, C.; Wang, P.J.; Wang, X.T.; Li, Y.S.; Xu, F.H.; Wang, Y.; Dai, Y.Q.; Zhang, L.; Li, X.P.; et al. Improvements of emergence and tuber yield of potato in a seasonal spring region using plastic film mulching only on the ridge. *Field Crop. Res.* **2018**, *223*, 57–65. [CrossRef]
9. Djaman, K.; Irmak, S.; Koudahe, K.; Allen, S. Irrigation Management in Potato (*Solanum tuberosum* L.) Production: A Review. *Sustainability* **2021**, *13*, 1504. [CrossRef]
10. Li, N.Y.; Qu, J.H.; Yang, J.Y. Microplastics distribution and microbial community characteristics of farmland soil under different mulch methods. *J. Hazard. Mater.* **2023**, *445*, 130408. [CrossRef]
11. Sun, D.B.; Li, H.G.; Wang, E.L.; He, W.Q.; Hao, W.P.; Yan, C.R.; Li, Y.Z.; Mei, X.R.; Zhang, Y.Q.; Sun, Z.X.; et al. An overview of the use of plastic-film mulching in China to increase crop yield and water-use efficiency. *Natl. Sci. Rev.* **2020**, *7*, 1523–1526. [CrossRef]
12. Wang, N.J.; Ding, D.Y.; Malone, R.W.; Chen, H.X.; Wei, Y.S.; Zhang, T.B.; Luo, X.Q.; Li, C.; Chu, X.S.; Feng, H. When does plastic-film mulching yield more for dryland maize in the Loess Plateau of China? A meta-analysis. *Agric. Water Manag.* **2020**, *240*, 106290. [CrossRef]
13. Ma, D.D.; Chen, L.; Qu, H.C.; Wang, Y.L.; Misselbrook, T.; Jiang, R. Impacts of plastic film mulching on crop yields, soil water, nitrate, and organic carbon in Northwestern China: A meta-analysis. *Agric. Water Manag.* **2018**, *202*, 166–173. [CrossRef]
14. Gao, H.H.; Yan, C.R.; Liu, Q.; Ding, W.L.; Chen, B.Q.; Li, Z. Effects of plasticmulching and plastic residue on agricultural production: A meta-analysis. *Sci. Total Environ.* **2019**, *651*, 484–492. [CrossRef] [PubMed]
15. Gu, X.B.; Cai, H.J.; Fang, H.; Li, Y.P.; Chen, P.P.; Li, Y.N. Effects of degradable film mulching on crop yield and water use efficiency in China: A meta-analysis. *Soil Till. Res.* **2020**, *202*, 104676. [CrossRef]
16. Zhang, S.H.; Wang, H.D.; Sun, X.; Fan, J.L.; Zhang, F.C.; Zheng, J.; Li, Y.P. Effects of farming practices on yield and crop water productivity of wheat, maize and potato in China: A meta-analysis. *Agric. Water Manag.* **2021**, *243*, 106444. [CrossRef]
17. He, G.; Wang, Z.H.; Hui, X.L.; Huang, T.M.; Luo, L.C. Black film mulching can replace transparent film mulching in crop production. *Field Crop. Res.* **2021**, *261*, 108026. [CrossRef]
18. Li, H.; Chang, W.Y. Exploring optimal film mulching to enhance potato yield in China: A meta-analysis. *Agron. J.* **2021**, *113*, 4099–4115. [CrossRef]
19. Mo, F.; Yu, K.L.; Crowther, T.W.; Wang, J.Y.; Zhao, H.; Xiong, Y.C.; Liao, Y.C. How plastic mulching affects net primary productivity, soil C fluxes and organic carbon balance in dry agroecosystems in China. *J. Clean. Prod.* **2020**, *263*, 121470. [CrossRef]
20. Wang, L.L.; Coulter, J.A.; Palta, J.A.; Xie, J.H.; Luo, Z.Z.; Li, L.L.; Carberry, P.; Li, Q.; Deng, X.P. Mulching-Induced Changes in Tuber Yield and Nitrogen Use Efficiency in Potato in China: A Meta-Analysis. *Agronomy* **2019**, *9*, 793. [CrossRef]
21. Hedges, L.V.; Gurevitch, J.; Curtis, P.S. The meta-analysis of response ratios in experimental ecology. *Ecology* **1999**, *80*, 1150–1156. [CrossRef]
22. Gurevitch, J.; Hedges, L.V. Statistical issues in ecological meta-analyses. *Ecology* **1999**, *80*, 1142–1149. [CrossRef]
23. Yu, Y.Y.; Turner, N.C.; Gong, Y.H.; Li, F.M.; Fang, C.; Ge, L.J.; Ye, J.S. Benefits and limitations to straw- and plastic-film mulch on maize yield and water use efficiency: A meta-analysis across hydrothermal gradients. *Eur. J. Agron.* **2018**, *99*, 138–147. [CrossRef]
24. He, G.; Wang, Z.H.; Li, S.X.; Malhi, S.S. Plastic mulch: Tradeoffs between productivity and greenhouse gas emissions. *J. Clean. Prod.* **2018**, *172*, 1311–1318. [CrossRef]
25. Li, Y.Z.; Yang, J.B.; Shi, Z.; Pan, W.H.; Liao, Y.C.; Li, T.; Qin, X.L. Response of root traits to plastic film mulch and its effects on yield. *Soil Till. Res.* **2021**, *209*, 104930. [CrossRef]
26. Cutler, D.R.; Edwards, T.C.; Beard, K.H.; Cutler, A.; Hess, K.T. Random forests for classification in ecology. *Ecology* **2007**, *88*, 2783–2792. [CrossRef] [PubMed]

27. Grömping, U. Variable Importance Assessment in Regression: Linear Regression versus Random Forest. *Am. Stat.* **2009**, *63*, 308–319. [CrossRef]
28. Tseng, M.-C.; Roel, A.; Macedo, I.; Marella, M.; Terra, J.; Zorrilla, G.; Pittelkow, C.M. Field-level factors for closing yield gaps in high-yielding rice systems of Uruguay. *Field Crop. Res.* **2021**, *264*, 108097. [CrossRef]
29. Smidt, E.R.; Conley, S.P.; Zhu, J.; Arriaga, F.J. Identifying Field Attributes that Predict Soybean Yield Using Random Forest Analysis. *Agron. J.* **2016**, *108*, 637–646. [CrossRef]
30. Philibert, A.; Loyce, C.; Makowski, D. Prediction of N2O emission from local information with Random Forest. *Environ. Pollut.* **2013**, *177*, 156–163. [CrossRef]
31. Dang, C.R.; Kong, F.L.; Li, Y.; Jiang, Z.X.; Xi, M. Soil inorganic carbon dynamic change mediated by anthropogenic activities: An integrated study using meta-analysis and random forest model. *Sci. Total Environ.* **2022**, *835*, 155463. [CrossRef]
32. Liu, X.T.; Tan, S.W.; Song, X.J.; Wu, X.P.; Zhao, G.; Li, S.P.; Liang, G.P. Response of soil organic carbon content to crop rotation and its controls: A global synthesis. *Agric. Ecosyst. Environ.* **2022**, *335*, 108017. [CrossRef]
33. Munyasya, A.N.; Koskei, K.; Zhou, R.; Liu, S.T.; Indoshi, S.N.; Wang, W.; Zhang, X.C.; Cheruiyot, W.K.; Mburu, D.M.; Nyende, A.B.; et al. Integrated on-site & off-site rainwater-harvesting system boosts rainfed maize production for better adaptation to climate change. *Agric. Water Manag.* **2022**, *269*, 107672.
34. Tuck, S.L.; Winqvist, C.; Mota, F.; Ahnström, J.; Turnbull, L.A.; Bengtsson, J. Land-use intensity and the effects of organic farming on biodiversity: A hierarchical meta-analysis. *J. Appl. Ecol.* **2014**, *51*, 746–755. [CrossRef] [PubMed]
35. Rossetti, M.R.; Tscharntke, T.; Aguilar, R.; Batáry, P. Responses of insect herbivores and herbivory to habitat fragmentation: A hierarchical meta-analysis. *Ecol. Lett.* **2017**, *20*, 264–272. [CrossRef] [PubMed]
36. Pittelkow, C.M.; Liang, X.Q.; Linquist, B.A.; van Groenigen, K.J.; Lee, J.; Lundy, M.E.; van Gestel, N.; Six, J.; Venterea, R.T.; van Kessel, C. Productivity limits and potentials of the principles of conservation agriculture. *Nature* **2015**, *517*, 365–368. [CrossRef]
37. Terrer, C.; Phillips, R.P.; Hungate, B.A.; Rosende, J.; Pett-Ridge, J.; Craig, M.E.; van Groenigen, K.J.; Keenan, T.F.; Sulman, B.N.; Stocker, B.D.; et al. A trade-off between plant and soil carbon storage under elevated CO_2. *Nature* **2021**, *591*, 599–603. [CrossRef]
38. Hou, X.Y.; Wang, F.X.; Han, J.J.; Kang, S.Z.; Feng, S.Y. Duration of plastic mulch for potato growth under drip irrigation in an arid region of Northwest China. *Agric. For. Meteorol.* **2010**, *150*, 115–121. [CrossRef]
39. Zhang, X.D.; Yang, L.C.; Xue, X.K.; Kamran, M.; Ahmad, I.; Dong, Z.Y.; Liu, T.N.; Jia, Z.K.; Zhang, P.; Han, Q.F. Plastic film mulching stimulates soil wet-dry alternation and stomatal behavior to improve maize yield and resource use efficiency in a semi-arid region. *Field Crop. Res.* **2019**, *233*, 101–113. [CrossRef]
40. Ren, A.T.; Zhou, R.; Mo, F.; Liu, S.T.; Li, J.Y.; Chen, Y.L.; Zhao, L.; Xiong, Y.C. Soil water balance dynamics under plastic mulching in dryland rainfed agroecosystem across the Loess Plateau. *Agric. Ecosyst. Environ.* **2021**, *312*, 107354. [CrossRef]
41. Chen, N.; Li, X.Y.; Simunek, J.; Shi, H.B.; Hu, Q.; Zhang, Y.H. Evaluating the effects of biodegradable and plastic film mulching on soil temperature in a drip-irrigated field. *Soil Till. Res.* **2021**, *213*, 105116. [CrossRef]
42. Wang, H.; Fan, J.; Fu, W.; Du, M.G.; Zhou, G.; Zhou, M.X.; Hao, M.D.; Shao, M.A. Good harvests of winter wheat from stored soil water and improved temperature during fallow period by plastic film mulching. *Agric. Water Manag.* **2022**, *274*, 107910. [CrossRef]
43. Wang, X.K.; Xing, Y.Y. Effects of Mulching and Nitrogen on Soil Nitrate-N Distribution, Leaching and Nitrogen Use Efficiency of Maize (*Zea mays* L.). *PLoS ONE* **2016**, *11*, e0161612. [CrossRef] [PubMed]
44. Zheng, J.; Fan, J.L.; Zhou, M.H.; Zhang, F.C.; Liao, Z.Q.; Lai, Z.L.; Yan, S.C.; Guo, J.J.; Li, Z.J.; Xiang, Y.Z. Ridge-furrow plastic film mulching enhances grain yield and yield stability of rainfed maize by improving resources capture and use efficiency in a semi-humid drought-prone region. *Agric. Water Manag.* **2022**, *269*, 107654. [CrossRef]
45. Zhao, H.; Wang, R.Y.; Ma, B.L.; Xiong, Y.C.; Qiang, S.C.; Wang, C.L.; Liu, C.A.; Li, F.M. Ridge-furrow with full plastic film mulching improves water use efficiency and tuber yields of potato in a semiarid rainfed ecosystem. *Field Crop. Res.* **2014**, *161*, 137–148. [CrossRef]
46. Biswal, P.; Swain, D.K.; Jha, M.K. Straw mulch with limited drip irrigation influenced soil microclimate in improving tuber yield and water productivity of potato in subtropical India. *Soil Till. Res.* **2022**, *223*, 105484. [CrossRef]
47. Xing, Z.S.; Toner, P.; Chow, L.; Rees, H.W.; Li, S.; Meng, F.R. Effects of Hay Mulch on Soil Properties and Potato Tuber Yield under Irrigation and Nonirrigation in New Brunswick, Canada. *J. Irrig. Drain. Eng.* **2012**, *138*, 703–714. [CrossRef]
48. Zhang, G.X.; Zhang, Y.; Zhao, D.H.; Liu, S.J.; Wen, X.X.; Han, J.; Liao, Y.C. Quantifying the impacts of agricultural management practices on the water use efficiency for sustainable production in the Loess Plateau region: A meta-analysis. *Field Crop. Res.* **2023**, *291*, 108787. [CrossRef]
49. Xiong, L.; Liang, C.; Ma, B.L.; Shah, F.; Wu, W. Carbon footprint and yield performance assessment under plastic film mulching for winter wheat production. *J. Clean. Prod.* **2020**, *270*, 122468. [CrossRef]
50. Li, Y.Z.; Xie, H.X.; Ren, Z.H.; Ding, Y.P.; Long, M.; Zhang, G.X.; Qin, X.L.; Siddique, K.H.M.; Liao, Y.C. Response of soil microbial community parameters to plastic film mulch: A meta-analysis. *Geoderma* **2022**, *418*, 115851. [CrossRef]
51. Ding, F.; Ji, D.C.; Yan, K.; Dijkstra, F.A.; Bao, X.L.; Li, S.Y.; Kuzyakov, Y.; Wang, J.K. Increased soil organic matter after 28 years of nitrogen fertilization only with plastic film mulching is controlled by maize root biomass. *Sci. Total Environ.* **2022**, *810*, 152244. [CrossRef]
52. Hou, X.Q.; Li, R.; He, W.S.; Ma, K. Effects of planting density on potato growth, yield, and water use efficiency during years with variable rainfall on the Loess Plateau, China. *Agric. Water Manag.* **2020**, *230*, 105982. [CrossRef]

53. Xiong, L.; Wu, W. Can additional agricultural resource inputs improve maize yield, resource use efficiencies and emergy based system efficiency under ridge-furrow with plastic film mulching? *J. Clean. Prod.* **2022**, *379*, 134711. [CrossRef]
54. Zhao, H.; Xiong, Y.C.; Li, F.M.; Wang, R.Y.; Qiang, S.C.; Yao, T.F.; Mo, F. Plastic film mulch for half growing-season maximized WUE and yield of potato via moisture-temperature improvement in a semi-arid agroecosystem. *Agric. Water Manag.* **2012**, *104*, 68–78. [CrossRef]
55. Gao, H.H.; Yan, C.R.; Liu, Q.; Li, Z.; Yang, X.; Qi, R.M. Exploring optimal soil mulching to enhance yield and water use efficiency in maize cropping in China: A meta-analysis. *Agric. Water Manag.* **2019**, *225*, 105741. [CrossRef]
56. Zhang, Y.L.; Wang, F.X.; Shock, C.C.; Yang, K.J.; Kang, S.Z.; Qin, J.T.; Li, S.E. Effects of plastic mulch on the radiative and thermal conditions and potato growth under drip irrigation in arid Northwest China. *Soil Till. Res.* **2017**, *172*, 1–11. [CrossRef]
57. Qin, X.L.; Li, Y.Z.; Han, Y.L.; Hu, Y.C.; Li, Y.J.; Wen, X.X.; Liao, Y.C.; Siddique, K.H.M. Ridge-furrow mulching with black plastic film improves maize yield more than white plastic film in dry areas with adequate accumulated temperature. *Agric. For. Meteorol.* **2018**, *262*, 206–214. [CrossRef]
58. Ma, J.T.; Cheng, H.B.; Chai, S.X.; Wang, S.E.; Gao, T.T.; Chang, L.; Ding, W. Influence of different mulching patterns on soil moisture and yield of potato cultivars with different maturities in dryland farming. *J. Gansu Agric. Univ.* **2019**, *54*, 55–64.
59. Zhang, Q.Q.; Ma, Z.R.; Cai, Y.Y.; Li, H.R.; Ying, G.G. Agricultural Plastic Pollution in China: Generation of Plastic Debris and Emission of Phthalic Acid Esters from Agricultural Films. *Environ. Sci. Technol.* **2021**, *55*, 12459–12470. [CrossRef]
60. Zhu, D.; Ma, J.; Li, G.; Rillig, M.C.; Zhu, Y.G. Soil plastispheres as hotpots of antibiotic resistance genes and potential pathogens (Aug, 10.1038/s41396-021-01137-z, 2021). *ISME J.* **2022**, *16*, 615. [CrossRef]
61. Qi, Y.; Ossowicki, A.; Yergeau, É.; Vigani, G.; Geissen, V.; Garbeva, P. Plastic mulch film residues in agriculture: Impact on soil suppressiveness, plant growth, and microbial communities. *FEMS Microbiol. Ecol.* **2022**, *98*, fiac017. [CrossRef]
62. Wang, D.; Xi, Y.; Shi, X.Y.; Zhong, Y.J.; Guo, C.L.; Han, Y.N.; Li, F.M. Effect of plastic film mulching and film residues on phthalate esters concentrations in soil and plants, and its risk assessment. *Environ. Pollut.* **2021**, *286*, 117546. [CrossRef]

Disclaimer/Publisher's Note: The statements, opinions and data contained in all publications are solely those of the individual author(s) and contributor(s) and not of MDPI and/or the editor(s). MDPI and/or the editor(s) disclaim responsibility for any injury to people or property resulting from any ideas, methods, instructions or products referred to in the content.

Article

Optimizing Well Placement for Sustainable Irrigation: A Two-Stage Stochastic Mixed Integer Programming Approach

Wanru Li [1,*], Mekuanent Muluneh Finsa [2,3], Kathryn Blackmond Laskey [1], Paul Houser [4], Rupert Douglas-Bate [5] and Kryštof Verner [6,7]

1. Department of Systems Engineering and Operations Research, George Mason University, Fairfax, VA 22030, USA; klaskey@gmu.edu
2. Institute of Hydrogeology, Engineering Geology and Applied Geophysics, Charles University, 128 44 Prague, Czech Republic; finsam@natur.cuni.cz or mekuanent.muluneh@amu.edu.et
3. Water Resource Research Center, Arba Minch University, Arba Minch P.O. Box 21, Ethiopia
4. Department of Geography and Geoinformation Science, George Mason University, Fairfax, VA 22030, USA; phouser@gmu.edu
5. Global MapAid, Aylesbury HP17 8RZ, UK; rupertdouglasbate@globalmapaid.org
6. Czech Geological Survey, Klárov 3, 118 21 Prague, Czech Republic; krystof.verner@geology.cz
7. Institute of Petrology and Structural Geology, Faculty of Science, Charles University, Albertov 6, Prague 2, 128 00 Prague, Czech Republic
* Correspondence: wli15@gmu.edu

Citation: Li, W.; Finsa, M.M.; Laskey, K.B.; Houser, P.; Douglas-Bate, R.; Verner, K. Optimizing Well Placement for Sustainable Irrigation: A Two-Stage Stochastic Mixed Integer Programming Approach. *Water* **2024**, *16*, 2715. https://doi.org/10.3390/w16192715

Academic Editors: Chenglong Zhang and Xiaojie Li

Received: 18 August 2024
Revised: 14 September 2024
Accepted: 21 September 2024
Published: 24 September 2024

Copyright: © 2024 by the authors. Licensee MDPI, Basel, Switzerland. This article is an open access article distributed under the terms and conditions of the Creative Commons Attribution (CC BY) license (https://creativecommons.org/licenses/by/4.0/).

Abstract: Utilizing groundwater offers a promising solution to alleviate water stress in Ethiopia, providing a dependable and sustainable water source, particularly in regions with limited or unreliable surface water availability. However, effective decision-making regarding well drilling and placement is essential to maximize groundwater resource potential, enhancing agricultural productivity, reducing hunger, and bolstering food security in Ethiopia. This study concentrates on the development of two-stage stochastic mixed integer programming (SMIP) models to optimize well placement for sustainable agricultural irrigation, considering uncertain demand scenarios. Additionally, a deterministic mixed integer programming model is formulated for comparison with the two-stage SMIP. Experiments are conducted to explore various demand scenario distributions, revealing that the optimized total cost for the two-stage SMIP generally exceeds that of a deterministic setting, aligning with the two-stage SMIP's focus on long-term benefits. Moreover, slight differences are observed in well layouts under different assumption scenarios. The study also examines the impact of selected parameters, such as fixed construction costs, per-meter drilling costs, and demand scenarios. The out-of-sample performance shows that the stochastic model is more flexible and resilient, with 11% and 4% lower costs than deterministic cases 1 and 3, respectively. This flexibility provides a more robust long-term strategy for well placement and resource allocation in groundwater management.

Keywords: well placement optimization; well layout optimization; mixed integer programming (MIP); two-stage stochastic mixed integer programming (SMIP); Bilate watershed; southern Ethiopia; sustainable irrigation

1. Introduction

Ethiopia is characterized by a complex geological structure and rugged topography, with extremely diverse hydrogeological conditions. These include (a) Neoproterozoic, low- to high-grade metamorphic rocks that are part of the Arabian-Nubian Shield, overlain by (b) platform sedimentary units ranging from the Lower Paleozoic to the Paleogene, as seen in the large Ogaden Basin, Blue Nile Basin, and Mekele Basin, and (c) Eocene to Holocene volcanic rocks and volcaniclastic deposits associated with the active NNE-SSV trending Main Ethiopian Rift [1,2]. Especially in light of these specific natural factors, Ethiopia faces a pressing challenge in ensuring access to clean and sustainable water sources for its burgeoning population. With over 60 million people lacking access to clean water, Ethiopia ranks among the countries with the lowest water supply coverage in sub-Saharan Africa [3]. Compounded by climate variability, recurrent droughts, and geological risks associated with slope deformations, the country's water resources are under significant strain, threatening the livelihoods and well-being of millions of Ethiopians [4]. Inadequate infrastructure and poor water management exacerbate the situation, further deepening the water crisis in both rural and urban areas [5]. Utilizing groundwater presents a promising avenue for mitigating water stress and ensuring sustainable water access in Ethiopia. Groundwater resources, if managed effectively, offer a reliable and resilient water source, particularly in regions where surface water availability is limited or erratic [6]. However, unlocking the potential of groundwater requires strategic decision-making in well drilling and placement.

Well drilling decision-making holds significant importance in optimizing groundwater utilization for sustainable development in Ethiopia. The strategic placement of wells with regard to hydrogeological conditions can enhance water access, improve agricultural productivity, and support rural livelihoods. However, managing groundwater resources involves uncertainties, particularly regarding future demand, recharge rates, and water table depth. Often, these characteristics only become fully known after drilling has occurred, requiring adaptive risk management strategies. By continuously adjusting management practices as new information becomes available, we can ensure more effective, long-term sustainability. Optimizing well placement and layout not only minimizes water wastage, reduces energy consumption, and mitigates the risk of groundwater depletion but also allows for flexible responses to these uncertainties. This adaptive, holistic approach to groundwater management benefits farming communities while contributing to broader societal goals such as poverty reduction, food security, and environmental sustainability [7].

There are two main optimization techniques that have been applied in well placement problems: conventional optimization methods and non-conventional methods. Conventional optimization methods have been extensively employed for well placement optimization in utilizing groundwater for irrigation. Drawing insights from a previous study [8], conventional approaches often utilize mathematical programming techniques such as linear programming (LP), nonlinear programming (NLP), integer programming (IP), mixed-integer programming (MIP), stochastic mixed integer programming (SMIP), and adjoint methods. In 2014, Tafteh et al. formulated a LP model for the optimization of irrigation water distribution [9]. Studies by authors such as Ma et al. (2019) [10] and Kuvichko and Ermolaev (2020) [11] have demonstrated the efficacy of MIP in optimizing well locations. Also, research by Liu et al. (2015) has optimized the layout of pumping wells based on NLP models for irrigation planning and management [12]. Additionally, Halilovic et al. (2022) introduced a new gradient-based optimization method using the adjoint approach for well layouts in groundwater heat pumps, facilitating continuous well locations and accommodating a large number of wells. Their method demonstrates versatility in optimizing layouts for multiple neighboring or single systems, as evidenced by a real case study involving 10 systems, showcasing its practical applicability [13]. Furthermore, the study by Bayer et al. in 2008 introduced a novel approach to solving stochastic optimization problems with multiple equally probable realizations of uncertain parameters, utilizing dynamically ordered stacks of realizations during the search process. By applying this technique to a water supply well field design problem, the study demonstrates that simple stack ordering can significantly reduce computational effort by up to 97% without compromising optimization results, offering promise for similar water management and reliability-based design challenges [14]. However, despite their ability to ensure optimality under certain conditions, conventional methods may encounter challenges, such as getting trapped in local optima.

In contrast, gradient-free optimization methods as non-conventional methods present promising alternatives for well placement optimization in irrigation systems, offering robustness against complex objective functions and solution spaces. Researchers such as Emerick et al. (2009) implemented a genetic algorithm to optimize the number and location of the wells [15]. Investigations by researchers such as Sharifipour et al. (2021) have explored the application of the shuffled frog leaping algorithm, particle swarm optimization, and genetic algorithm in optimizing well placement problems. They found that the shuffled frog leaping algorithm outperformed other algorithms, demonstrating superior objective function values and well spacing in both intermediate and late optimization stages [16]. These gradient-free optimization techniques, as highlighted in [8], do not rely on gradient information and employ diverse search strategies to explore solution spaces iteratively. While they provide flexibility and scalability, a notable disadvantage of gradient-free optimization methods is their potential for slower convergence compared to gradient-based methods, which can result in longer computational times, particularly for high-dimensional optimization problems and large-scale irrigation systems AlQahtani et al. (2013) conducted a comparison between evolutionary metaheuristics and mathematical optimization methods for the well placement problem. Their study revealed that mathematical optimization outperformed genetic algorithms by yielding superior solutions in a shorter computational timeframe [17].

Many previous research studies have investigated the optimization of pumping well layouts to enhance groundwater extraction efficiency. However, some of the studies focus on deterministic models that may not adequately account for the uncertain water demand and the need for sustainable groundwater development [9–12,18]. Additionally, some papers have relied on simplified models that assume an infinite aquifer, overlooking the potential risks of groundwater depletion and adverse environmental impacts [10,12]. To ensure the sustainable development of groundwater resources, it is essential to incorporate groundwater level and recharge parameters into the optimization model. By considering the variability of groundwater levels and recharge rates over different spaces, as well as the finite nature of aquifers, these parameters can provide valuable insights into the long-term viability of well placement strategies. Moreover, integrating sustainability constraints into the optimization framework can help balance water extraction with aquifer replenishment, thus safeguarding groundwater resources for future generations.

In this study, we chose to use mixed-integer programming (MIP) and two-stage stochastic mixed-integer programming (SMIP) over non-conventional methods such as genetic algorithms or shuffled frog leaping. The primary reason for this choice is the deterministic nature of MIP and its proven effectiveness in handling well placement optimization problems with precise constraints and objectives. Additionally, two-stage SMIP allows us to incorporate uncertainty in demand scenarios, providing a more robust solution under varying conditions. While non-conventional methods like genetic algorithms and shuffled frog leaping offer flexibility and robustness, they often require more computational effort and may not guarantee optimal solutions. Our initial focus on MIP and two-stage SMIP was due to their structured approach and computational efficiency, making them suitable for the preliminary phase of this research.

This study aims to model and optimize optimal drilling locations to minimize costs while meeting farm demand and ensuring sustainable groundwater development for subsistence agriculture in water-scarce regions. To achieve this objective, MIP and two-stage SMIP models are constructed. The two-stage stochastic model offers the benefit of long-term planning by considering uncertain demand scenarios. Therefore, our focus in this study is on the two-stage SMIP. These models are developed using predicted groundwater levels [19] and estimated groundwater recharge [20] from our previous studies conducted in a larger region that encompasses the current study area. The specific study area selected for analysis is a small region in the western flank of the Bilate watershed (westernmost part of the Sidama Region) in southern Ethiopia. This study also explores various uncertain demand scenarios for the two-stage SMIP by employing different distributions and comparing the outcomes. Additionally, we present the well layout maps for each experiment and analyze the impact of key parameters.

The remainder of this paper is organized as follows: Section 2 delineates the problem formulation for both the deterministic and stochastic models and provides an overview of the study area and the data utilized; Section 3 offers a comprehensive presentation of the main results from the optimization models and shows optimal well layouts; Section 4 discusses the impact of different parameters, provides an analysis on out-of-sample performance, discusses the limitations of this study, and suggests potential avenues for future research; finally, Section 5 concludes the paper and summarizes key findings.

2. Materials and Methods

2.1. Problem Formulation and Reformulation

In this section, we provide problem formulation and reformulation for the deterministic MIP and two-stage SMIP in Sections 2.1.1 and 2.1.2, respectively.

2.1.1. Deterministic Mixed Integer Programming Model

Mixed-integer programming (MIP) is an optimization technique that involves decision variables that can be both continuous and discrete, allowing for complex modeling of real-world problems. It is widely used in various fields to find optimal solutions under given constraints, such as scheduling [21,22] and portfolio optimization [23,24]. In this section, we will introduce the sets, parameters, decision variables, formulation, and reformulation of the deterministic MIP.

Sets:

- I: set of farm sites, indexed by i.
- J: set of potential well sites, indexed by j.

Parameters:

- c_j: construction fixed cost at site j.
- u_j: unit cost for construction at site j.
- t_{ij}: transportation cost from j to i.
- d_i: demand at farm i.
- l_{swl_j}: static water level of the well at site j.
- r: average recharge quantity.
- s: area of the study region.
- w_{max}: the deepest depth the well can be drilled.
- l_{min}: the minimum difference between the depth well w_j and static water level l_{swl_j}.

Decision variables:

- x_j: binary, equal to 1 if the well at location j is placed, 0 otherwise.
- y_{ij}: quantity of water that is transported from site j to farm i.
- w_j: depth of the well to drill at site j.
- K_j: capacity of the well at site j.

Deterministic MIP problem formulation:

Our objective is to minimize the summation of the total construction cost, the total transportation cost, and the cost for drilling different depths of wells. The objective function is determined in collaboration with experts in the Arba Minch Water Technology Institute. The objective function is defined as:

$$\min_{x} \sum_{j=1}^{J} c_j x_j + \sum_{i=1}^{I}\sum_{j=1}^{J} t_{ij} y_{ij} + \sum_{j=1}^{J} u_j w_j, \tag{1}$$

Constraint (2) ensures that all water demand from the farms is met. Constraint (3) requires that each constructed well j must have a capacity K_j. The total quantity of water transported from well j to all the farms must be less than or equal to K_j, the capacity of that specific well. Constraint (4) specifies that the capacity K_j is defined as the volume of groundwater above the deepest point of the well. Figure 1 presents a conceptual plot illustrating the capacity of a well. Constraint (5) mandates that the total amount of pumped-out water must be less than the quantity of recharged water. Constraint (6) requires that each constructed well j must have a depth less than the maximum allowable depth. Constraint (7) indicates that the lower bound of the well depth w_j should be greater than the static water level plus the minimum difference between the well depth and the static water level. Constraints (8) and (9) are standard binary and non-negative constraints.

$$\text{s.t.} \quad \sum_{j=1}^{J} y_{ij} = d_i, \ \forall i \tag{2}$$

$$\sum_{j=1}^{J} y_{ij} \leq K_j x_j, \ \forall i \tag{3}$$

$$K_j = s * \left(w_j - l_{swl_j}\right), \ \forall j \tag{4}$$

$$\sum_{i=1}^{I} \sum_{j=1}^{J} y_{ij} \leq r \tag{5}$$

$$w_j \leq w_{max}, \ \forall j \tag{6}$$

$$w_j \geq \left(l_{swl_j} + l_{min}\right) x_j, \ \forall j \tag{7}$$

$$y_{ij} \geq 0, \ \forall i, \forall j \tag{8}$$

$$x_j \in \{0, 1\}, \ \forall j \tag{9}$$

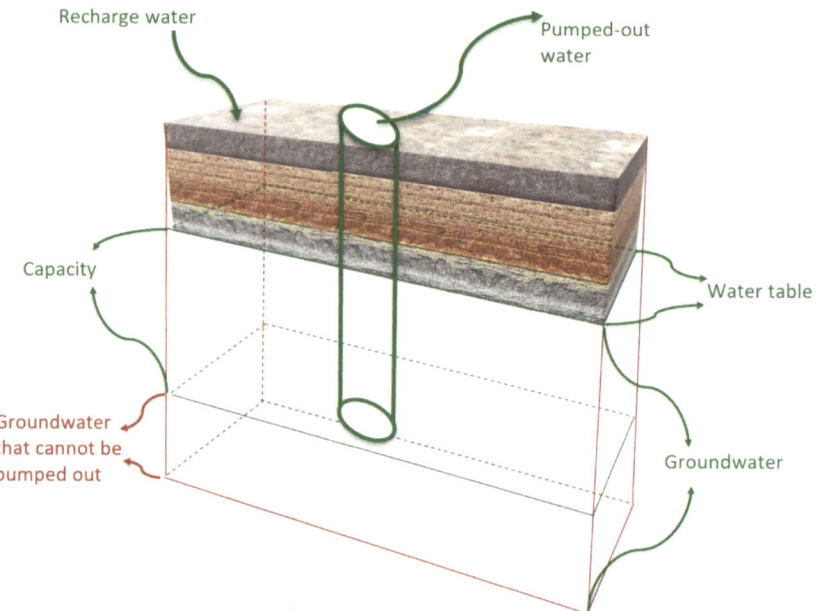

Figure 1. Conceptual plot illustrating the capacity of a well (Constraints (3) and (4)). The volume of groundwater below the deepest point of the well cannot be pumped out for drinking or irrigation.

Deterministic MIP problem reformulation:

The constraints (3) and (4) can be written into $\sum_{j=1}^{J} y_{ij} \leq s*\left(w_j x_j - l_{swl_j} x_j\right)$, where $w_j\, x_j$ is a bilinear term. To linearize the bilinear term, we reformulate the problem using McCormick Envelope, which is a relaxation technique for bilinear non-convex nonlinear programming problems [25]. The reformulation can be performed by replacing the bilinear term into a new variable (letting $w_j\, x_j = v_j$) and adding four sets of constraints. Below are the reformulated problems:

$$\min_{x} \sum_{j=1}^{J} c_j x_j + \sum_{i=1}^{I}\sum_{j=1}^{J} t_{ij} y_{ij} + \sum_{j=1}^{J} u_j w_j, \tag{10}$$

$$\text{s.t.} \quad \sum_{j=1}^{J} y_{ij} = d_i, \quad \forall i \tag{11}$$

$$\sum_{j=1}^{J} y_{ij} \leq s(v_j - l_{swl_j} x_j), \quad \forall i \tag{12}$$

$$v_j \geq \left(l_{swl_j} + l_{min}\right) x_j, \quad \forall j \tag{13}$$

$$v_j \geq w_j + w_{max} x_j - w_{max}, \quad \forall j \tag{14}$$

$$v_j \leq w_j + \left(l_{swl_j} + l_{min}\right) x_j - \left(l_{swl_j} + l_{min}\right), \quad \forall j \tag{15}$$

$$v_j \leq w_{max} x_j, \quad \forall j \tag{16}$$

$$\sum_{i=1}^{I}\sum_{j=1}^{J} y_{ij} \leq r \tag{17}$$

$$w_j \leq w_{max}, \quad \forall j \tag{18}$$

$$w_j \geq \left(l_{swl_j} + l_{min}\right) x_j, \quad \forall j \tag{19}$$

$$y_{ij} \geq 0, \quad \forall i, \forall j \tag{20}$$

$$x_j \in \{0, 1\}, \quad \forall j \tag{21}$$

2.1.2. Two-Stage Stochastic Mixed Integer Programming Model

Two-stage mixed-integer programming is an advanced optimization technique used to make decisions under uncertainty. In the first stage, decisions are made based on initial information before the uncertainty is revealed. In the second stage, additional decisions are adjusted after the uncertainty is known, allowing the model to respond dynamically to new information [26]. This approach is particularly useful for complex systems where future conditions are uncertain. In this section, the sets, parameters, decision variables, formulation, and reformulation of the two-stage SMIP are described.

Sets:
- I: set of farm sites, indexed by i.
- J: set of potential well sites, indexed by j.
- M: set of farm demand scenarios, indexed by m.

Parameters:
- c_j: construction fixed cost at site j.

- u_j: unit cost for drilling one meter at site j.
- p_m: probability of the demand scenario m.
- d_{im}: demand at farm i for scenario m.
- t_{ij}: transportation cost from j to i.
- l_{swl_j}: static water level of the well at site j.
- r: average recharge quantity
- s: area of the study region
- w_{max}: the deepest depth the well can be drilled.
- l_{min}: the minimum difference between the depth of the well w_j and static water level l_{swl_j}.

Decision variables:

- x_j: binary, equal to 1 if the well at location j is placed, otherwise 0.
- y_{ijm}: quantity of water that is transported from site j to farm i for demand scenario m.
- w_j: depth of the well to drill at site j.
- K_j: capacity of the well at site j.

Two-stage SMIP problem formulation:

We formulate the problem into a two-stage stochastic integer programming problem. The first stage decision variables are x_j denoting whether a well at location j is placed, w_j, denoting depth of the well to drill at site j, and K_j denoting the capacity of the well at site j. The second stage decision variable is y_{ij}, denoting the quantity of water that is transported from site j to farm i for demand scenario m. For the first stage, the objective is to minimize the total fixed construction costs and total drilling depth costs plus the expected cost of the second stage decision on transportation costs.

The first-stage constraints (23), (24), (26), and (27) are identical to the constraints in the deterministic mixed-integer programming model (constraints (6), (7), (8), and (9), respectively). Constraint (25) defines K_j as the volume of groundwater above the deepest point of the well. The second-stage constraints are also similar to their corresponding constraints in the deterministic model, with the key difference being the incorporation of uncertain demand scenarios m.

Specifically, constraint (29) ensures that the demand for all scenarios is satisfied Constraint (30) requires that each constructed well j has a capacity K_j, and the total quantity of water transported from well j to all farms for each demand scenario m must be less than or equal to K_j. Constraint (31) mandates that the total pumped-out water for each demand scenario m must be less than the quantity of recharged water. Finally, constraint (32) is the standard non-negativity constraint. The two-stage stochastic integer program is defined as

$$\min_x \sum_{j=1}^{J} c_j x_j + \sum_{j=1}^{J} u_j w_j + E_m Q(x, m), \qquad (22)$$

$$\text{s.t.} \quad w_j \leq w_{max}, \ \forall j \qquad (23)$$

$$w_j \geq \left(l_{swlj} + l_{min}\right) x_j, \ \forall j \qquad (24)$$

$$K_j = s * \left(w_j - l_{swl_j}\right), \ \forall j \qquad (25)$$

$$w_j \geq 0, \ \forall j \qquad (26)$$

$$x_j \in \{0, 1\}, \ \forall j \qquad (27)$$

where E_m denote mathematical expectation with respect to m. $Q(x, m)$ is the value of the second-stage for a given realization of the random vector m.

$$Q(x, m) = \min \sum_{m=1}^{M} (p_m(\sum_{i=1}^{I} \sum_{j=1}^{J} t_{ij} y_{ijm})), \tag{28}$$

$$s.t. \quad \sum_{j=1}^{J} y_{ijm} = d_{im}, \quad \forall i, \forall m \tag{29}$$

$$\sum_{j=1}^{J} y_{ijm} \leq K_j x_j, \quad \forall i, \forall m \tag{30}$$

$$\sum_{i=1}^{I} \sum_{j=1}^{J} y_{ijm} \leq r, \quad \forall m \tag{31}$$

$$y_{ij} \geq 0, \quad \forall i, \forall j \tag{32}$$

Two-stage SMIP reformulation and deterministic equivalent form:

The constraints (25) and (30) can be written into $\sum_{j=1}^{J} y_{ijm} \leq s * \left(w_j x_j - l_{swl_j} x_j \right)$, where $w_j x_j$ is a bilinear term. To linearize the bilinear term, we reformulate the problem using McCormick Envolope, which is a relaxation technique for bilinear non-convex nonlinear programming problems. The reformulation can be performed by replacing the bilinear term into a new variable (letting $w_j x_j = v_j$) and adding four sets of constraints. The objective is to minimize the summation of the total fixed cost for construction, the total transportation cost, and the cost for digging different depths of wells. We reformulate and define the problem in a deterministic equivalent form:

$$\min \sum_{j=1}^{J} c_j x_j + \sum_{j=1}^{J} u_j w_j + \sum_{m=1}^{M} (p_m(\sum_{i=1}^{I} \sum_{j=1}^{J} t_{ij} y_{ijm})), \tag{33}$$

$$s.t. \quad \sum_{j=1}^{J} y_{ijm} = d_{im}, \quad \forall i, \forall m \tag{34}$$

$$\sum_{j=1}^{J} y_{ijm} \leq s(v_j - l_{swl_j} x_j), \quad \forall j, \forall m \tag{35}$$

$$v_j \geq \left(l_{swl_j} + l_{min} \right) x_j, \quad \forall j \tag{36}$$

$$v_j \geq w_j + w_{max} x_j - w_{max}, \quad \forall j \tag{37}$$

$$v_j \leq w_j + \left(l_{swl_j} + l_{min} \right) x_j - \left(l_{swl_j} + l_{min} \right), \quad \forall j \tag{38}$$

$$v_j \leq w_{max} x_j, \quad \forall j \tag{39}$$

$$\sum_{i=1}^{I} \sum_{j=1}^{J} y_{ijm} \leq r, \quad \forall m \tag{40}$$

$$w_j \leq w_{max}, \quad \forall j \tag{41}$$

$$w_j \geq \left(l_{swl_j} + l_{min} \right) x_j, \quad \forall j \tag{42}$$

$$y_{ij} \geq 0, \quad \forall i, \forall j \tag{43}$$

$$x_j \in \{0, 1\}, \quad \forall j \tag{44}$$

2.2. Study Area

From the regional point of view, the study area is located in the northeastern part of the Omo Basin, southern Main Ethiopian Rift, in the western flank of the Bilate watershed (Figure 2a,b). The Bilate watershed spans a latitude of 6°34′ to 8°6′ N and a longitude of 37°46′ to 38°18′ E, with a total area of 5276.25 km². The study area in the western flank of the Bilate watershed spans a latitude of 7°17′ to 7°23′ N and a longitude of 37°46′ to 37°49′ E, covering approximately 43.6 km². Based on the Digital Elevation Model (DEM), the elevation ranges from 2255 m to 2803 m (Figure 2c). This study area is characterized by high elevation and belongs to a semi-arid climate zone.

The diverse geological features of the Bilate watershed further influence the movement and availability of groundwater in this region. The geological features of the upper Bilate watershed are quite varied, playing a major role in shaping the area's underground water sources. This region is mainly made up of rocks dating back to the Cenozoic era, particularly basalts from the Oligocene to Miocene periods, which form the geological foundation. In the Rift Valley and the highlands are rhyolites and trachyte (acidic volcanic rocks) resulting from recent Quaternary volcanic activities, which add to the intricate geological composition of the area. Moreover, a considerable number of sedimentary deposits, such as lacustrine sediments, are vital for storing and facilitating the movement of groundwater resources [27]. The geological formations in the area significantly affect how groundwater moves due to faults and fractures. These structures can impede groundwater flow depending on how permeable they are. One distinctive characteristic of the Rift Valley is its graben zones running from northeast to southwest. These zones were shaped by forces, and they are filled with a mix of volcanic and sedimentary materials [28]. These geological formations have implications for groundwater systems. The mix of sedimentary rocks forms an aquifer network where fractured volcanic rocks act as efficient aquifers because of their high permeability, while sedimentary layers can serve as both aquifers and aquitards. Faults and fractures play a role in groundwater replenishment and release by creating pathways for water to flow through the layers [27].

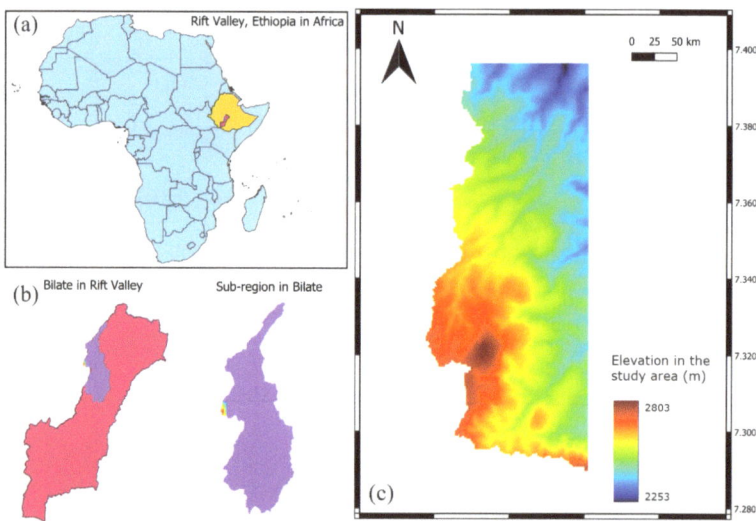

Figure 2. (**a**) Location of the southern part of the Main Ethiopian Rift in Ethiopia, Africa; (**b**) Bilate watershed (violet polygon) within the Southern Main Ethiopian Rift Valley (left) and the location of the study area within the Bilate watershed (right); (**c**) Simplified digital elevation model (DEM) of the area of interest (on the western edge of the Bilate watershed).

To determine the farm locations for the optimization model, a grid with a resolution of 1 km by 1 km was created. The center points of each grid cell were then generated and designated as farm locations, resulting in a total of 43 farms. This is for purposes of demonstration; in actual application, we would want data on where the farms really are. Previously, we created a grid with a resolution of 100 m by 100 m to predict groundwater levels within the Bilate region [19]. These 100 m resolution grid points are considered potential well locations, totaling 4264 potential sites. Figure 3 illustrates the farm and well locations on the map.

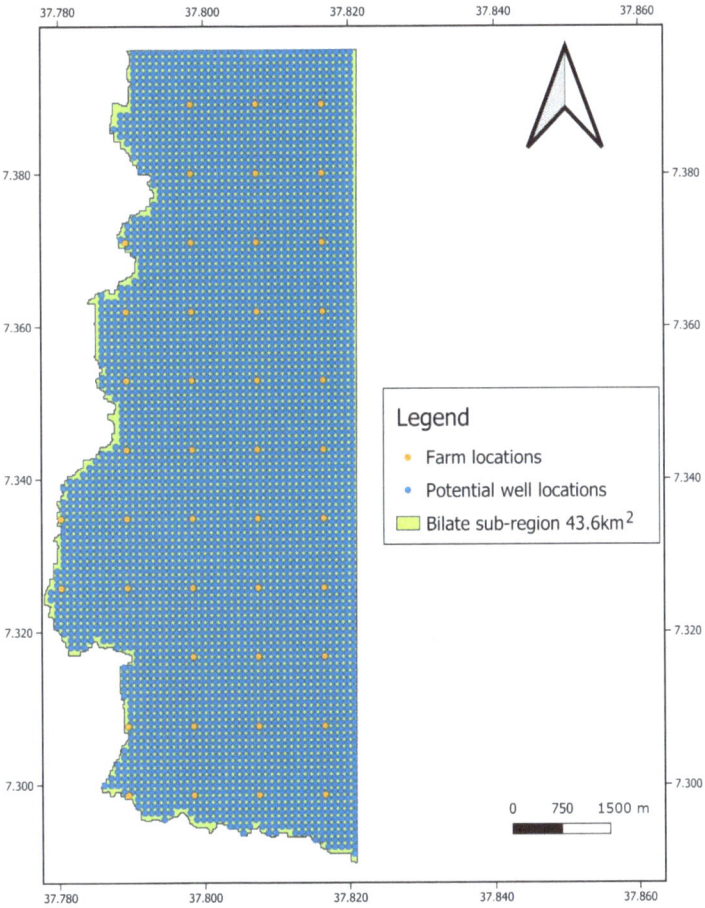

Figure 3. Farm and potential well locations in the study area (western flank of Bilate watershed).

2.3. Data

In this section, the initial settings of all the parameters are introduced. These settings were determined in consultation with Ethiopian hydrology experts who have extensive field experience.

2.3.1. Construction Costs and Per-Meter Drilling Costs

In this study, we initially assume a fixed construction cost of 5000 USD for each well site and a unit cost of 100 USD per meter drilled to solve the optimization problems. To ensure the robustness of our findings, we conduct a sensitivity analysis by considering a range of fixed cost values. The details and results of this sensitivity analysis will be presented in the next section.

2.3.2. Transportation Costs

We consider two scenarios for calculating the transportation costs from well site j to farm i. One is that the farm elevation is greater than the elevation of the well site, so that water needs to be pumped uphill. In this case, we assume that the transportation costs are very large, set at 9999 USD, when the pipe length between the farm and well is greater than 1000 m or the elevation difference between the farm and well is greater than 200 m. For the case where the pipe length is less than 1000 m and the elevation difference is less than 200 m, we use the formula below to calculate the transportation costs:

We consider two scenarios for calculating the transportation costs from well site j to farm i. In the first scenario, the farm elevation is greater than the elevation of the well site, requiring water to be pumped uphill. In this case, we assume that the transportation cost is prohibitively high, set at 9999 USD, if the pipe length between the farm and well exceeds 1000 m or if the elevation difference between the farm and well exceeds 200 m. In the other case, where the pipe length is less than 1000 m and the elevation difference is less than 200 m, we calculate the transportation costs using the following formula:

$$t_{ij} = u_c * E = u_c * \left(H_{ij} + h_{f1} * L_{ij} \right), \text{ for } H_{ij} <= 200 \text{ and } L_{ij} <= 1000$$

where u_c is the unit cost, E is the total energy, H is the elevation difference between the farm and well, h_{f1} is the friction loss for PVC pipe for scenario 1, and L is the pipe length

The second scenario considers the situation where the farm elevation is lower than the elevation of the well, allowing water to flow downhill by gravity. In this case, we assume there is no cost for pumping. However, if the pipe length exceeds 1000 m, we assume a very high transportation cost. If the pipe length is less than 1000 m, the transportation costs are calculated using the following formula:

$$t_{ij} = u_c * E = u_c * h_{f2} * L_{ij}, \text{ for } L_{ij} \leq 1000$$

where u_c is the unit cost, E is the total energy, h_{f2} is the friction loss for PVC pipe for scenario 2 and L is the distance (pipe length).

To estimate the friction loss, the Hazen–Williams equation is used for this study. The equation is defined as:

$$h_f = \frac{10.67}{d^{4.8704}} \left(\frac{Q}{C} \right)^{1.85}$$

where h_f denotes the friction head loss in meters over the water pipe, d is the inside diameter in meters, Q is the flow rate in cubic meters per second, and C is the roughness coefficient. For PVC pipe, the standard C value is 150 [29]. We assume a 3-inch PVC pipe is used, giving d = 0.0762 m. The flow rate Q is assumed to be 280 gallons per minute (gpm) for scenario 1, and 140 gpm for scenario 2 [30]. Converting these flow rates to cubic meters per second, we have approximately Q = 0.0177 m^3/s for scenario 1 and Q = 0.0088 m^3/s for scenario 2. Plugging these values into the equation, we obtain a friction loss of 0.1603 m for scenario 1 and 0.0445 m for scenario 2.

2.3.3. Water Demand

For the deterministic mixed-integer programming problem, a single demand scenario (m = 1) has been considered, with the demand set to, for example, 1000 kg/km^2 per year This simplifies the problem by assuming a constant demand across the entire study area. In this study, deterministic MIP models are constructed using demand values ranging from 800 to 1200, with increments of 100 (Section 3.1). These demand values were recommended by local water resource management experts based on regional water usage patterns and anticipated demand.

In contrast, the two-stage stochastic mixed-integer programming problem considers multiple demand scenarios (m = 10), which are assumed to follow a uniform distribution, each with a probability of 0.1. This approach accounts for the uncertainty in demand, allowing for a more robust optimization model that can adapt to various possible demand levels. To enhance the robustness and reliability of the model, different distribution assumptions are explored in the sensitivity analysis. This helps to understand the impact of varying demand patterns on the optimization outcomes and ensures that the proposed solutions are effective under a range of possible future scenarios.

2.3.4. Average Groundwater Recharge Quantity

The groundwater recharge has been estimated in a previous study within this dissertation using an observation-constrained land surface model (LSM). This model provides straightforward recharge estimates for the data-limited, water-scarce Rift Valley basin in Ethiopia. It primarily utilizes publicly available data from NASA's GLDAS Noah model, covering the years 2000 to 2022, offering a cost-effective method for estimating groundwater recharge over large regions. To validate this approach, we conducted a comprehensive analysis that integrates groundwater recharge, precipitation, and land cover and land use data, alongside temporal and spatial variability assessments. We also compared our results with existing literature on similar geographical and climatic regions. The findings suggest that our model aligns well with other methods of estimating groundwater recharge. According to the results, the study area used for the optimization model has an approximate yearly average recharge quantity of 323,000 kg/km^2.

2.3.5. Static Water Level

The static water level for each potential well location has been predicted in a previous study within this dissertation using machine learning algorithms. Models including multiple linear regression, multivariate adaptive regression splines, artificial neural networks, random forest regression, and gradient boosting regression (GBR) have been developed to forecast the static water level using 75 borehole data in Bilate watershed, a sub-basin in Rift Valley. The study incorporated 20 independent variables, such as elevation, soil type, and seasonal data for precipitation, specific humidity, wind speed, day and night land surface temperatures, and the Normalized Difference Vegetation Index (NDVI). Among these models, GBR demonstrated the highest performance, achieving an average R-squared value of 0.77 and a median absolute error of 19 m on the testing data. The static water level for the potential well locations in the optimization model has been predicted using the best-performing GBR model. The predicted static water levels in this sub-region range from 60 to 137 m.

2.3.6. Other Data

To formulate the optimization problem, the area of the study region (s) is required. As previously described, the study region covers 43.6 km^2. According to Kebede, groundwater depths are categorized as follows: very shallow (0–30 m), shallow (30–100 m), deep (100–250 m), and very deep (over 250 m). Our study focuses on finding shallow groundwater for irrigation. However, the results from the GBR model show that the predicted static water levels range from 1.6 to 245.9 m in the Bilate watershed and from 60 to 137 m within the study area for the optimization model. Considering the cost per meter of drilling, the maximum depth for well drilling (w_{max}) in this optimization model is set at 140 m. Additionally, the minimum difference between the depth of the well and the static water level (l_{min}) is set to be 1 m.

3. Results

In this study, the Gurobi solver was used to address the optimization problems, ensuring efficient and accurate solutions [31]. The software tools employed included Python 3.12 for scripting and algorithm implementation [32] and QGIS 3.24.1 for geographic data processing and visualization [33]. These tools provided a robust framework for developing and analyzing the well placement optimization problem.

3.1. Model Optimization Results

Three models were employed in this study: a deterministic mixed-integer programming (MIP) model with demand values set at 800, 900, 1000, 1100, and 1200; a two-stage stochastic mixed-integer programming (SMIP) model with demand scenarios randomly generated from different uniform distributions (U (400, 1200), U (500, 1300), U (600, 1400), U (700, 1500), and U (800, 1600)); and another two-stage SMIP model with demand scenarios randomly generated from five different normal distributions with means of 800, 900, 1000, 1100, and 1200, each with a standard deviation of 300. The means of the demand scenarios generated from each uniform distribution are 800, 900, 1000, 1100, and 1200, which match the settings of the deterministic MIP model. A similar rationale applies to the two-stage SMIP model with the normal distribution. The results show that the average total costs for the three models were 42,264,421, 42,298,103, and 42,317,641, respectively (Table 1). The demand values considered in this study range from 800 to 1200 kg/km^2, with increments of 100. As mentioned in Section 2.3.3, these values were recommended by local water resource management experts, reflecting regional water usage patterns and anticipated demand. Considering different scenarios allows us to account for the inherent uncertainty in demand patterns, which can significantly influence well placement and resource allocation decisions.

Table 1. Results for the optimization models.

	Deterministic MIP with Fixed Demand					
Demand (kg/km^2)	800	900	1000	1100	1200	Average
Optimized number of wells	43	44	45	46	46	45
Total costs (USD)	42,177,807	42,224,273	42,265,578	42,306,933	42,347,516	42,264,421
Total fixed construction costs (USD)	215,000	220,000	225,000	230,000	230,000	224,000
Total drilling costs (USD)	41,719,176	41,728,938	41,738,701	41,748,463	41,758,326	41,738,721
Total transportation costs (USD)	243,631	275,335	301,877	328,470	359,190	301,700
	Two-stage SMIP (Demand scenarios: uniform distribution)					
Demand Instance	U (400, 1200)	U (500, 1300)	U (600, 1400)	U (700, 1500)	U (800, 1600)	Average
Optimized number of wells	45	45	46	46	46	45.6
Total costs (USD)	42,215,743	42,256,368	42,297,266	42,339,950	42,381,187	42,298,103
Total fixed construction costs (USD)	225,000	225,000	230,000	230,000	230,000	228,000
Total drilling costs (USD)	41,748,893	41,758,756	41,768,351	41,778,241	41,787,793	41,768,406
Total transportation costs (USD)	241,850	272,612	298,915	331,709	363,394	301,696
	Two-stage SMIP (Demand scenarios: normal distribution)					
Demand Instance	N (800, 300)	N (900, 300)	N (1000, 300)	N (1100, 300)	N (1200, 300)	Average
Optimized number of wells	44	44	46	46	46	45.2
Total costs (USD)	42,233,743	42,275,352	42,316,817	42,360,022	42,402,269	42,317,641
Total fixed construction costs (USD)	220,000	220,000	230,000	230,000	230,000	226,000
Total drilling costs (USD)	41,762,470	41,772,055	41,783,204	41,793,020	41,802,142	41,782,578
Total transportation costs (USD)	251,273	283,297	303,613	337,002	370,127	309,062

3.2. Optimal Well Layouts

As demand values increased in the deterministic MIP model, the number of wells placed also increased, leading to higher total costs. This is because higher demand requires additional groundwater extraction to meet the increased water needs. As a result, more wells are necessary to ensure sufficient water supply across the region. For uniform distribution demand scenarios, slightly more wells were placed, but the total costs were lower than those from models with demand scenarios generated from a normal distribution, possibly due to deeper wells being placed, leading to higher drilling costs. Table 1 summarizes the optimized number of wells, total costs, and cost components for the three models.

Maps were created to illustrate the optimized well placement for each experiment and each model (Figures 4–6). In the deterministic MIP model experiments, well placements/layouts were largely similar, with differences observed in the placement of wells 419, 420, 506, 3666, 3780, and 3781 across experiments based on assumed water demand. For the two-stage SMIP model using different uniform distributions, wells 276, 419, 420, 3551, 3666, 3781, 3629, and 3744 were placed slightly differently across experiments. With an increase in the mean demand, the number of wells placed slightly increased. In the two-stage SMIP model using a normal distribution, decisions regarding the placement of wells 338, 419, 420, 506, 1457, 1458, 2170, 2398, 2285, 3551, 3629, 3666, 3744, 3780, and 3781 varied slightly across models.

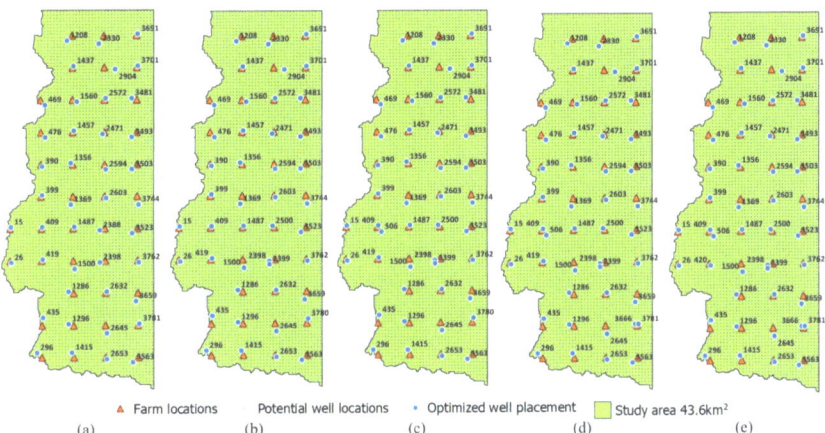

Figure 4. (a–e) Five deterministic MIP experiments assuming water demands of 800, 900, 1000, 1100, and 1200 kg/km², respectively. (Note: The numbers on the map represent the well ID). The layout of a few wells is different across different models, including well ID 419, 420, 506, 3666, 3780, and 3781, which are all located in the southern part of the study area.

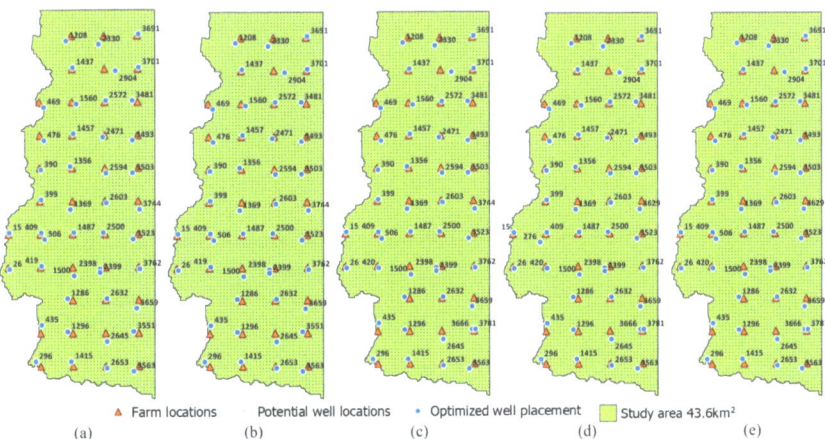

Figure 5. (a–e) Illustration of five two-stage SMIP experiments depicting water demand scenarios randomly generated from five uniform distributions: U (400, 1200), U (500, 1300), U (600, 1400), U (700, 1500), and U (800, 1600), respectively. (Note: The numbers on the map represent the well ID). The layout of a few wells is different across different models, including well ID 276, 419, 420, 3551, 3666, 3781, 3629, and 3744. Well 276 is located in the eastern part of the study area in map (**d**). Wells 3629 and 3744 are situated in the mid-eastern region, while the remaining wells are located in the southern part of the study area.

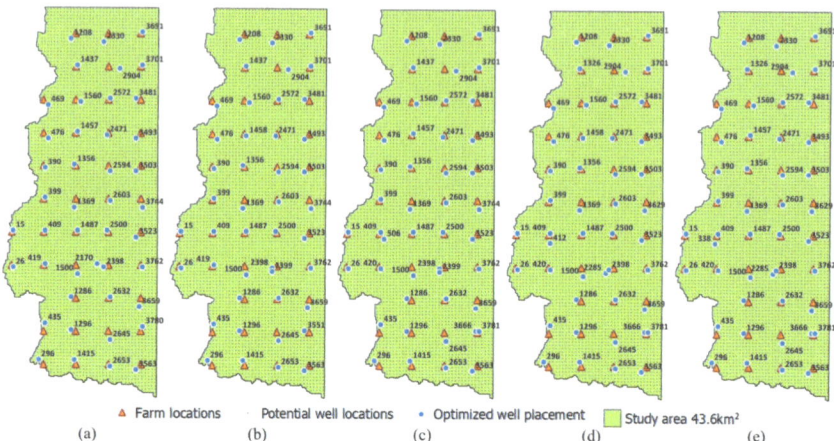

Figure 6. (**a–e**) Visualization of five two-stage SMIP experiments depicting water demand scenarios randomly generated from five normal distributions: N (800, 300), N (900, 300), N (1000, 300), N (1100, 300), and N (1200, 300), respectively. (Note: The numbers on the map represent the well ID). The layout of a few wells is different across different models, including well ID 338, 419, 420, 506, 1457, 1458, 2170, 2398, 2285, 3551, 3629, 3666, 3744, 3780, and 3781. Well 338 is located in the eastern part of the study area in map (**e**). Wells 1457 and 1458 are located in the northwestern part. Wells 3629 and 3744 are situated in the mid-eastern region, while the remaining wells are located in the southern part of the study area.

4. Discussion

4.1. Impact of Different Parameters

In this study, we consider the experiment for the uniform distribution U (600, 1400) for the demand scenarios of the two-stage SMIP as the base model. The impact of different parameters demonstrated in this section is created based on this model.

4.1.1. Impact of Different Fixed Construction Costs

The fixed construction costs are the coefficients of the decision variable x_j (whether to place a well at location j). The objective total costs include three components: total fixed construction costs, total per-meter drilling costs (also known as unit costs), and total transportation costs. The range of fixed construction costs is set from 3000 to 18,000, with an increment of 1000. As the fixed construction costs increase (Figure 7), the total costs also rise.

The differences between consecutive objective values are calculated and plotted in Figure 8. From this plot, we can see that when the fixed cost increases from 5000 to 6000, the difference between the objective values is 46,000. When the fixed cost increases from 6000 to 7000, the difference is approximately 44,000. This clear drop from 46,000 to 44,000 occurs because, when the fixed cost increases from 6000 to 7000, the total number of wells placed decreases from 46 to 45. Additionally, the well layout changes, indicating that the model chooses to drill deeper wells with higher drilling costs rather than maintaining the same number of wells with higher construction costs to minimize the total costs. Another observation is an increase in total cost from 44,000 to 44,700. This occurs because, when the fixed costs increase from 7000 to 8000, the number of wells decreases from 45 to 44. Consequently, the well layout changes to satisfy the water demand, leading to approximately 7600 higher transportation costs and 100 higher per-meter drilling costs at the 8000 fixed cost compared to the 7000 fixed cost. Additionally, there is a notable drop in total cost from 44,000 to 41,454. This decrease happens because, when the fixed costs increase from 11,000 to 12,000, the number of wells to be placed decreases from 44 to 43. Subsequently, the number of wells placed remains constant at 43, resulting in a consistent difference of 43,000 in objective values as the fixed construction cost increases by increments of 1000.

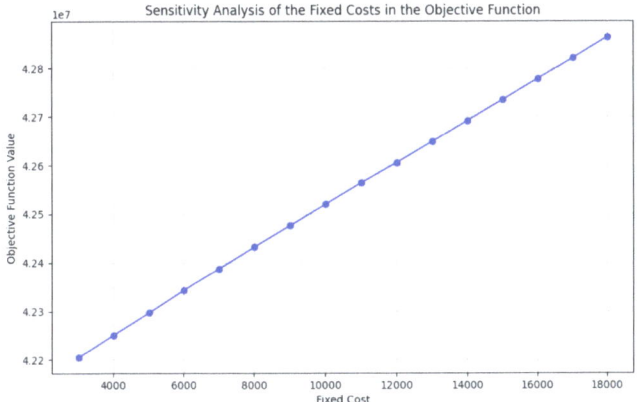

Figure 7. Impact of the fixed construction costs on the objective values.

Figure 8. Increment of objective values vs. fixed construction costs.

4.1.2. Impact of Different Per-Meter Drilling Costs

The per-meter drilling costs are the coefficient of the decision variable w_j (depth of the well placed at location j). Similarly, the range of the per-meter drilling costs is set from 30 to 150, with an increment of 10. As the per-meter drilling costs increase, the total costs also rise (Figure 9).

Figure 10 shows the differences between consecutive objective values as the per-meter drilling cost increases. From this plot, we can see a clear drop in the increment of the objective when the per-meter drilling cost increases from 110 to 120. This occurs because, as the per-meter drilling cost becomes higher, the model tends to select shallower wells that are slightly farther from the farm to minimize the total costs.

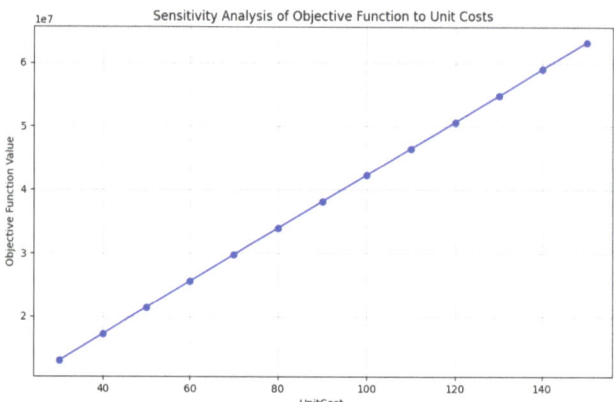

Figure 9. Impact of the different per-meter drilling costs on the objective values.

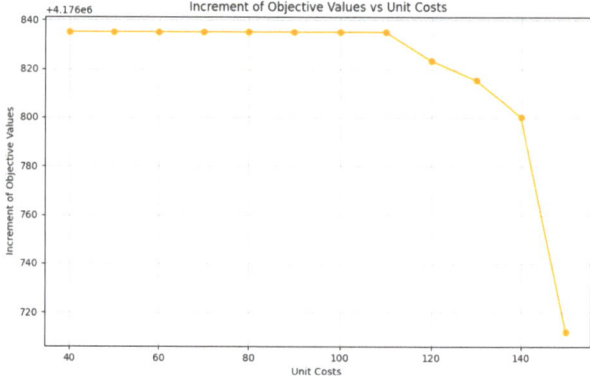

Figure 10. Increment of objective values vs. per-meter drilling costs (unit costs).

4.1.3. Impact of Different Demand Scenarios

Since the demand scenarios are generated randomly, the increments in the objective values between each pair of demand scenarios are not constant. Some increments are big and some are small. Two big increment examples are described next.

First, in Figure 11b, there is an obvious increase in the objective values when moving from U (1200, 2000) to U (1300, 2100). This can be explained by the fact that as the mean demand increases from 1600 to 1700, the number of open wells increases from 47 to 49. This increase in the number of open wells is the main contributor to the substantial increment in the objective values (Figure 12a).

When increasing the demand range from U (600, 1400) to U (1100, 1900), the number of open wells remains the same (Figure 12a). However, the objective values see a greater increase when moving from U (800, 1600) to U (900, 1700); see Figure 11a. This phenomenon can be explained by the fact that as the mean demand increases, different well layouts are likely to be required to meet the rising demand. This leads to higher transportation costs and necessitates deeper wells, which increases capacity and results in higher total drilling costs. Note that the capacity directly impacts the total quantity of water transported from the well to the farm, which in turn affects the demand (Figure 12).

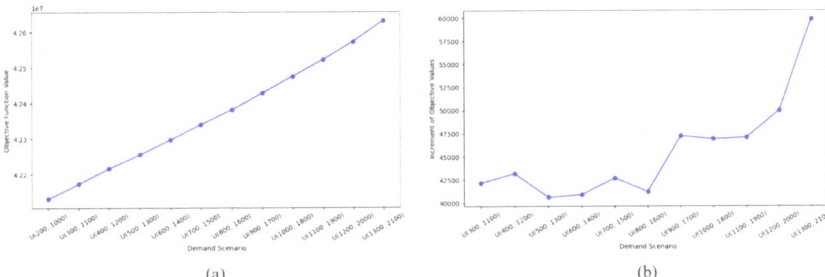

Figure 11. (**a**) Objective values vs. different demand scenarios (**b**) Increment of the objective values vs. different demand scenarios.

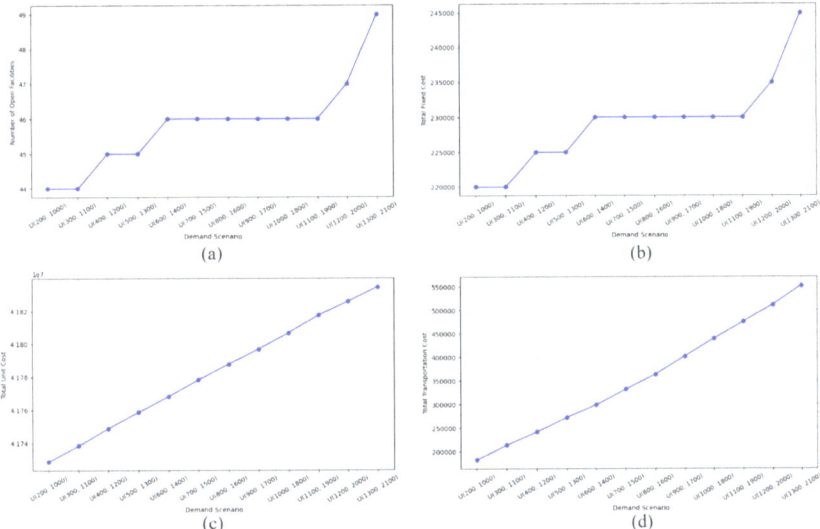

Figure 12. (**a**) Number of open wells vs. demand scenarios; (**b**) Total fixed costs vs. demand scenarios; (**c**) Total unit costs vs. demand scenarios; (**d**) Total transportation costs vs. demand scenarios.

4.2. Out-of-Sample Performance

To evaluate out-of-sample performance, we first formulate the problem with a deterministic demand (in-sample demand data) and solve it. The solution of the first-stage decision variables, whether to place a well at site j (x_j) and the depth of the well to drill at site j (w_j), is then fixed. Using these fixed first-stage decision variables, we solve a new problem with stochastic demand scenarios (out-of-sample demand data) to obtain the objective values. By fixing the same first-stage decision variable results, the second-stage demand scenarios can be randomly generated from the same distribution multiple times, producing a set of objective values.

Next, we formulate the problem with a stochastic demand scenario (in-sample demand data) and solve it. The solutions of the first-stage decision variables are then fixed to solve a problem with the same stochastic demand scenario (out-of-sample demand), but with newly randomly generated demand scenarios, iterating this process multiple times to obtain objective values. Finally, the two cases—the deterministic case and the stochastic case—can be compared by analyzing the mean and standard deviation of their objective values to evaluate out-of-sample performance.

In this study, four cases were investigated, each iterated 50 times. The uniform distribution was chosen to range from 600 to 1400, with a mean of 1000. For the normal distribution, a mean of 1000 and a standard deviation of 155 were selected. This choice was made to ensure that 99% of the values fall between 600 and 1400 in a normal distribution, as the standard deviation is approximately calculated as $400/2.576 = 155$. The deterministic demand was set at 1100. This value was chosen instead of 1000 to ensure the feasibility of the model.

It is important to note that the number of wells to place is 45 for the deterministic MIP with a demand of 1000. In contrast, the SMIP with demand scenarios generated from a uniform distribution U (600, 1400) indicates that 46 wells should be placed. Therefore, if we use an in-sample demand with a deterministic value of 1000, the placement of 45 wells may not satisfy the out-of-sample demand scenarios generated from U (600, 1400).

The mean and standard deviation of the objective values from experiments using the out-of-sample demand data for the four cases are summarized in Table 2. We compared Case 1 with Case 2 and Case 3 with Case 4, and created boxplots in Figures 13 and 14. The results indicate that the deterministic case 1 achieves 11% higher average out-of-sample objective costs compared to the stochastic case 2. Additionally, the average costs for deterministic case 3 are 4% higher than those for stochastic case 4. This demonstrates that the stochastic approach yields lower average costs and is more effective than the deterministic approach. The boxplots also show that the stochastic cases perform better than the deterministic cases, with tighter distributions and lower variability in the objective values. This indicates that stochastic models consistently perform better across a wider range of out-of-sample scenarios.

Table 2. Results of the out-of-sample performance.

	Case 1	Case 2	Case 3	Case 4
In-sample demand	Deterministic with demand = 1100	Uniform distribution U (600, 1400)	Deterministic with demand = 1100	Normal distribution N (1000, 155)
Out-of-sample demand	Uniform distribution U (600, 1400)	Uniform distribution U (600, 1400)	Normal distribution N (1000, 155)	Normal distribution N (1000, 155)
Mean	47,124,183	42,494,243	44,200,128	42,625,237
Standard deviation	938,970	107,608	437,315	172,968

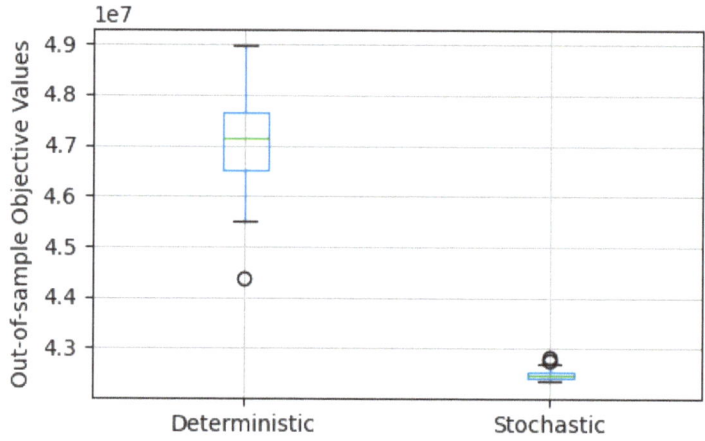

Figure 13. Comparison of the out-of-sample performance between a deterministic in-sample demand and a stochastic out-of-sample demand generated from a U (600, 1400).

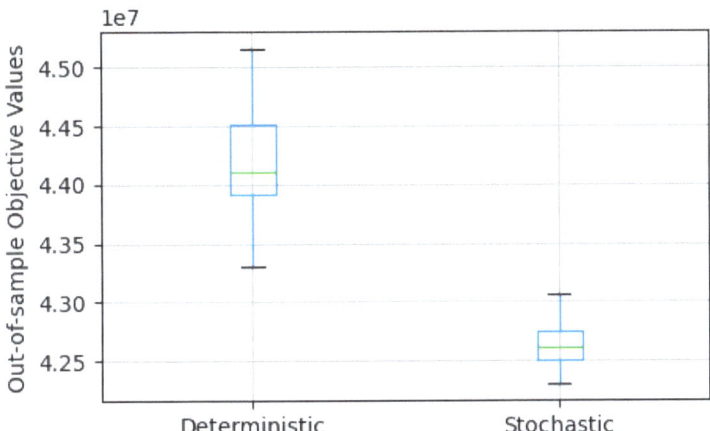

Figure 14. Comparison of the out-of-sample performance between a deterministic in-sample demand and a stochastic out-of-sample demand generated from a N (1000, 155).

These results emphasize that incorporating demand uncertainty through stochastic optimization (Case 2 and Case 4) enables more adaptive and responsive well placement strategies, reducing the risk of inefficiencies or resource shortages that may occur with deterministic models (Case 1 and Case 3), which tend to result in higher objective costs. Stochastic models are better equipped to handle variability, offering consistent performance across various scenarios, which is essential for long-term groundwater management. Their flexibility and resilience to changing demand provide a more robust strategy for well placement and resource allocation, ensuring the sustainable management of groundwater resources.

4.3. Limitations and Future Research Directions

The optimization study has several limitations that should be acknowledged. Firstly, the model is restricted to a very small region due to computational constraints, which limits the generalizability of the findings to larger areas. This restriction means that the results may not fully capture the complexities and variances that could occur in a broader geographical context. However, it is important to recognize that applying these algorithms to larger regions or areas with greater hydrological variability presents its own set of challenges, even though the stochastic model shows clear advantages for well placement optimization. Secondly, the study does not incorporate data on the actual locations of farms, which is crucial for accurate well placement and optimizing irrigation strategies. Without this data, the model's applicability to real-world scenarios is limited. Additionally, while the study examines the uncertainty of demand scenarios, it does not consider other important uncertainties, such as variations in static water levels and groundwater recharge. These factors could significantly impact the effectiveness of the well placement optimization and should be included in future research to provide a more comprehensive analysis.

Future research could focus on conducting extensive scenario analyses to explore the impacts of different demand scenarios on well placement optimization, which could provide valuable insights into the robustness and reliability of the model under various geological and hydrogeological conditions. Examining the effect of uncertainty in static water levels and groundwater recharge should be considered, also with regard to the progressing climate change. Additionally, exploring and comparing different optimization methods could provide a broader perspective on the most effective approaches for well placement optimization. This development could enable optimizing well placement over a larger region, potentially providing a broader impact. Furthermore, collecting data on the actual locations of farms is crucial for more accurate well placement. Another future research direction could be developing adaptive optimization techniques that can dynamically update well placement decisions based on real-time data and changing conditions, which could enhance the flexibility and responsiveness of the decision support system.

5. Conclusions

In this study, we formulated both deterministic MIP and two-stage SMIP problems to optimize well placement with minimal total costs while satisfying farm demand and ensuring sustainable groundwater extraction. Our study area is located in the northeastern part of the Omo Basin in the southern Main Ethiopian Rift, on the western flank of the Bilate watershed. Experiments were conducted with different distributions of demand scenarios for the two-stage SMIP models. The results show that the average total costs for the MIP are slightly higher than those for the SMIP models. This is intuitive because the two-stage stochastic model incorporates the benefit of long-term planning by considering uncertain demand scenarios. By accounting for these uncertainties, the SMIP can optimize well placement more effectively over time, potentially avoiding costly misallocations and adjustments that a deterministic MIP model might incur when it does not account for demand variability. Consequently, the SMIP's more comprehensive approach can lead to more cost-effective solutions in the long run.

Maps of well placement for each experiment have been created, revealing slight changes in the layout of the wells near a few farms in the south across different experiments with varying demand scenarios. In practice, this could guide drilling decisions by incorporating expert judgment based on soil conditions and geographic features. The analysis of the impact of varying fixed construction costs, per-meter drilling costs, and demand scenarios on the objective values shows that total costs increase with these costs and demand. The increments in objective values relative to these three parameters have also been examined and explained. The results of the out-of-sample performance show that the stochastic cases perform better than the deterministic cases, with tighter distributions and lower variability in the objective values (11% and 4% lower costs than deterministic cases 1 and 3, respectively). This indicates that the solutions from the stochastic models are more flexible and resilient to changes in demand.

Author Contributions: Conceptualization, W.L. and K.B.L.; methodology, W.L.; software, W.L.; validation, W.L., M.M.F., K.B.L., P.H., R.D.-B. and K.V.; formal analysis, W.L.; investigation, W.L. and K.B.L.; resources, M.M.F., R.D.-B. and K.V.; data curation, W.L.; writing—original draft preparation, W.L.; writing—review and editing, W.L., M.M.F., K.B.L., P.H., R.D.-B. and K.V.; visualization, W.L.; supervision, K.B.L., P.H. and R.D.-B.; project administration, K.B.L., P.H., and R.D.-B.; funding acquisition, K.B.L., P.H., R.D.-B. and K.V. All authors have read and agreed to the published version of the manuscript.

Funding: This research was partially supported by a graduate research fellowship to W.L. from Czech Geological Survey development aid project No. ET-2023-006-RO-43040 and George Mason University's Center for Resilient and Sustainable Communities.

Data Availability Statement: The data presented in this study will be available on interested request from the corresponding author.

Acknowledgments: This work represents a collaboration among George Mason University, Arba Minch University, and Global Map Aid with support by the Czech Geological Survey. This elaboration was processed as a part of the development aid project by the Czech Geological Survey No. ET-2023-006-RO-43040 (to K. Verner) entitled "Improving the quality of life by ensuring availability and sustainable management of water resources in Sidama Region and Gamo and Gofa Zones (Ethiopia)" financed by the Czech Republic through the Czech Development Agency.

Conflicts of Interest: Author Rupert Douglas-Bate was employed by the company Global MapAid. The remaining authors declare that the research was conducted in the absence of any commercial or financial relationships that could be construed as a potential conflict of interest.

References

1. Williams, F.M. *Understanding Ethiopia*; Wolfgang Eder AV; Springer International Publishing: New York, NY, USA, 2016.
2. Verner, K.; Šíma, J.; Megerssa, L. (Eds.) *A Synopsis of Regional Geology and Hydrogeology of Ethiopia*; Czech Geological Survey: Staré Brno, Czech Republic, 2024.
3. UNICEF. Ethiopia For Every Child, Clean Water! Available online: https://www.unicef.org/ethiopia/every-child-clean-water (accessed on 30 May 2024).
4. UNHCR. Horn of Africa Food Crisis Explained. Available online: https://www.unrefugees.org/news/horn-of-africa-food-crisis-explained/ (accessed on 24 May 2024).
5. The World Bank. The World Bank in Ethiopia. Available online: https://www.worldbank.org/en/country/ethiopia/overview (accessed on 30 May 2024).
6. Scanlon, B.R.; Fakhreddine, S.; Rateb, A.; de Graaf, I.; Famiglietti, J.; Gleeson, T.; Grafton, R.Q.; Jobbagy, E.; Kebede, S.; Kolusu, S.R.; et al. Global Water Resources and the Role of Groundwater in a Resilient Water Future. *Nat. Rev. Earth Environ.* **2023**, *4*, 87–101. [CrossRef]
7. Carter, R.C.; Bevan, J.E. Groundwater Development for Poverty Alleviation in Sub-Saharan Africa. In *Applied Groundwater Studies in Africa*; CRC Press: Boca Raton, FL, USA, 2008; ISBN 978-0-429-20736-5.
8. Mahmood, H.A.; Al-Fatlawi, O. Well Placement Optimization: A Review. *AIP Conf. Proc.* **2022**, *2443*, 030009. [CrossRef]
9. Tafteh, A.; Babazadeh, H.; Ebrahimipak, N.; Kaveh, F. Optimization of Irrigation Water Distribution Using the Mga Method and Comparison with a Linear Programming Method. *Irrig. Drain.* **2014**, *63*, 590–598. [CrossRef]
10. Ma, T.; Wang, J.; Liu, Y.; Sun, H.; Gui, D.; Xue, J. A Mixed Integer Linear Programming Method for Optimizing Layout of Irrigated Pumping Well in Oasis. *Water* **2019**, *11*, 1185. [CrossRef]
11. Kuvichko, A.; Ermolaev, A. *Mixed-Integer Programming for Optimizing Well Positions*; OnePetro: Richardson, TX, USA, 2020.
12. Liu, X.; Wang, S.; Huo, Z.; Li, F.; Hao, X. Optimizing Layout of Pumping Well in Irrigation District for Groundwater Sustainable Use in Northwest China. *Hydrol. Process.* **2015**, *29*, 4188–4198. [CrossRef]
13. Halilovic, S.; Böttcher, F.; Kramer, S.C.; Piggott, M.D.; Zosseder, K.; Hamacher, T. Well Layout Optimization for Groundwater Heat Pump Systems Using the Adjoint Approach. *Energy Convers. Manag.* **2022**, *268*, 116033. [CrossRef]
14. Bayer, P.; Bürger, C.M.; Finkel, M. Computationally Efficient Stochastic Optimization Using Multiple Realizations. *Adv. Water Resour.* **2008**, *31*, 399–417. [CrossRef]
15. Emerick, A.A.; Silva, E.; Messer, B.; Almeida, L.F.; Szwarcman, D.; Pacheco, M.A.C.; Vellasco, M.M.B.R. *Well Placement Optimization Using a Genetic Algorithm with Nonlinear Constraints*; OnePetro: Richardson, TX, USA, 2009.
16. Sharifipour, M.; Nakhaee, A.; Yousefzadeh, R.; Gohari, M. Well Placement Optimization Using Shuffled Frog Leaping Algorithm. *Comput. Geosci.* **2021**, *25*, 1939–1956. [CrossRef]
17. AlQahtani, G.; Alzahabi, A.; Kozyreff, E.; de Farias, I.R., Jr.; Soliman, M. A Comparison between Evolutionary Metaheuristics and Mathematical Optimization to Solve the Wells Placement Problem. *Adv. Chem. Eng. Sci.* **2013**, *3*, 30–36. [CrossRef]
18. Yin, J.; Pham, H.V.; Tsai, F.T.-C. Multiobjective Spatial Pumping Optimization for Groundwater Management in a Multiaquifer System. *J. Water Resour. Plan. Manag.* **2020**, *146*, 04020013. [CrossRef]
19. Li, W.; Finsa, M.M.; Laskey, K.B.; Houser, P.; Douglas-Bate, R. Groundwater Level Prediction with Machine Learning to Support Sustainable Irrigation in Water Scarcity Regions. *Water* **2023**, *15*, 3473. [CrossRef]
20. Li, W.; Finsa, M.M.; Houser, P.; Laskey, K.B.; Douglas-Bate, R. Groundwater Recharge Estimation in Data-Limited Water-Scarce Regions. *Hydrol. Res.* **2024**. Submitted.
21. Fan, Z.; Ji, R.; Chang, S.-C.; Chang, K.-C. Novel Integer L-Shaped Method for Parallel Machine Scheduling Problem under Uncertain Sequence-Dependent Setups. *Comput. Ind. Eng.* **2024**, *193*, 110282. [CrossRef]
22. Fan, Z.; Chang, K.-C.; Ji, R.; Chen, G. Data Fusion for Optimal Condition-Based Aircraft Fleet Maintenance with Predictive Analytics. *J. Adv. Inf. Fusion (JAIF)* **2023**, *18*, 102.
23. Fan, Z.; Ji, R.; Lejeune, M.A. Distributionally Robust Portfolio Optimization under Marginal and Copula Ambiguity. *J. Optim. Theory Appl.* **2024**. [CrossRef]
24. Ji, R.; Lejeune, M.A.; Fan, Z. Distributionally Robust Portfolio Optimization with Linearized STARR Performance Measure. *Quant. Financ.* **2022**, *22*, 113–127. [CrossRef]
25. McCormick, G.P. Computability of Global Solutions to Factorable Nonconvex Programs: Part I—Convex Underestimating Problems. *Math. Program.* **1976**, *10*, 147–175. [CrossRef]
26. Ahmed, S. Two-Stage Stochastic Integer Programming: A Brief Introduction. In *Wiley Encyclopedia of Operations Research and Management Science*; John Wiley & Sons, Ltd.: Hoboken, NJ, USA, 2011; ISBN 978-0-470-40053-1.
27. Cholo, B.E.; Tolossa, J.G. Identification of Groundwater Recharge and Flow Processes Inferred from Stable Water Isotopes and Hydraulic Data in Bilate River Watershed, Ethiopia. *Hydrogeol. J.* **2023**, *31*, 2307–2321. [CrossRef]
28. Deribew, K.T.; Arega, E.; Moisa, M.B. Prediction of the Topo-Hydrologic Effects of Soil Loss Using Morphometric Analysis in the Upper Bilate Watershed. *Bull. Eng. Geol. Environ.* **2024**, *83*, 162. [CrossRef]
29. Powers, P.J.; Corwin, A.B.; Schmall, P.C.; Kaeck, W.E.; Herridge, C.J.; Morris, M.D. Appendix A: Friction Losses for Water Flow Through Pipe. In *Construction Dewatering and Groundwater Control*; John Wiley & Sons, Ltd.: Hoboken, NJ, USA, 2007; pp. 597–602, ISBN 978-0-470-16810-3.

30. FlexPVC. Water Flow Charts Based on Pipe Size. Available online: https://flexpvc.com/Reference/WaterFlowBasedOnPipeSize shtml (accessed on 25 May 2024).
31. Gurobi Optimization, LLC. Gurobi Optimizer Reference Manual 2024. Available online: https://www.gurobi.com/documentation/11.0/refman/index.html (accessed on 25 May 2024).
32. Python Software Foundation. The Python Language Reference, Version 3.12. Available online: https://docs.python.org/3/reference/index.html (accessed on 25 May 2024).
33. QGIS. Development Team QGIS Geographic Information System. Available online: https://qgis.org/en/site/ (accessed on 1 August 2023).

Disclaimer/Publisher's Note: The statements, opinions and data contained in all publications are solely those of the individual author(s) and contributor(s) and not of MDPI and/or the editor(s). MDPI and/or the editor(s) disclaim responsibility for any injury to people or property resulting from any ideas, methods, instructions or products referred to in the content.

Article

Evaluation of the Effect of Agricultural Return Flow on Water Quality, Water Quantity and Aquatic Ecology in Downstream Rivers

Taeseong Kang [1], Yongchul Shin [2], Minhwan Shin [1], Dongjun Lee [3], Kyoung Jae Lim [4] and Jonggun Kim [4,*]

1. EM Research Institute, Chuncheon-si 24408, Republic of Korea; kangstar2003@naver.com (T.K.)
2. Department of Agricultural Civil Engineering, Kyungpook National University, Daegu 41566, Republic of Korea
3. College of Forestry, Wildlife and Environment, Auburn University, Auburn, AL 36849, USA
4. Department of Regional Infrastructure Engineering, Kangwon National Univiersity, Daegu 41566, Republic of Korea
* Correspondence: jgkim@kangwon.ac.kr

Abstract: Agricultural water serves various functions, including public interest purposes, beyond its primary role in agricultural production. In order to evaluate the various public interest purposes of agricultural water, a quantified study of the effect of agricultural water on river flow, water quality, and aquatic ecosystems is needed. Therefore, this study quantified the impact of agricultural water on the environmental and ecological maintenance function of downstream rivers, taking into account the return flow of agricultural water in rural areas. To this end, first, the effect of agricultural return flow on river maintenance function was evaluated by comparing the return flow quantity calculated using the reservoir supply data with the simulated river flow rate through the SWAT model. Second, the effect of the agricultural return flow on the downstream river environmental ecological function was analyzed using the optimal flow rate results calculated through the PHABSIM model. The lastly, the effect of agricultural water by farming period on the water quality of downstream rivers was analyzed. As a result of the analysis, it was found that the return flow of agricultural water had a large effect on the river flow rate in the case of the non-rainy season, but the optimal ecological flow rate was not satisfied. In the case of river water quality, it was confirmed that the effect of agricultural water (mainly considered as a pollutant) was not significant, except for the drainage duration of rice paddies. Therefore, it can be understood that agricultural water is not only used for the purpose of production but can also have a positive impact on the aquatic ecology of downstream rivers.

Keywords: agricultural return flow; SWAT model; PHABSIM model; water quality; aquatic ecology

1. Introduction

Agricultural water holds significant importance not merely in terms of its primary function for crop production and irrigation but also plays a crucial role from a broader public interest standpoint. This encompasses aspects such as maintaining the ecosystem balance, supporting biodiversity, and contributing to the landscape's aesthetic and recreational value. Additionally, it serves as a vital resource for sustaining rural communities and their economies, thereby underlining its multifaceted value beyond mere agricultural productivity. From this perspective, it is necessary to secure agricultural water and re-evaluate its value (in terms of water quantity, quality, and aquatic ecology). In Korea, the demand for agricultural, domestic, and industrial water is continuously increasing [1]. However, the development of new water sources for securing the water supply is facing considerable restrictions due to environmental protection concerns, leading to a growing interest in identifying alternative water resources [2,3]. In addition, efficient water resource use is subject to integrated project management policies and the demands of the times [4,5].

Recently, changes in farming practices and the industrialization of rural areas have led to an increased demand for agricultural water in these regions [6]. Additionally, the frequency of droughts and floods, exacerbated by climate change, has heightened the need for the rational and efficient management of agricultural water to ensure a stable water supply [7–9]. Therefore, the reasonable and efficient management of agricultural water is required to ensure a stable water supply [10]. Agricultural water, which accounts for 41% of Korea's total water resources, is broadly defined as water used for agricultural activities ranging from livestock production to irrigation water for crop cultivation and has social and environmental functions, such as atmospheric circulation, groundwater recharge, provision of river maintenance water, and amenity and ecosystem conservation [11–13]. In particular, the agricultural return water drained from rice paddies flows into nearby rivers (rapid return flow) and groundwater (delayed return water), greatly affecting the river ecosystem and performing a multidisciplinary function in the agricultural environment [14,15].

Recently, there has been an increasing interest in the multifunctional aspects of agricultural water [16,17]. The perception of agricultural water, previously understood solely as irrigation water necessary for crop growth, is shifting towards a more inclusive concept that encompasses various environmental and ecological uses, such as improving rural living environments and supporting the environmental flow for ecosystems. Return flows from agricultural water play a crucial role in efficient water use and the maintenance of the environmental flow for ecosystems. They are significantly important for watershed water supply planning, predicting river flow conditions, determining irrigation water usage, preventing river dry-up, protecting aquatic ecosystems, and ensuring biodiversity. These processes underscore the importance of managing agricultural water not just for agricultural productivity but also for broader environmental and ecological health [18,19].

In Korea, according to previous studies [20], the return rate of agricultural water greatly varies from 38.1% to 70.5%, depending on the regional characteristics and crop cultivation methods used. Accurate calculation of the return rate of agricultural water is required for rational and economical water resource use and water management. Song et al. [21] calculated the amount of irrigation return during 2011–2012 for the irrigation district of the Idong Reservoir. They found that, the higher the supply of agricultural water, the higher the irrigation return rate and the return rate of agricultural water and that agricultural return water was the main component contributing to the river flow rate. Also, as part of a survey on river water use, a systematic survey of the return flow rate of agricultural water was conducted; based on the results, approximately 35% of the supply of agricultural water is expected to return to rivers.

Recent changes in rivers, including the installation of hydraulic structures for water diversion and flood control, water quality deterioration from various pollution sources, and alterations to the water cycle system due to industrialization and urbanization, have led to river contamination. Therefore, restoring river functions to conserve river ecosystems and create environmentally stable rivers requires the efficient management of environmental flows [22,23]. To recover the self-purification capability and normal functions of rivers, various conditions are needed, including creating habitats for aquatic organisms, blocking sources of pollution, and maintaining appropriate river flow levels, with the maintenance of ecological flows in rivers serving as a fundamental component for other habitat formations. The Ministry of Environment in Korea defines environmental ecological flow as the minimum flow necessary to maintain the health of aquatic ecosystems. The Water Environment Conservation Act mandates consideration of the environmental ecological flow when announcing river maintenance flows, allowing for the announcement of such flows at representative points in small streams, dried-up tributaries, or branches [24]. For larger rivers, like national rivers, maintenance flows have often been established based on drought flows rather than calculated flows considering the aquatic ecosystem, necessitating the determination and development of necessary flow measures to protect habitats of endangered species in these rivers [25].

Although recent studies on calculating the return flows of agricultural water and estimating ecological flows have been actively carried out [26–28], research on the impact of agricultural water on the quantity and quality of water in downstream rivers and its effects on aquatic ecosystems remains insufficient. In the past, the management of water resources in rivers focused on quantity and quality [29], but as interest in the environment has increased, the aspect of aquatic ecology has been emphasized in river flow management [30,31]. Furthermore, agricultural water is often perceived as a potential source of pollution in downstream rivers from an environmental perspective. However, there is still a lack of seasonal and quantitative studies conducted on this matter. From this view, the calculation of the environmental and ecological flow rates of rivers can be of great significance, and the integrated management of river quantity, water quality, and aquatic ecology is needed.

Therefore, in this study, the effect of agricultural water on the water quantity and quality and the ecology in downstream rivers were evaluated considering agricultural water return in a rural basin. This research assessing the effective impact of agricultural water return flows not only in terms of quantity and quality but also from an aquatic ecology perspective can serve as fundamental data for quantitatively assessing the public benefits of agricultural water. This information can be foundational for decision-making in policy and institutional development aimed at the efficient management of agricultural water.

2. Materials and Methods

In this study, the river flow and water quality were monitored in the main stream and inflow streams of the reservoir irrigation district (this study site), and aquatic health was evaluated using data from the National Biometric Network and previous studies [32,33]. In addition, the return flow for agricultural water was calculated using the reservoir supply data, and the long-term flow rate of downstream rivers was simulated using the Soil and Water Assessment Tool (SWAT) model to evaluate the impact of the return water on river function maintenance. Using the simulation results of the optimal ecological flow, the impact of return water for agricultural use on the environmental ecological function of the downstream rivers was evaluated (Figure 1).

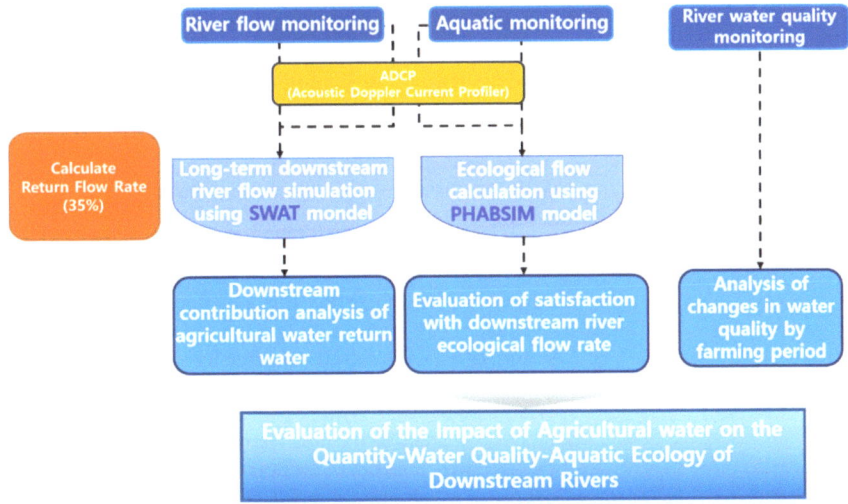

Figure 1. Flowchart of this study.

2.1. Study Area

This study was conducted in the Heungeop Reservoir irrigation district located in Maeji-ri, Heungeop-myeon, Wonju-si, Gangwon-do, Korea, where it is possible to analyze the contribution of agricultural water to downstream river pollution. The Heungeop Reservoir was built to supply agricultural water, and the irrigation area is approximately 165 ha. The water volume of the reservoir is 1,098,000 tons, the basin area is 1750 ha, and the full water area is 25 ha. The reservoir is managed by the Wonju Branch of the Rural Community Corporation.

There are six weather stations in Wonju, where the Heungeop Reservoir is located. The Wonju weather station was selected as a representative observatory for this study. The average annual precipitation (2011–2020 years) was 1209.9 mm, which is less than the national average precipitation (1397.7 mm). The total annual precipitation in Wonju-si shows a decreasing trend; it decreased significantly from 2011 to 2014, then gradually increased until 2020, except in 2019. In particular, in 2020, the precipitation was approximately 545 mm higher than in 2019. The total annual precipitation was 1063.6 mm in 2017, 1229.2 mm in 2018, and 771.9 mm in 2019. Rainfall occurrences of 30 mm or more occurred 10 times in 2017, 14 times in 2018, and 3 times in 2019 and, thus, were the most prevalent in 2018.

2.2. Monitoring for Analysis of the Contribution of Agricultural Return Flow to Downstream Rivers

2.2.1. River Flow Monitoring

To assess the quantitative and aquatic ecological impact of agricultural water on downstream rivers, precise investigations into the river flow are essential. This necessitates accurate cross-sectional surveys and ongoing flow measurements. In this study, the water level gauges were installed at monitoring points selected during a preliminary field survey to measure the river flow rate and to monitor the long-term water-level and flow rate changes (Figure 2a). River flow monitoring was conducted at two points: Bonghyeon Bridge (H1), located downstream in the benefiting area, and Heongeop 2 Bridge (H2), where the Majicheon (MJ) and Seogokcheon (SG) tributaries converge. For flow monitoring at each point, a float level meter and river gauge were installed on the bridge piers, and flow rates were measured using the velocity-area method. In addition, tributaries inflowing from downstream areas other than the Seogok Stream were identified, and at the inflow point (H3), regular on-site flow measurements were conducted instead of using a water gauge. The primary aim of this study is to analyze the contribution of agricultural return flows to the flow rates in the downstream rivers at the extents of the Heungeop Reservoir beneficiary areas. Consequently, monitoring sites were selected to assess the proportion of agricultural return flows compared to the total flow originating from the Heungeop Reservoir watershed, including the beneficiary regions. The final downstream location, H1, was chosen, and other specific points such as Seogok Stream (H2) entering at the middle part of the Heungeop Reservoir beneficiary area and a downstream tributary (H3) were selected to exclude their flows. In addition, the flow rate change before the Seogok Stream was measured by monitoring the flow rate in the Maeji Stream at point H4. The river flow monitoring data were used as correction data for the basin model used for the long-term river flow simulation. The river flow rate was measured using the velocity-area method, in which the average flow rate and the flow rate in a cross-section of a small section are calculated by dividing the flow channel into several small sections and measuring the depth of the water in each small section using a flow meter. In this study, the water flow and water quality were monitored at 11 time points during rainfall and non-rainfall periods.

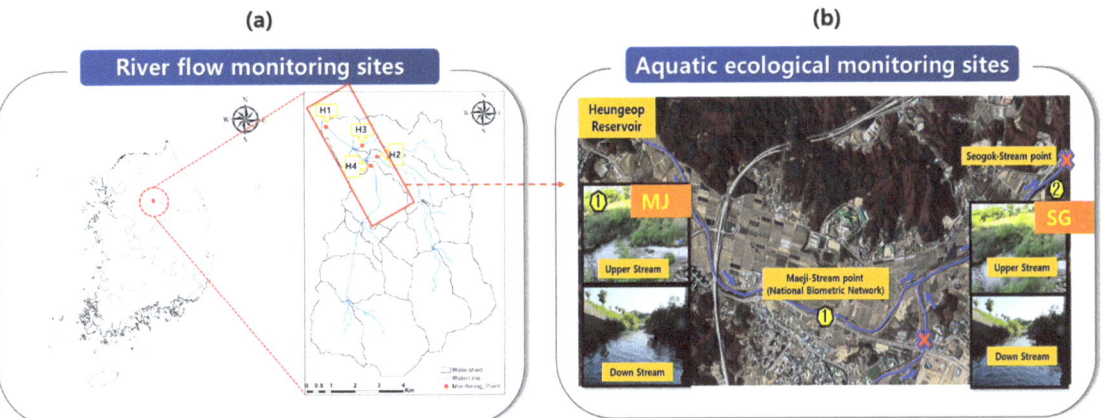

Figure 2. (**a**) River flow, water quality, and (**b**) aquatic ecological monitoring points. Red box means the irrigated area as shown in (**b**). Arrows mean the flow direction. H1~H3 means flow monitoring sites. MJ(MaeJi) and SG(SeoGok) are the aquatic ecological monitoring sites.

2.2.2. Aquatic Ecological Monitoring

In this study, monitoring was conducted to assess the health of aquatic ecosystems in a downstream river of a district receiving agricultural water, utilizing methods such as the Trophic Diatom Index (TDI), Benthic Macroinvertebrate Index (BMI), and Fish Assemblage Index (FAI) at selected aquatic ecological monitoring sites. Additionally, representative fish species were selected based on fish surveys, and their optimal ecological flow was calculated to evaluate the impact of agricultural water on the downstream river's aquatic ecology. Monitoring was conducted on downstream rivers directly affected by agricultural return flows from the Heungeop Reservoir (Figure 2b), and the investigation point was selected to reflect the current operating aquatic ecological monitoring sites (operated by the Ministry of Environment) to verify aquatic ecosystem monitoring data. The SG monitoring site was selected for its suitability for aquatic ecological monitoring, including fish surveys, due to adequate space available near the downstream flow and water quality monitoring sites conducted in this study. For the MJ site, located upstream, a currently operating national biometric network site was chosen, allowing for comparative validation of the aquatic monitoring performed in this study. Therefore, we selected point MJ, which was, to some extent, secured from the upstream flow, and we selected the point SG near the Bonghyeon Bridge located at the end of the beneficiary area as the branch.

Considering the physical water system and seasonal characteristics, the major habitat characteristics and cluster structures (including species diversity, wind patterns, and uniformity) were analyzed, and aquatic health was analyzed and evaluated. The monitoring was conducted twice at two points (MJ and SG) on 30 May and 21 September 2020 (once a month for each site), before and after the flood season, to exclude the effects of rainfall and typhoons in the summer. The field survey followed the "Guidelines for Surveying and Evaluating River Ecosystem Health" by the Ministry of Environment. As outlined in the guidelines, surveys were conducted once each in the spring and autumn annually. Monitoring was conducted in a relatively uniform condition, without any issues such as rainfall, and all biotas were investigated considering the hydraulic and hydrologic parameters, such as river width, water depth, and flow rate. In addition, data of the site from the National Biometric Network were collected and used to evaluate the aquatic ecology health. The health of the river aquatic ecosystems at each branch was evaluated using three indices:

- Fish Assessment Index (FAI),
- Tropic Diatom Index (TDI),
- Benthic Macroinvertebrate Index (BMI).

In the case of fish, the species composition, population, share, dominant species, and least dominant species were analyzed for the fish collected through monitoring. In this study, fish communities were collected along a 50–100 m stretch of the water system around the survey points using a throw net (mesh size 7 × 7 mm) and a dip net (mesh size 5 × 5 mm). To ensure the quantification of fish communities between locations, collections with the throw net were repeated 10 times, and those with the dip net were conducted over 50 min. The collected fish were preserved in a 10% formalin solution. In addition, the distribution of legally protected species, such as rare and endangered species of classes I and II and foreign, introduced species, was analyzed. Tolerance guilds and feeding guilds of fish were analyzed. Tolerance guilds were classified into sensitive species, intermediate species, and tolerant species, respectively, and feeding guilds were classified into insectivores, omnivores, carnivores, and herbivores, according to their feeding characteristics. The FAI was calculated based on the scores for eight metrics (total number of domestic species; number of riffle-benthic species, sensitive species; proportion of the population of tolerant species; proportion of omnivores; proportion of domestic insectivores; total number of domestic species collected; and proportion of individuals with disease, tumors, fin damage, and other anomalies), and the end score (0–100) was classified as very good, good, normal, bad, or very bad. Cluster analysis was used to calculate the likelihood, variety, uniformity, and abundance of species based on the species and populations collected quantitatively at each survey point [34,35].

In the case of epilithic diatoms, the total taxonomic groups, species composition, density, dominant species, subdominant species, and cumulative dominant frequency of epilithic diatoms were analyzed, and cluster analysis was conducted. The dominant species was determined as the species with the largest population in each survey section. The epilithic algae collection was conducted in riffle areas of the streambed, ideally located near the center of the river with average flow rates between 10 and 50 cm/s and composed primarily of substrates larger than gravel size. The substrate chosen for algae collection was the most stable and solid natural material available within the river, specifically flat-surfaced rocks. Approximately 250 cm^2 of the collection area was brushed off to gather samples. The collected material was preserved on-site in 10% formalin and later identified in the laboratory. The health of the aquatic ecosystem at each point was evaluated using the TDI.

In the case of benthic macroinvertebrates, the total taxonomic groups, species composition, density, dominant species, and subdominant species of benthic macroinvertebrates were analyzed, and the dominance, diversity, abundance, and evenness indices of small benthic macroinvertebrates were calculated for each monitoring point. Collection was conducted at each survey site using a Surber net (30 × 30 cm^2, 1 mm mesh size), taking into account the flow and environment of each location. Quantitative collection was performed once at riffle and pool areas at each site. The collected samples were preserved on-site in 10% formalin, transported to the laboratory, and then sorted using a sieve (1 mm mesh). Finally, the samples were stored in 75% ethanol. Aquatic ecosystem health was evaluated by calculating the BMI based on benthic macroinvertebrates for each survey section and calculating scores based on the evaluation criteria. In addition, we calculated the ecological score of the benthic macroinvertebrate community (ESB), which is another biological index widely used for rivers, wetlands, and lakes. The ESB is used to evaluate the water environment by assigning an environmental quality score (Q_i) to each benthic macroinvertebrate species according to the National Natural Environment Survey Guidelines issued by the Ministry of Environment. The calculation method of each index is shown in Table 1.

Table 1. Calculation method for each index.

Index	Formula for Calculation	Indicators	Reference
Species diversity index (H')	$H' = -\sum_{i=1}^{s} p_i \ln p_i$	H': diversity S: total number of appearances Pi: percentage of the i-th species	[35]
Evenness index (J)	$J = H'/\ln(S)$	J: evenness H': diversity S: total number of appearances	[36]
Richness index (R)	$R = (S-1)/\ln(N)$	R: richness S: total number of appearances N: total cover degree	[37]
Dominance index (D)	$D = (n1 + n2)/N$	D: dominance $n1$: cover degree of first dominant species $n2$: cover degree of second dominant species	[34]
Ecological score of benthic macroinvertebrate community (ESB)	$ESB = \sum_{i=1}^{s} Q_i$	ESB: ecological score of benthic macroinvertebrate community S: total number of species Qi: environmental quality score of i-th species	[38]
Trophic diatom index (TDI)	$TDI = 100 - [(WMS \times 25) - 25]$	Aj: abundance of j species in sample (%) Sj: pollution sensitivity of j species Vj: indicator value of j species	[39]
Benthic macroinvertebrate index (BMI)	$BMI = \left(4 - \frac{\sum_{i=1}^{n} s_i \cdot h_i \cdot g_i}{\sum_{i=1}^{n} h_i \cdot g_i}\right) \times 25$	BMI: benthic macroinvertebrates index i: the number assigned to the species n: the number of species si: the saprobic value of the species i hi: the relative abundance of the species i gi: the indicator weight value of the species i	[40]
Fish assessment index (FAI)	$FAI = M1 + M2 + M3 + M4 + M5 + M6 + M7 + M8$	M1: total number of native fish species M2: number of riffle-benthic species M3: total number of sensitive species M4: proportion of tolerant species individuals M5: proportion of omnivorous individuals M6: proportion of native insectivore individuals M7: total number of individuals M8: proportion of abnormal individuals	[41]

2.2.3. Monitoring of River Water Quality

To assess the impact of agricultural water discharge on downstream river water quality during the farming season, additional monitoring of the river water quality was conducted at point H1 (Bonghyeon Bridge) located at the outlet of the basin. In this study, biological oxygen demand (BOD5) and total phosphorus (T-P) were chosen as indicators to evaluate the quality of agricultural water. BOD5 was selected because it measures the level of organic matter contamination and is widely used in water pollution assessments. T-P was chosen as it represents non-point source pollutants such as agricultural and livestock wastewater and serves as an indicator of eutrophication in rivers and lakes. Moreover, in Korea, the current Total Maximum Daily Load (TMDL) program sets and manages water quality targets for the BOD5 and T-P. The BOD5 was measured using the method that involves collecting a water sample, measuring its initial dissolved oxygen (DO) level, incubating it at 20 °C in the dark for five days, remeasuring the DO, and calculating the oxygen depletion to determine the organic pollution level. Total phosphorus (TP) was measured by digesting the water sample with acid to convert all phosphorus forms to orthophosphate, which was then quantified using a spectrophotometer after reacting with a color reagent, typically resulting in a blue color. Water quality monitoring was conducted on the same dates as the river flow monitoring, including 14 May, 16 June, 25 June, 14 July, 29 July, 10 August, 27 August,

11 September, 25 September, 8 October, and 26 October, totaling 11 sessions. Water quality analysis was performed by collecting samples and transporting them to a laboratory, where they were analyzed according to the water pollution testing standards. According to the management manual for each crop (agricultural day management schedule) proposed by the Rural Development Administration, the management schedule varies depending on the climate characteristics and temperature of the year, but the rice field is usually drained completely for 30–35 days from the end of September to October. From November to December, the rice straw is ploughed to prepare for farming activities in the following year. In this study, the effect of agricultural water discharge on the water quality of downstream rivers during rainy periods was evaluated in different farming management periods.

2.3. Modeling for the Analysis of the Contribution of Agricultural Water to Downstream River Parameters

2.3.1. Determining Long-Term Flow Rate Fluctuations Using the SWAT Model

Long-term river flow was simulated using the SWAT model to identify long-term flow rate fluctuations in the basin. The SWAT model can simulate the behavior of agricultural chemicals based on soil properties, land use, and land management status, as well as the water quality of basins with complex characteristics, using similar principles [42]. The predictive ability of the SWAT model was assessed based on the coefficient of determination (R^2) and Nash–Sutcliffe efficiency (NSE) between the simulated and measured results. R^2 is a statistical measure that indicates the correlation between simulated and measured values and takes a value between 0 and 1, with 1 indicating a completely linear relationship. NSE is a normalized statistic that indicates the relative magnitude of the residual variance between measured and simulated values. It takes a value from $-\infty$ to 1, with 1 indicating a perfect match between measured and simulated values [43]. The interpretation of these parameters for model evaluation is presented in Table 2.

Table 2. Statistic interpretation guidance for model assessment.

Statistic	Measurement Frequency	Very Good	Good	Fair	Poor	Reference
R^2	Daily	$0.80 < R^2 \leq 1$	$0.70 < R^2 \leq 0.80$	$0.60 < R^2 \leq 0.70$	$R^2 \leq 0.60$	[44]
	Monthly	$0.86 < R^2 \leq 1$	$0.75 < R^2 \leq 86$	$0.65 < R^2 \leq 0.75$	$R^2 \leq 0.65$	
NSE	Monthly	$0.75 < NSE \leq 1$	$0.65 < NSE \leq 0.75$	$0.50 < NSE \leq 0.65$	$NSE \leq 0.50$	[45]

The parameters of the SWAT model for long-term river flow simulation were optimized based on the measured results of the river flow rate monitoring at points H1 and H2, using the SWAT-CUP calibration and uncertainty program. SWAT-CUP was developed for model calibration and validation and sensitivity and uncertainty analyses of SWAT models. In this study, the parameters were optimized using the SUFI2 algorithm, which can quantify and represent parameter uncertainty. The parameters calibrated in this study are described in Table 3.

Table 3. Definition and scope of the parameters for SWAT model calibration.

Num	Parameter	Description	Min	Max
1	CN2	Initial SCS runoff curve number for moisture condition 2	35	95
2	ALPHA_BF	Baseflow alpha factor (1/days)	0	1
3	GWQMN	Threshold depth of water in the shallow aquifer required for return flow to occur (mm H_2O)	0	5000
4	LAT_TTIME	Lateral flow travel time (days)	0	180
5	CH_K2	Effective hydraulic conductivity in main channel alluvium	−0.01	500
6	SOL_AWC	Available water capacity of the soil layer (mm H_2O/mm soil)	0	1

Table 3. Cont.

Num	Parameter	Description	Min	Max
7	ESCO	Soil evaporation compensation factor	0	1
8	SURLAG	Surface runoff lag coefficient	0.05	24
9	SOL_K	Saturated hydraulic conductivity (mm/h)	0	2000
10	SLSOIL	Slope length for lateral subsurface flow (m)	0	150
11	EPCO	Plant uptake compensation factor	0	1
12	GW_DELAY	Groundwater delay time (days)	0	500
13	RCHRG_DP	Deep aquifer percolation fraction	0	1
14	REVAPMN	Threshold depth of water in the shallow aquifer for "revap" or percolation to the deep aquifer to occur (mm H_2O)	0	500
15	GW_REVAP	Groundwater "revap" coefficient	0.02	0.2
16	HRU_SLP	Average slope steepness (m/m)	0	1

2.3.2. Evaluation of the Contribution of Return Water to Downstream River Parameters

The agricultural return water quality varies greatly depending on the natural environment, crop cultivation method, and regional characteristics. According to long-term comprehensive water resource planning, rationalization of rural water use planning, and the Ministry of Land, Infrastructure, and Transport in Korea, approximately 35% of the agricultural water supply is expected to return to rivers (rapid and delayed returns) in Korea [24]. In this study, the environmental preservation function of agricultural water return was indirectly evaluated using the available data, and the return was quantified by setting the return rate of agricultural water to 35%. The effect of the agricultural water return quantity on the river was evaluated using the calculated return quantity results and the river flow rate data simulated with the SWAT model.

2.3.3. Environmental Ecological Flow Estimation Using the Physical Habitat Simulation System (PHABSIM) Model

The ecological flow rate in rivers downstream of the Heungeop Reservoir was calculated using the PHABSIM model, which was developed to evaluate the effect of agricultural water return on the aquatic ecology of downstream rivers. The PHABSIM model predicts the physical habitat of aquatic species according to their growth stage based on changes in the flow characteristics (e.g., flow rate and water depth) and calculates the optimal flow rate required for the aquatic species of interest in relation to the available habitat area (weighted available area). This model can calculate the ecological flow rate for representative fish species in the simulated area and was used to quantitatively evaluate the environmental conservation function of agricultural water return based on the environmental ecological flow rate (Figure 3). The PHABSIM model uses river cross-section data, flow rate data by depth, and flow rate data by water level as the input data. In this study, we utilized Habitat Suitability Indices (HSI) for the depth and velocity for key fish species as presented through field surveys in the existing literature [46]. Additionally, river cross-section data for the targeted research area were sourced from the River Maintenance Basic Plan for point H1. River sections and depth-specific velocities were also measured on-site using Acoustic Doppler Current Profiler (ADCP) equipment and incorporated into the model (Figure 4). Following the collection of this baseline data, the PHABSIM model was employed to calculate the relationship curves between Weighted Usable Area (WUA) and the flow rate based on the Instream Flow Incremental Methodology (IFIM). These curves were then used to estimate environmental ecological flows corresponding to the physical habitats. The effect of agricultural water on downstream rivers was evaluated by calculating the optimal ecological flow rate for representative fish species in the Seogok Stream, a terminal river in the study area. Ecological flow satisfaction was analyzed by comparing the simulation results of the long-term outflow of the rivers at the end of the beneficiary area downstream of the Heungeop Reservoir obtained with the SWAT model and the optimal ecological flow results calculated using the PHABSIM model.

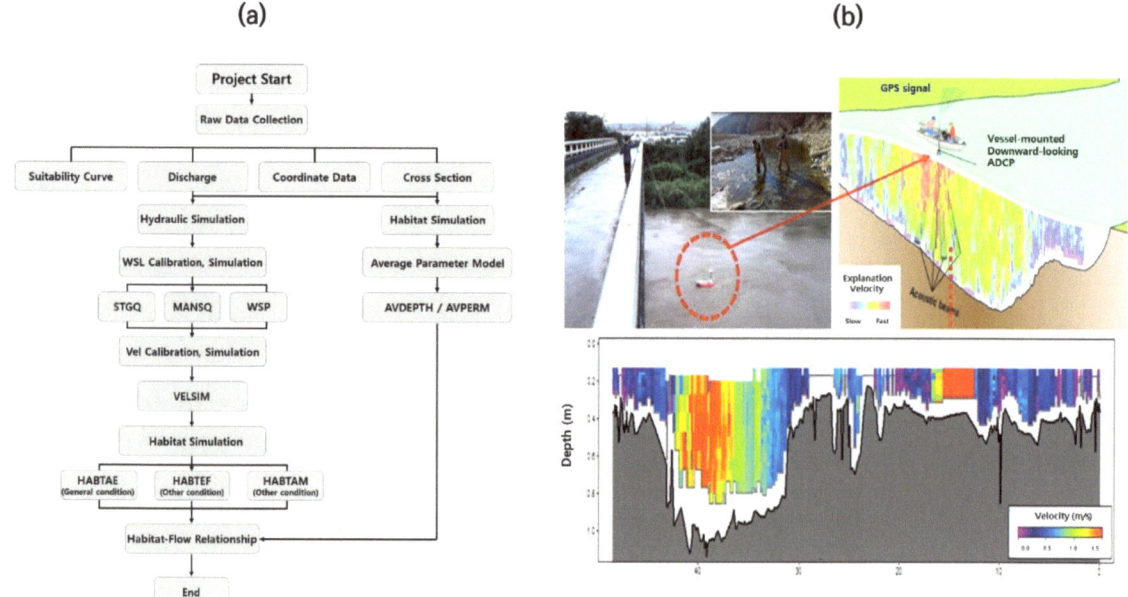

Figure 3. PHABSIM modeling process and river sections/depth-specific velocity measurements using ADCP (**a**,**b**). Red dotted circle means the ADCP equipment in operation on site.

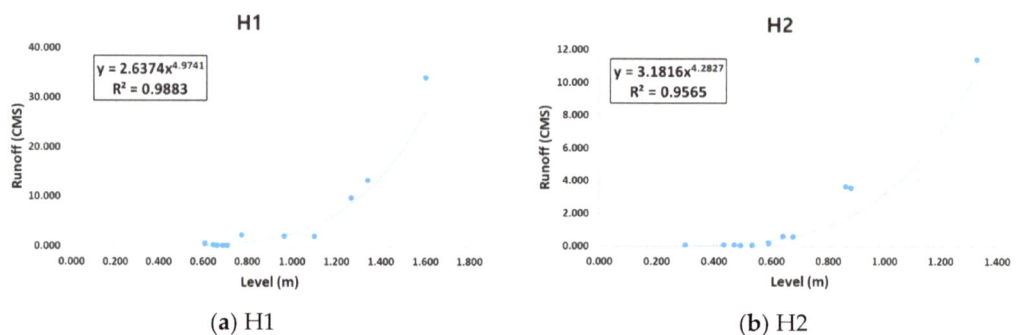

(**a**) H1 (**b**) H2

Figure 4. Water level-flow curves for monitoring sites H1 (Bonghyeon Bridge) and H2 (Heongeop 2 Bridge).

3. Results and Discussion

3.1. Water Quantity and Water Quality Monitoring Results

3.1.1. River Flow Monitoring Results

The flow rate at point H1 showed the largest change after rainfall and was the lowest, at 0.170 m^3/s, in May, the dry season. Similarly, the flow rate at point H2 was the highest, at 3.641 m^3/s, in July, and the lowest, at 0.059 m^3/s, in May. At point H3, the basin area was small, and the flow rate was 1.206 m^3/s in August, the flood season. At point H4, the flow rate was 0.527 m^3/s at the 9th measuring time point but only 0.055 m^3/s at the 11th measuring time point after a relatively long period of rainless days (26 days). The water level flow curves for H1 and H2 were calculated to be y = 2.6374 × 4.9741 (R^2 = 0.988) and y = 3.1816 × 4.2827 (R^2 = 0.956), respectively, using flow rate data obtained from river flow monitoring and water level data at H1 and H2, where water level gauges were installed (Figure 4).

3.1.2. River Water Quality Monitoring Results

To evaluate the effect of agricultural water return on the aquatic ecological health of downstream rivers, which was the purpose of this study, long-term aquatic ecology monitoring at the Seogok Stream point at the end of the beneficiary area downstream of the Heungeop Reservoir was conducted. River water quality monitoring at point H1 (at the outlet of the basin) revealed an average BOD5 value of 1.3 mg/L (0.6–2.6 mg/L), indicating a river living environment standard of Ia (very good) to II (slightly good) (Table 4). The average T-P value was 0.067 mg/L (0.034–0.135 mg/L), representing Ib (good) to III (normal).

Table 4. River living environment standards.

Grade	Standard	BOD	T-P
Very good	Ia	<1	<0.02
Good	Ib	<2	<0.04
Slightly good	II	<3	<0.1
Normal	III	<5	<0.2
Slightly bad	IV	<8	<0.3
Bad	V	<10	<0.5
Very bad	VI	>10	>0.05

The factors that affect river water quality vary widely depending on the characteristics of the watershed and of the pollution source. In the rural basin, we considered that pollutants running off from crop fields flow into the river depending on farming activities and affect the river water quality; therefore, the river water quality was analyzed in different farming periods. In the rice transplanting period in May, the BOD5 value was 1.7 mg/L, and in the active tillering stage in early June, it was 2.3 mg/L. In late June and early July, the water quality was good, as indicated by BOD5 values of 0.7 mg/L and 1.2 mg/L, respectively. At the end of July, when the water was drained, the water pollutant concentration was the highest, with a BOD5 value of 2.6 mg/L. After the water was drained, the water pollutant concentration decreased, and it increased again in October.

The T-P value was 0.088 mg/L at the end of May, and it decreased from the end of June to mid-July but was the highest, at 0.135 mg/L, around the end of July, when the water was drained. Then, the T-P value decreased, but it increased again during the complete draining period from late September to October, showing a similar tendency to the BOD5 parameter. The average BOD5 value during the water draining period was 1.5 mg/L, which was 1.6 times higher than that in other periods (1.0 mg/L). The average T-P value during the water draining period was 0.081 mg/L, which was 1.9 times higher than the average value in other periods (0.043 mg/L) (Figure 5).

The agricultural land area in the irrigation district of the Heungeop Reservoir is 815.2 ha, accounting for 14.5% of the total basin area (Figure 6). The fertilizers used for soil amendment and agricultural activities during rainy periods, including the end of May, the end of July to early August, and the end of September to October, were found to affect water pollution in the river. The effect of agricultural water discharge on the river water quality was limited, except for run-off during intensive farming periods. Therefore, appropriate measures should be taken to reduce pollutant runoff during water drainage periods. In addition, the target water quality for small basins in the future should be set according to the characteristics of the basin and in agricultural areas, considering the characteristics of the different farming periods. However, as this study did not account for the effect of prior precipitation and analyzed only short-term monitoring data, it is necessary to analyze long-term river water flow and quality and account for prior rainfall in future studies.

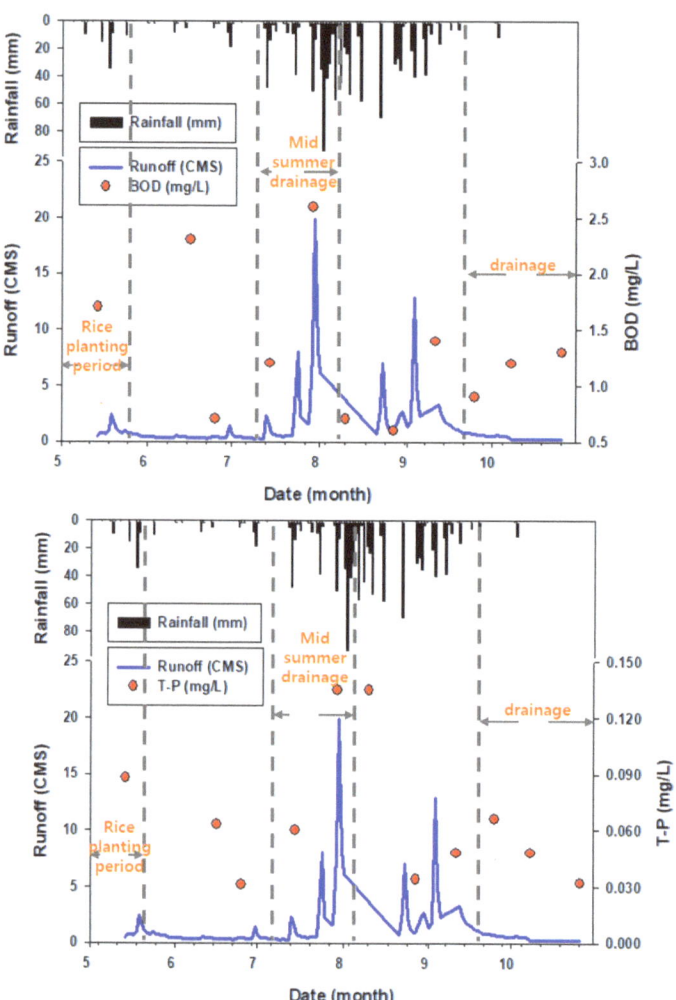

Figure 5. Analysis of the water quality(BOD and T-P) according to the farming period at point H1.

3.2. Aquatic Ecology Monitoring Results

In the case of fish, a total of six families and 10 species of freshwater fish were surveyed at the aquatic ecological monitoring sites (Table 5). A total of 100 freshwater fish from three families and six species were detected at the MJ site, and 188 fish from five families and eight species were detected at the SG site. No legally protected species (rare and endangered species) were detected. One specimen of *Odontobutis interrupta* was found in each of the MJ and SG sites, respectively, and *Micropterus salmoides*, an ecosystem disturbance species, was detected at the SG site in the autumn. In terms of comparative abundance according to family, at the MJ site, *carps* were the most abundant, at 66.6%, and the *Korean dark sleeper* and *Gobiidae* each had an abundance of 16.7%. At the SG site, *carps* were the most abundant, at 50.0%, while *loach, catfish, black rockfish,* and *Korean dark sleeper* each had an abundance of 12.5%. These results indicate that *carps* were predominant in both streams. At the species level, *Chinese minnow* was the most abundant, at 45.0%, at the MJ site, followed by the *striped shiner*, *pale chub*, *Korean spotted sleeper*, and *Amur goby*. At the SG site, the *striped shiner* was the most abundant, at 40.4%, followed by the *pale chub*, *Chinese minnow*, *Korean spotted sleeper*, and *dojo loach*. Thus, *striped shiner* was the most abundant, followed by *pale chub*, *Chinese minnow*, *Korean spotted sleeper*, and *Amur goby*.

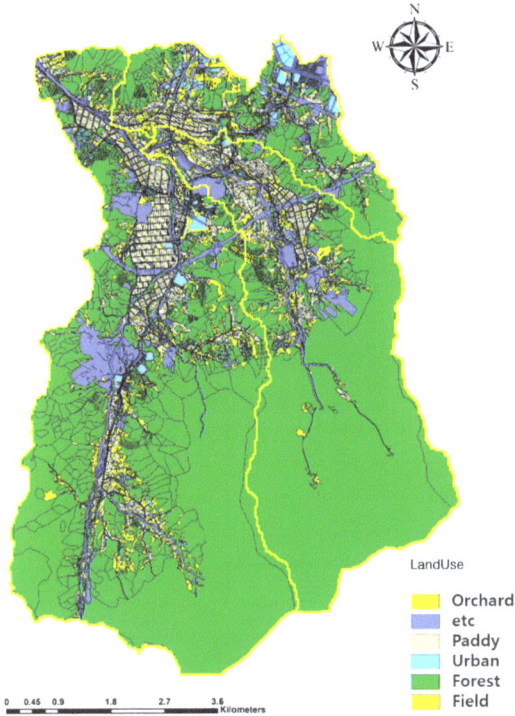

Figure 6. Land use in the study area.

We next calculated the clustering index for each stream. The dominance index was 0.36–1.00 for the MJ site and 0.71–0.86 for the SG site, and the diversity index was 0.44–1.49 for the MJ site and 1.06–1.40 for the SG site. The abundance index was higher for the SG site (0.88–1.34) than for the MJ site (0.34–1.09). The evenness index for the MJ site was 0.63–0.83 and, for the SG site, was 0.66–0.78. The FAI for freshwater fish was evaluated as normal (grade C), with an average value of 46.9 for the MJ site and 53.2 for the SG site. In the first survey conducted in May, the FAI for the MJ site was 68.8, which is considered good (grade B), whereas, in the second survey conducted in September, it was 25.0, which is interpreted as bad (grade D). The MJ site has various flow rates and various riverbed structures; however, numerous fish were lost due to flooding caused by excessive rainfall, and the FAI was low because the fish population did not recover in the second survey. In contrast, the SG site was located downstream, and due to the influence of various pollutants, such as drainage of the surrounding agricultural land and soil, the FAI was evaluated as normal (grade C).

In the case of epilithic diatoms, a total of 2 order, 3 suborders, 8 families, 20 genera, and 54 species were identified. For the MJ site, the average cluster index was 0.39, the diversity index (H') was 1.93, the abundance index (R) was 2.31, and the evenness index (J) was 0.58. In the first survey, the diversity, abundance, and evenness indices were found to be high, and in the second survey, the dominance index was high. For the SG site, the dominance index was 0.44, the diversity index was 3.51, the abundance index was 1.96, and the evenness index was 1.10. In the first survey, the dominant species, diversity, abundance, and evenness indices were all high. The abundance index for the MJ site was higher than that for the SG site, whereas the dominance, diversity, and evenness indices for the SG site were higher than those for the MJ site. The TDI for epilithic diatoms was normal (grade C) for the MJ site, with an average of 58.7, and bad (grade D), based on a value of 35.8, for the SG site.

Table 5. Fish species results collected in this study for each site.

Name	MJ			SG			Total
	May	Sep.	Sum	May	Sep.	Sum	
CYPRINIDAE							
Carassius auratus	3		3				3
Pseudogobio esocinus				1		1	1
Pungtungia herzi (Striped shiner)	18		18	25	51	76	94
Rhynchocypris oxycephalus	45		45	26		26	71
Zacco platypus		16	16	32	30	62	78
COBITIDAE							
Misgurnus anguillicaudatus				4		4	4
SILURIDAE							
Silurus asotus					1	1	1
CENTRACHIDAE							
Micropterus salmoides					2	2	2
ODONTOBUTIDAE							
Odontobutis interrupta	9	3	12	6	10	16	28
GOBIIDAE							
Rhinogobius brunneus	6		6				

In the case of benthic macroinvertebrates, a total of 5 phyla, 6 classes, 13 orders, 31 families, and 41 species were identified at each of the two points at the MJ and SG sites. The analysis according to season revealed that relatively many species were present in both rivers in autumn. The average population density per unit area was 137 for the MJ site and 188, which was relatively high, for the SG site. On average, insects accounted for 69.8% of the population of benthic macroinvertebrates, with percentages of 56.2% at the MJ site and 79.7% at the SG site. In the cluster index analysis, the dominance index was 0.52–0.59 for the MJ site and 0.49–0.66 for the SG site, the diversity index was 2.00–2.15 for the MJ site and 1.85–2.26 for the SG site, the abundance index was 2.60–3.72 for the MJ site and 3.23–4.22, which was higher, for the SG site, and the evenness index was 0.68–0.82 for the MJ site and 0.64–0.72 for the SG site.

The BMI values of 63.2 for the MJ site and 59.4 for the SG site in the first survey were evaluated as normal (grade C). In the second survey at the MJ site in autumn, the BMI of 65.3 was evaluated as good (grade B). The BMI values for the MJ site were higher than those for the SG site, likely because the MJ site has various habitat environments, various flow rates, and various riverbed structures, whereas, at the MJ site, land use, including agricultural land use and soil sedimentation, is higher. An environmental quality evaluation based on the ESB revealed that, in the first survey, the MJ site showed a grade II, whereas, in other surveys, the MJ site, and the SG site in all the surveys, showed a grade II.

The TDI was evaluated as normal (grade C), based on an average value of 58.7, for the MJ site and as bad (grade D), based on an average value of 35.8, for the SG site. We believe this is because the proportion of saproxenous taxa, which are more tolerant to relatively polluted water, was higher at the SG site than at the MJ site.

As a result of the two times of monitoring, the average BMI and FAI at both points were evaluated as normal (grade C). However, in the autumn survey, the FAI for the MJ site was evaluated as bad (grade D). We believe this temporal low was observed because numerous fish were lost due to flooding caused by excessive rainfall and the fish population had not yet recovered.

According to the analysis of the data from 2010 to the present, at the MJ site, the TDI had little effect on flow fluctuations and tended to gradually improve (Figure 7). The BMI tended to decrease, but we reason that this was a temporal phenomenon due to low flow rates because of drought damage in 2015 and 2016. The FAI has been improving, and in 2020, we reasoned that the lower rating was because numerous fish were lost due to flooding caused by excessive rainfall.

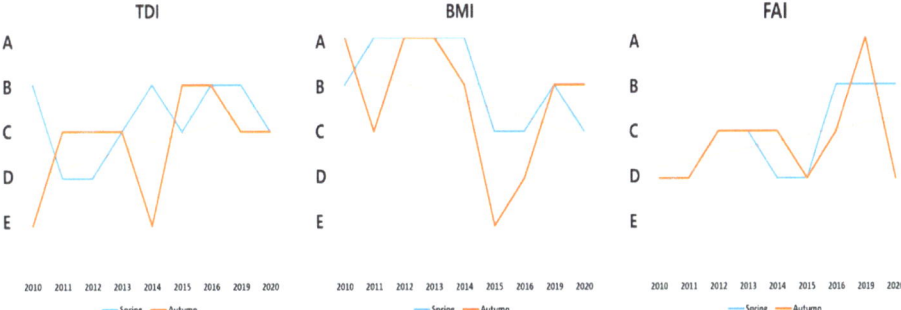

Figure 7. Analysis of past aquatic health evaluation data (TDI, BMI, FAI) of the MJ site.

3.3. Analysis of the Agricultural Return Flow Contribution to Downstream Rivers through Stream Flow and Ecological Modeling

3.3.1. Identification of Long-Term Flow Rate Variation Using the SWAT Model

Simulation of the river flow rate using the SWAT model revealed $R^2 = 0.737$, NSE = 0.706 at point H1 and $R^2 = 0.749$, NSE = 0.673 at point H2, all of which were rated as good, indicating that the model well simulated reality (Figure 8). Although model optimization was limited by a slight difference between the precipitation data collected at the weather station and flow measurement data obtained through monitoring, the results of the river flow measurements and simulation showed similar trends, as shown in Figure 9.

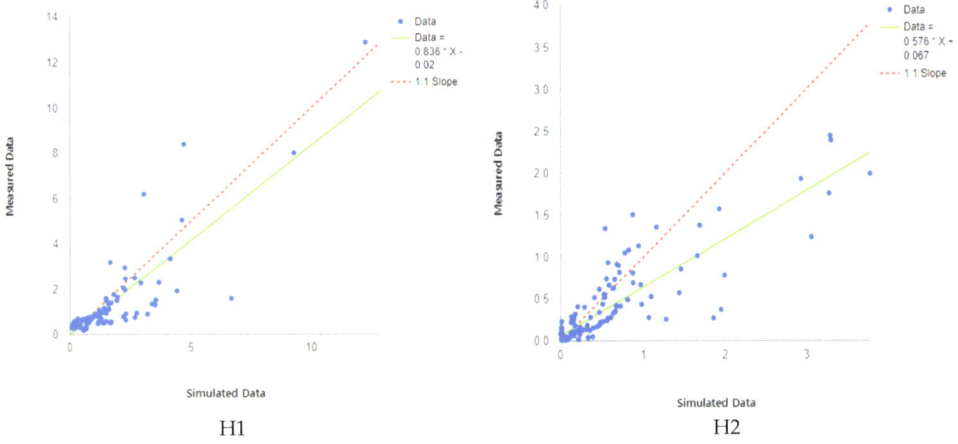

Figure 8. Correlation between river monitoring data and SWAT model simulation results (* mark means multiplication).

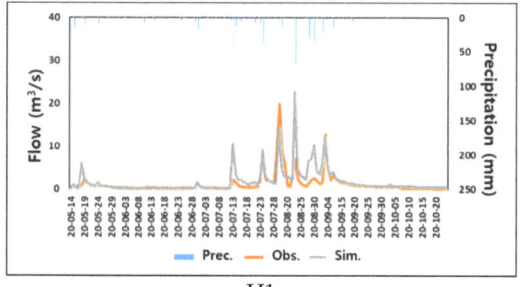 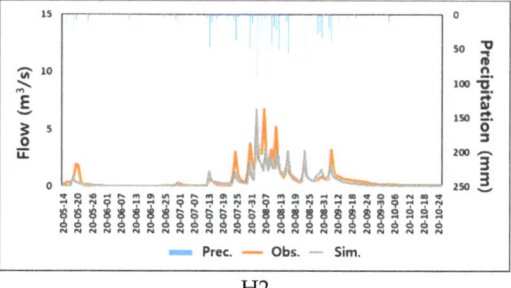

H1   H2

Figure 9. Comparison of river monitoring results and simulation results.

3.3.2. Evaluation of River Contribution of the Return Water

To evaluate the effect of agricultural water return on downstream rivers, we used data on the supply volumes from Rural Community Corporation reports to calculate the return quantity. The return quantity was calculated setting the return rate of agricultural water to 35%. The return quantity in the downstream beneficiary area of the Heungeop Reservoir was calculated to be approximately 45,000 tons (0.02 m^3/s) to 498,000 tons (0.19 m^3/s) from April to October 2017, 46,000 tons (0.02 m^3/s) to 465,000 tons (0.17 m^3/s) in 2018, and 63,000 tons (0.02 m^3/s) to 754,000 tons (0.28 m^3/s) in 2019. The flow rate of the main stream at the end of the beneficiary area downstream of the Heungeop Reservoir calculated by the hydrological model was calculated to be 0.17–7.66 m^3/s from April to October 2017, 1.20–3.24 m^3/s in 2018, 0.16–2.33 m^3/s in 2019, and 0.23–2.28 m^3/s in 2020 (Figure 10).

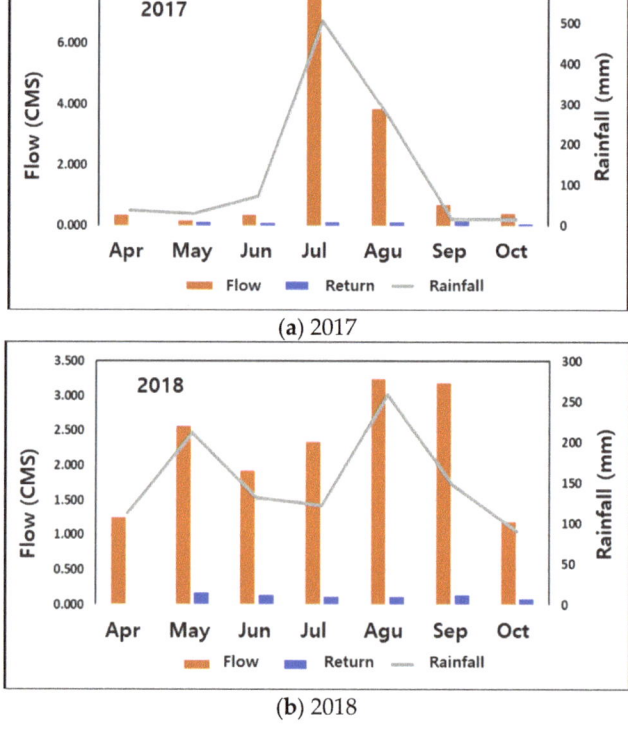

(a) 2017

(b) 2018

Figure 10. *Cont.*

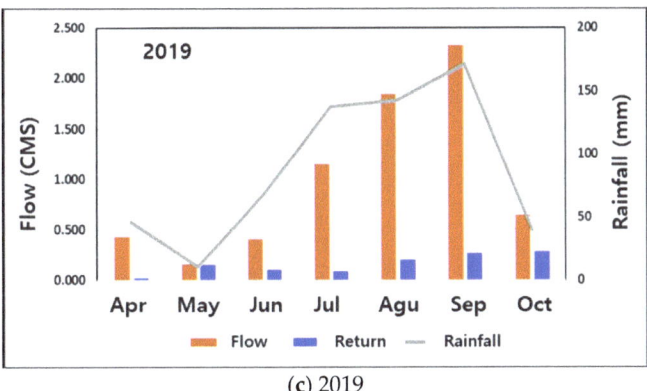

(c) 2019

Figure 10. Comparative analysis of the total flow and return flow in rivers at the outlet of the watershed.

The ratio of return water to main stream flow was 1.5–77.8% in 2017, 1.4–7.0% in 2018, and 5.7–96.5% in 2019. Therefore, it differed depending on the rainfall, ranging from a minimum of 1.4% to a maximum of 96.5%. The ratio of return water to main stream flow, excluding the inflow tributary flow at the end of the beneficiary area downstream of the Heungeop Reservoir, was 1.4–100.0% in 2017, 2.3–15.4% in 2018, 10.7–100.0% in 2019, and 6.5–35.9% in 2020. This ratio also differed depending on the rainfall, ranging from a minimum of 1.4% to a maximum of 100.0%. Comparison of the flow rate and return water by season for each year revealed that the fiver flow significantly increased during summer rainfall; therefore, the effect of the return water was generally small, but during the rainy season (April–June), it was significant.

3.3.3. Results of Environmental Ecological Flow Rate Calculation Using the PHABSIM Model

The optimal ecological flow rate for a representative fish species, striped shiner, was calculated considering the performance of river and aquatic ecology monitoring using the PHABSIM model and was approximately 1.2 m^3/s, as shown in Figure 11. Furthermore, according to the literature review [46], the optimal ecological flow for striped shiner was found to be 1.0 m^3/s, which is similar to the results of this study. It also indicates that the optimal ecological flow rate for striped shiner is 1.2 m^3/s, which implies that the habitat area for striped shiner is the most widely distributed

A comparison of the long-term runoff simulation results with the optimal ecological flow rate of the river at the outlet of the basin revealed that it differed by year depending on the difference in rainfall. In general, in the summer months of July–September—except in 2018, when the rainfall was higher than normal—the flow rate of the main stream was in line with the optimal ecological flow rate (1.2 CMS). In contrast, in April–June, which are non-rainy months, the optimal ecological flow rate was not reached. Therefore, the ratio of return water for agricultural use is not small during rainy periods, and the return water can play an important role in securing the optimal ecological flow rate of the river.

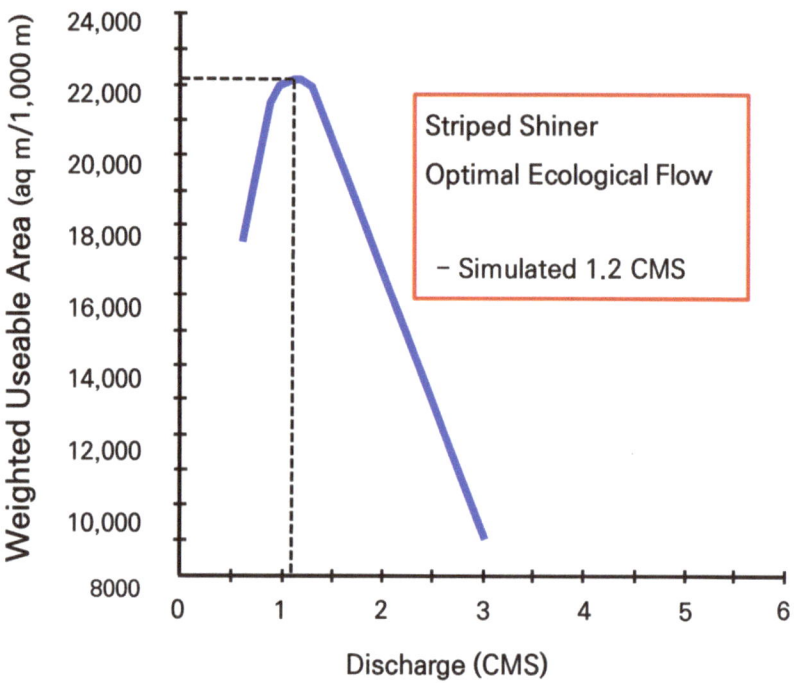

Figure 11. Result of the calculation of the optimal ecological flow for striped shiner.

4. Conclusions

In this study, the Heungeop Reservoir located in Wonju-si, Gangwon-do, was monitored to evaluate the impact of agricultural water on the downstream river quantity, quality, and ecology, considering the agricultural return water quantity in the rural watershed. Analysis of the aquatic ecology monitoring results and data from 2010 to the present for the Maeji Stream Branch revealed that the TDI was not substantially affected by flow rate fluctuations and gradually improved. The BMI tended to decrease, but we reasoned this was because of a temporal low due to a decrease in the flow rate because of drought in 2015 and 2016. The FAI showed an improving trend, except in 2020, when there was a temporary loss of fish due to floods caused by heavy rainfall. The results revealed that the average water pollutant concentration as reflected by the BOD5 and T-P was 1.6–1.9 times higher than the average during autumn, when paddy water is drained to prepare for farming activities in the following year. The effect of discharged agricultural water on the water quality of downstream rivers was found to be very limited, except during the autumn farming season. As a result of calculating the optimal ecological flow rate for the representative fish species (striped shiner) using the PHABSIM model, it was found to be about 1.2 m^3/s. As a result of comparing the optimal ecological flow rate calculation with the stream flow rate simulated by the SWAT model and comparing the flow rate and return water for each period, the river flow significantly increased during the summer rainy period. The effect of return water was small, but during the rainy period (April–June), it was significant. Comparison of the flow rate of long-term runoff simulation results and the optimal ecological flow rate (1.2 m^3/s) revealed that, at the outlet of the watershed, the optimal ecological flow rate was achieved in July–September, except in 2018, when the rainfall was higher than usual. In contrast, the optimal ecological flow rate was not achieved in April–June, which are non-rainy months. Thus, the ratio of return water for agricultural use was not small during rainstorms, and return water may play an important role in securing the optimal ecological flow rate of the river.

According to the results of this study, it is necessary to reduce the loss of agricultural water supplied from agricultural reservoirs and to secure river maintenance water by managing the ecological flow rate in downstream rivers. To secure the ecological flow rate in the dry season, it is necessary to secure the reservoir water volume and create a sufficient flow rate, and to secure the minimum flow rate for normal river function and river ecosystem conservation, the agricultural return water volume has to be secured. The optimal ecological flow rate calculated in this study can be used to evaluate the environmental function of agricultural return water and can be expected to reevaluate agricultural water.

Author Contributions: Conceptualization, J.K. and K.J.L.; methodology, Y.S.; investigation, M.S. and T.K.; writing—original draft preparation, T.K.; writing—review and editing, Y.S., D.L. and K.J.L.; supervision, J.K. All authors have read and agreed to the published version of the manuscript.

Funding: This work was supported by the Korea Institute of Planning and Evaluation for Technology in Food, Agriculture and Forestry (IPET) through the Agricultural Foundation and Disaster Response Technology Development Program, funded by the Ministry of Agriculture, Food and Rural Affairs (MAFRA) (322081-3).

Data Availability Statement: The data presented in this study are available on request from the corresponding author.

Conflicts of Interest: Author Taesung Kang was employed by the company EM Research Institute. The remaining authors declare that the research was conducted in the absence of any commercial or financial relationships that could be construed as a potential conflict of interest.

References

1. Li, Y.; Yang, W.; Shen, X.; Yuan, G.; Wang, J. Water environment management and performance evaluation in central China: A research based on comprehensive evaluation system. *Water* **2019**, *11*, 2472. [CrossRef]
2. Pachova, N.I.; Nakayama, M.; Jansky, L. *International Water Security*; United Nations University Press: Tokyo, Japan; New York, NY, USA; Paris, France, 2008.
3. Cook, C.; Bakker, K. Water security: Debating an emerging paradigm. *Glob. Environ. Chang.* **2012**, *22*, 94–102. [CrossRef]
4. Falkenmark, M.; Molden, D. Wake up to realities of river basin closure. *Int. J. Water Resour. Dev.* **2008**, *24*, 201–215. [CrossRef]
5. Ali, R.; Kuriqi, A.; Abubaker, S.; Kisi, O. Hydrologic alteration at the upper and middle part of the Yangtze river, China: Towards sustainable water resource management under increasing water exploitation. *Sustainability* **2019**, *11*, 5176. [CrossRef]
6. De Fraiture, C.; Wichelns, D. Satisfying future water demands for agriculture. *Agric. Water Manag.* **2010**, *97*, 502–511. [CrossRef]
7. Alcamo, J.; Dronin, N.; Endejan, M.; Golubev, G.; Kirilenko, A. A new assessment of climate change impacts on food production shortfalls and water availability in Russia. *Glob. Environ. Chang.* **2007**, *17*, 429–444. [CrossRef]
8. Hanjra, M.A.; Qureshi, M.E. Global water crisis and future food security in an era of climate change. *Food Policy* **2010**, *35*, 365–377. [CrossRef]
9. Turral, H.; Burke, J.; Faurès, J.-M. *Climate Change, Water and Food Security*; FAO Water Reports 36; FAO: Rome, Italy, 2011.
10. Nam, W.H.; Choi, J.Y.; Hong, E.M.; Kim, J.T. Assessment of irrigation efficiencies using smarter water management. *J. Korean Soc. Agric. Eng.* **2013**, *55*, 45–53.
11. Brown, D.M.; Reeder, R.J. *Farm-Based Recreation: A Statistical Profile*; Economic Research Service ERR-53: Washington, DC, USA, 2007.
12. Brinkley, C. Evaluating the benefits of peri-urban agriculture. *J. Plan. Lit.* **2012**, *27*, 259–269. [CrossRef]
13. Hellerstein, D.; Nickerson, C.; Cooper, J.C.; Feather, P.; Gadsby, D.; Mullarkey, D.; Tegene, A. *Farmland Protection: The Role of Public Preferences for Rural Amenities*; Economic Research Service AER-815: Washington, DC, USA, 2002; pp. 1–36.
14. Zoebl, D. Is water productivity a useful concept in agricultural water management? *Agric. Water Manag.* **2006**, *84*, 265–273. [CrossRef]
15. Swinton, S.M.; Lupi, F.; Robertson, G.P.; Hamilton, S.K. Ecosystem services and agriculture: Cultivating agricultural ecosystems for diverse benefits. *Ecol. Econ.* **2007**, *64*, 245–252. [CrossRef]
16. Kim, T.-C.; Gim, U.-S.; Kim, J.S.; Kim, D.-S. The multi-functionality of paddy farming in Korea. *Paddy Water Environ.* **2006**, *4*, 169–179. [CrossRef]
17. Kim, I.; Kwon, H.; Kim, S.; Jun, B. Identification of landscape multifunctionality along urban-rural gradient of coastal cities in South Korea. *Urban Ecosyst.* **2020**, *23*, 1153–1163. [CrossRef]

18. Dagnino, M.; Ward, F.A. Economics of agricultural water conservation: Empirical analysis and policy implications. *Int. J. Water Resour. Dev.* **2012**, *28*, 577–600. [CrossRef]
19. Kim, H.-Y.; Nam, W.-H.; Mun, Y.-S.; Bang, N.-K.; Kim, H.-J. Estimation of irrigation return flow on agricultural watershed in Madun reservoir. *J. Korean Soc. Agric. Eng.* **2021**, *63*, 85–96.
20. Kim, T.-C.; Lee, H.-C.; Moon, J.-P. Estimation of return flow rate of irrigation water in Daepyeong pumping district. *J. Korean Soc. Agric. Eng.* **2010**, *52*, 41–49.
21. Song, J.H.; Song, I.; Kim, J.-T.; Kang, M.S. Characteristics of irrigation return flow in a reservoir irrigated district. *J. Korean Soc. Agric. Eng.* **2015**, *57*, 69–78.
22. Kim, J.-K.; Kim, G.-H.; Ko, I.-H.; Park, S.-Y.; Seo, J.-W.; Jang, C.-L. Environmental Flow Assesment for Sustainable River Management in Guem River. In Proceedings of the Korea Water Resources Association Conference, Pyeongchang, Republic of Korea, 17 May 2007; pp. 622–627.
23. Hayes, D.S.; Brändle, J.M.; Seliger, C.; Zeiringer, B.; Ferreira, T.; Schmutz, S. Advancing towards functional environmental flows for temperate floodplain rivers. *Sci. Total Environ.* **2018**, *633*, 1089–1104. [CrossRef] [PubMed]
24. European Commission. *Ecological Flows in the Implementation of the Water Framework Directive*; CIS Guidance Document, No. 31; European Commission: Brussels, Belgium, 2014.
25. Lee, J.J.; Hur, J.W. Comparative Analysis of Environmental Ecological Flow Based on Habitat Suitability Index (HSI) in Miho stream of Geum river system. *Ecol. Resilient Infrastruct.* **2022**, *9*, 68–76.
26. Poff, N.L.; Richter, B.D.; Arthington, A.H.; Bunn, S.E.; Naiman, R.J.; Kendy, E.; Acreman, M.; Apse, C.; Bledsoe, B.P.; Freeman, M.C. The ecological limits of hydrologic alteration (ELOHA): A new framework for developing regional environmental flow standards. *Freshwater Biol.* **2010**, *55*, 147–170. [CrossRef]
27. Gerten, D.; Hoff, H.; Rockström, J.; Jägermeyr, J.; Kummu, M.; Pastor, A.V. Towards a revised planetary boundary for consumptive freshwater use: Role of environmental flow requirements. *Current Opinion in Environ. Sustainability* **2013**, *5*, 551–558.
28. Grantham, T.; Mezzatesta, M.; Newburn, D.; Merenlender, A. Evaluating tradeoffs between environmental flow protections and agricultural water security. *River Res. Appl.* **2014**, *30*, 315–328. [CrossRef]
29. Liu, J.; Liu, Q.; Yang, H. Assessing water scarcity by simultaneously considering environmental flow requirements, water quantity, and water quality. *Ecol. Indic.* **2016**, *60*, 434–441. [CrossRef]
30. Pal, S.; Singha, P. Linking river flow modification with wetland hydrological instability, habitat condition, and ecological responses. *Environ. Sci. Pollut. Res.* **2023**, *30*, 11634–11660. [CrossRef] [PubMed]
31. Luo, Z.; Zhang, S.; Liu, H.; Wang, L.; Wang, S.; Wang, L. Assessment of multiple dam-and sluice-induced alterations in hydrologic regime and ecological flow. *J. Hydrol.* **2023**, *617*, 128960. [CrossRef]
32. Kim, H.-Y.; Nam, W.-H.; Mun, Y.-S.; An, H.-U.; Kim, J.; Shin, Y.; Do, J.-W.; Lee, K.-Y. Estimation of irrigation return flow from paddy fields on agricultural watersheds. *J. Korea Water Resour. Assoc.* **2022**, *55*, 1–10. [CrossRef]
33. Cho, J.; Kim, C.; Lim, K.J.; Kim, J.; Ji, B.; Yeon, J. Web-based agricultural infrastructure digital twin system integrated with GIS and BIM concepts. *Comput. Electron. Agric.* **2023**, *215*, 108441. [CrossRef]
34. McNaughton, S.J. Relationships among functional properties of Californian grassland. *Nature* **1967**, *216*, 168–169. [CrossRef]
35. Pielou, E.C. Shannon's formula as a measure of specific diversity: Its use and misuse. *Am. Nat.* **1966**, *100*, 463–465. [CrossRef]
36. Ulfah, M.; Fajri, S.N.; Nasir, M.; Hamsah, K.; Purnawan, S. Diversity, evenness and dominance index reef fish in Krueng Raya Water, Aceh Besar. *Earth Environ. Sci.* **2019**, *348*, 012074. [CrossRef]
37. Goudarzian, P.; Erfanifard, S. The efficiency of indices of richness, evenness and biodiversity in the investigation of species diversity changes (case study: Migratory water birds of Parishan international wetland, Fars province, Iran). *Biodivers. Int. J.* **2017**, *1*, 41–45.
38. Kong, D.; Park, Y.; Jeon, Y. Revision of Ecological Score of Benthic Macroinvertebrates Community in Korea. *J. Korean Soc. Water Environ.* **2018**, *34*, 251–269.
39. Kelly, M. Use of the trophic diatom index to monitor eutrophication in rivers. *Water Res.* **1998**, *32*, 236–242. [CrossRef]
40. Zelinka, M.; Marvan, P. Zur präzisierung der biologischen klassifikation der reinheid fliessender grewässer. *Arch. Hydrobiol.* **1961**, *57*, 207–217.
41. Klemm, D.J. *Fish Field and Laboratory Methods for Evaluating the Biological Integrity of Surface Waters*; Environmental Monitoring Systems Laboratory, Office of Modeling, Monitoring Systems, and Quality Assurance, Office of Research and Development, US Environmental Protection Agency: Washington, DC, USA, 1993.
42. Arnold, J.G.; Srinivasan, R.; Muttlah, R. Large area hydrologic modeling and assessment part I: Model development. *J. Am. Water Resour. Assoc.* **1998**, *34*, 73. [CrossRef]
43. Seong, C.; Oh, C.; Hwang, S. Watershed-scale Hydrologic Modeling Considering a Detention Effect of Rice Paddy Fields using HSPF Surface-Ftable. *J. Korean Soc. Agric. Eng.* **2018**, *60*, 41–54.
44. Duda, P.B.; Hummel, P.R.; Donigian, A.S., Jr.; Imhoff, J.C. BASINS/HSPF: Model use, calibration, and validation. *Trans. ASABE* **2012**, *55*, 1523–1547. [CrossRef]

45. Moriasi, D.N.; Arnold, J.G.; Van Liew, M.W.; Bingner, R.L.; Harmel, R.D.; Veith, T.L. Model evaluation guidelines for systematic quantification of accuracy in watershed simulations. *Trans. ASABE* **2007**, *50*, 885–900. [CrossRef]
46. Kang, K.; Son, S.; Kim, J.; Kim, P.; Kwon, Y.; Kim, J.; Kim, Y.J.; Min, J.K.; Kim, A.R. Estimation of Habitat Suitability Index of Fish Species in the Gapyeong stream. *J. Korean Soc. Water Environ.* **2011**, *33*, 626–639.

Disclaimer/Publisher's Note: The statements, opinions and data contained in all publications are solely those of the individual author(s) and contributor(s) and not of MDPI and/or the editor(s). MDPI and/or the editor(s) disclaim responsibility for any injury to people or property resulting from any ideas, methods, instructions or products referred to in the content.

Article

Participatory Analysis of Impacts of Agricultural Production Systems in a Watershed Depicting Southern Brazilian Agriculture

Alexandre Troian [1,*], Mário Conill Gomes [2], Tales Tiecher [3], Marcos Botton Piccin [4], Danilo dos Santos Rheinheimer [1] and José Miguel Reichert [1,†]

1. Department of Soils, Federal University of Santa Maria (UFSM), Santa Maria 97105-900, Brazil
2. Eliseu Maciel Agronomy School, Federal University of Pelotas (UFPel), Pelotas 96010-610, Brazil
3. Soils Department, Federal University of Rio Grande do Sul (UFRGS), Porto Alegre 90010-150, Brazil
4. Department of Rural Extension, Federal University of Santa Maria (UFSM), Santa Maria 97105-900, Brazil
* Correspondence: xtroian@gmail.com
† Current address: Nuclear Energy Department, Federal University of Pernambuco (UFPE), Recife 50670-901, Brazil.

Abstract: The objective of this study was to propose a multidimensional model capable of evaluating, in a participatory method, the pressures agricultural production systems cause to aquatic ecosystems. The model was structured with information compiled from scientific articles, doctoral theses, public documents, and field research performed with the participation of stakeholders through interviews, questionnaires, and group evaluations. The evaluation matrix combines seven criteria and twenty-five sub-criteria with different weights to evaluate two main aspects: (i) land occupation and soil management and (ii) agricultural waste production and disposal. The model was tested in 14 agricultural farms, representing four productive arrangements, in a large watershed (2400 km^2) in southern Brazil. The geophysical characteristics of the site (18.3%), land use and occupation (28.2%), management practices (soil and water) (25.4%), manure and fertilizers (12.6%), pesticides (14.1%), agricultural waste and discards (1.4%) were the criteria and their respective weights used in the structure of the proposed evaluation model. The evaluation showed that the combination of the fragility of cultivated environments and the absence of conservation practices represented the greatest risks (72.9%) to maintaining the sound environmental conditions of aquatic ecosystems. For future research, it is recommended that a cost-effectiveness analysis be carried out to evaluate environmental conflicts.

Keywords: multicriteria analysis; farming systems; water resources

1. Introduction

The expansion of production of goods and services to meet growing human needs has compromised the natural regeneration of ecosystems, decreased biodiversity, and caused the extinction of many species [1]. The production models of contemporary society generate high emissions of greenhouse gases to the atmosphere [2], soil and water pollution [3], and compromise landscape and cultural values [4], among other negative implications for ecosystems [5,6].

The pressures exerted by human activities on the environment include changes in the natural state of aquatic ecosystems. The impacts monitored, both quantitative and qualitative, indicate modifications in hydrological cycles [7]. Changes in rainfall regime (temporal and spatial) and an increase in the volume of surface runoff immediately after rainfall are some of the observed variations related to water availability in terrestrial ecosystems [8–10]. In addition to physical changes in rivers, lakes and oceans, human activities (which include animal husbandry) accelerate the transfer of pollutants to aquatic ecosystems. Industrial chemical elements, organic compounds used in agriculture (pesticides and pharmaceuticals, especially), and even in human medical treatments, hygiene, and aesthetics have modified the quality of hydric resources [11,12].

While the planet's population has doubled in the last 60 years, water consumption has increased sevenfold [13]. By the year 2030, water withdrawal will be 25% greater than the current volume [14]. The increase in per capita consumption is mainly determined by changes in global food consumption habits—higher consumption of meat and dairy products. In addition to consuming more water, the current development model has not safely safeguarded natural areas from the effects of anthropogenic activities on aquatic ecosystems. For instance, wetlands across the planet have declined by 40% since the 1970s [15].

The concentration of urbanized areas and the expansion of agricultural areas in recent years have intensified the uncertainties around water resource management. Occupying about 50% of the planet's habitable land and using approximately 70% of the freshwater volume, agriculture has been the subject of several studies. Research conducted in distinct watersheds [3,12,16,17] has linked the degradation and pollution of aquatic ecosystems directly to agricultural activities. Agricultural land use has been indicated to be an important factor affecting the nutrient status and sedimentation of streams [18], imposing restrictions on fish and the human community [3,19].

Agricultural production systems, consisting of a combination of crop and livestock systems with distinct technical factors structured from different measures of agricultural area, knowledge, and capital [20,21] have put pressure on aquatic ecosystems at a level many times greater than their natural renewal capacity. Even systems characterized by smallholder production and the use of primarily family labor cause changes in the natural characteristics of water resources. Agricultural pressures are related to two main factors: (i) use of natural resources and (ii) waste generated, transported, or disposed of in the environment. The former refers to the different land uses and landscape changes, while the latter refers to the use of products and inputs. Isolated or accompanied by extreme natural events (climate oscillations), waste can have serious impacts on the environment and human well-being [22].

The identification and characterization of environmental problems is a challenge for scientists, starting with defining the methods used since not all techniques can adequately recognize agri–environmental interactions [23,24]. Because of this, our objective was to structure and test a multidimensional model to identify and measure the pressures that different agricultural production systems exert at the watershed level. To this end, we used the MCDA (Multicriteria Decision Aid) methodology, with the participation of stakeholders through interviews, questionnaires, and group evaluation. We developed a "multi-criteria aid evaluation model" [25], which took into account the perceptions of managers who make public decisions, as well as seeking to raise awareness among farmers who live in the area and use the natural resources evaluated. The hypothesis was that models structured through participatory processes have the potential to identify and solve complex problems such as the management of natural ecosystems, since these approaches shorten the distance between decision makers and the identification of problems, as well as bringing them closer to the formulation of alternatives.

2. Theoretical and Methodological Framework

The diagnosis and resolution of problems with a socio-environmental dimension are faced with a series of uncertainties that can be of a technical, epistemological, methodological, and even ethical nature. At the same time, the decisions at stake, in general, involve costs, benefits, compromise, and the distinct interests of the various agents involved [26].

A set of methods called Multicriteria Decision Aid (MCDA), originating from the European School, which is based on the scientific paradigm of constructivism, can be used as an alternative to ponder and minimize these uncertainties [27]. These methods integrate multiple dimensions into problems to incorporate the preferences of those interested in the phenomenon, to sort, rank, or elect the alternatives constructed for a given context [28].

The studies of [27,29–31] are theoretical-conceptual references of the principles and bases of the multicriteria methods. Operationally, such methods establish ways to model complex situations and can cover both qualitative (environmental, social, and organizational factors) and quantitative factors (costs involved, physical variables, among others) [31].

Multicriteria methods, unlike monocriteria methods in which alternatives are evaluated based on a single immediate criterion, bring together a considerable variety of aspects relevant to the problem, including taking into account the psychological aspects of the behavior of stakeholders [32]. More precisely, the multicriteria methods start from the rationality that underlies fuzzy set theories to compare preferences through a set of alternatives subjected to a series of criteria. A criterion is a "tool" that allows for comparing alternatives according to a particular "axis of meaning" or a "point of view" [33] (p. 59).

Relations between alternatives are classified into three properties: Strict preference, when an alternative is preferable to another (a P b); Indifference, when there is no preference between alternatives (a I b); and Incomparability, in this case it is not possible to compare two alternatives (a R b) [34]. The literature organizes multicriteria methods as of the following approaches: (i) interactive methods, (ii) outranking methods, and (iii) single synthesizing criterion [34]. The single synthesizing criterion make use of the Multi-Attribute Utility Theory (MAUT) to identify a marginal utility function in each criterion and subsequently group individual utility functions into a global utility function [33] (Table 1).

Table 1. Additive aggregation decision matrix.

		Criteria				
		$w_1 \cdot g_1(.)$	$w_2 \cdot g_2(.)$	...	$w_i \cdot g_i(.)$	G(.)
Alternative	a_1	$w_1 \cdot g_1(a_1)$	...	...	$w_i \cdot g_i(a_1)$	$v(a_1)$
	a_2	...	...	...	...	...
	...	...	...	...	...	...
	a_n	$w_1 \cdot g_1(a_n)$	...	...	$w_i \cdot g_i(a_n)$	$v(a_n)$

Note: a_1, a_2 e a_n—Alternatives; w_1, w_2 e w_i—Weights assigned to the criteria; g_1 and g_i—criteria; $v(a_1)$—Value of alternative a_1 in the i criterion; $v(a_n)$—Value of the n-th alternative in the i-th criterion; and G(.)—Value Global of the model.

According to the additive aggregation matrix presented in Table 1, the value of each alternative results from the weighted sum of the criteria. This value is obtained for each alternative, according to Equation (1):

$$v(a) = w_1 \cdot g_1(a) + w_2 \cdot g_2(a) \ldots + w_i \cdot g_i(a) \tag{1}$$

The global value of the multicriteria model—G(.)—is obtained for the i-th criterion in the n-th alternative, according to Equation (2):

$$G(.) = \sum_{i=a_1}^{a_n} i(w_i \cdot g_i) \tag{2}$$

The main implication of using this mathematical model is the compensation between criteria—the loss of value in one criterion may be mutually offset by the gain in another (pareto optimum). Compensations are also called trade-offs [35]. The model used in this research (which includes the criteria used) was built with the participation of stakeholders through interviews, questionnaires, and group evaluation. The criteria and weights were built from the knowledge and experience of experts who know the location and, consequently, the problems related to agricultural production and natural ecosystems in the region. In summary, such processes occur in three consecutive steps: (i) identification of the context and delimitation of the problem (Phase Exploratory); (ii) structuring of the multicriteria model (Phase Constructive); and (iii) evaluation of the different alternatives and model results (Phase the of Structuring and Evaluation) (Figure 1).

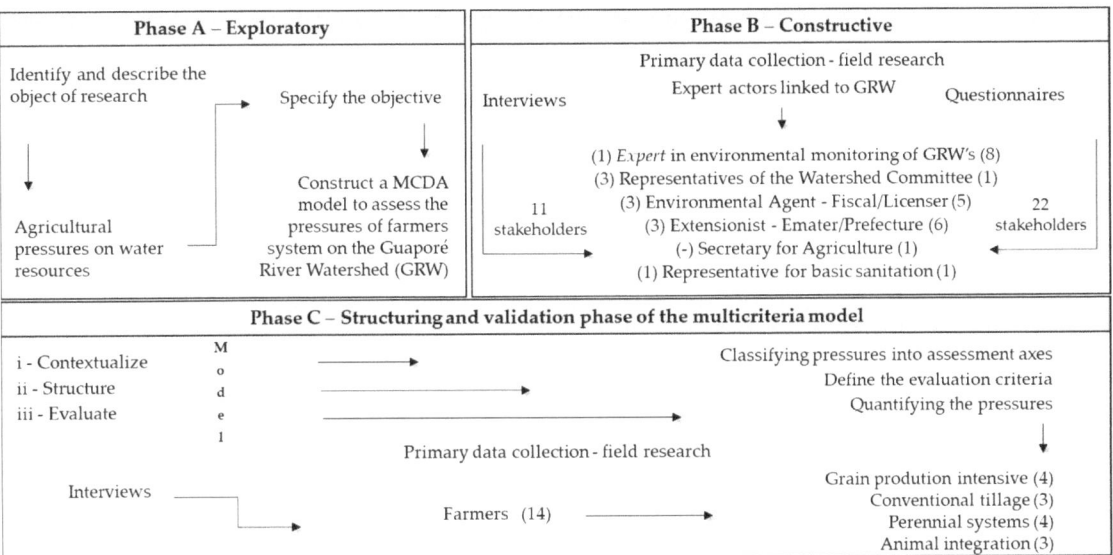

Figure 1. Research development phases.

2.1. Stages of the Investigation

Phase A—Exploratory. At this moment, we delimited the problem through the collection and analysis of secondary and primary data made available by the following units: Brazilian Institute of Geography and Statistics (IBGE), United States Geological Survey (USGS), National Water Agency (ANA, as in Portuguese abbreviation), Rio Grande do Sul Secretariat of Environment and Infrastructure (SEMA, as in Portuguese abbreviation), Scopus, Web of Science, and Science Direct.

In this stage, a regional zoning was also carried out to identify and characterize the different agrarian landscapes of the GRW. The process was carried out on the basis of the Agricultural Systems Theory, which has been developed since the 1960s in France (AgroParisTech (which includes the former National Agronomic Institute of Paris-Grignon—INA-PG) [36]. This analytical tool makes it possible to understand the complexity and classify—through organization and functioning—different forms of agriculture. The main stages were as follows: (i) retrieval of the historical formation of agriculture in the region; and (ii) characterization of agro-ecological and socio-economic conditions (based on the description of climatic, geological, hydro-sedimentological, relief, flora, and the main physical characteristics of the watershed) [20].

Phase B—Constructive. In this stage, major assessment areas to be included in the multicriteria model were identified and defined through primary data collection in field research. To obtain this information, 11 interviews were conducted and 22 questionnaires were applied, with social actors able to analyze the situation of water resources in the GRW in the period between July 2018 and October 2019.

The actors who made up the sample were chosen based on purposeful sampling criterion [37] according to the following profile: experts in the public supply and sanitary sewage, authorities in environmental licensing, enforcement and sanitary surveillance, rural development agents, and researchers in soil management, conservation, and environmental monitoring of watersheds.

The contact with the informants selected to participate in the research was carried out by the procedures of 'network systems' [38]. Operationally, geographic micro-regions of interest were defined within the GRW and, subsequently, the social agents with horizontal amplitude were framed in pole elements of the network of micro-regions of interest.

Phase C—Structuring and validation phase of the multicriteria model. The information collected from the social agents identified during the previous phase was organized into evaluation axes, also called Fundamental Viewpoints—FPV in the literature [39]. They were arranged in a tree structure using two techniques: (i) Cognitive mapping [40], designed to answer the following question: which aspects must be considered to preserve the natural characteristics of aquatic ecosystems? (ii) Frame mode of the decision-making context [41] to decompose the cognitive map into clusters with similar evaluation themes. After defining and organizing them hierarchically, the evaluation axes and each FPV was transformed into a criterion through an attribute (measurement scale) and a value function associated with this attribute [42], which enables measuring, in the least ambiguous manner possible, the performance of the available alternatives for each axis [32]. Complex FPVs were subdivided into two or more Elementary Viewpoints (EPVs), each generating a sub-criterion as well.

The attributes can be classified as direct, constructed, or indirect (proxy). They can also be qualitative or quantitative, and continuous or discrete. Once the attributes were defined, they were ordered from most attractive to least attractive. Subsequently, a value function was associated with each attribute's impact level through the Direct Rating method [33,35]. This function is obtained using interval scales estimated with arbitrary values 0 (zero) and 100 (one hundred). Respectively, the maximum level represents the most desirable situation; on the other hand, the minimum level represents the least desirable but possible situation. The values of the intermediate impacts are scaled to the minimum and maximum values. Table 2, for example, illustrates the value function for Elementary Points of View (sub-criterion). The structures of the other criteria is available in Appendix A.

Table 2. Sub-criterion "C1.3 Distance from ploughing to streams or springs".

Impact Levels	Reference Levels	Description	Original Value Function	Rescaled Value Function
Maximum		The distance between the cultivation and the stream(s) is more than 60 m, and between the cultivation and the springs is more than 50 m	100	150
Range of expectations	Good	The distance between cultivation and the stream(s) is between 50 and 60 m, and between the cultivation and springs is more than 50 m	80	100
		The distance between cultivation and the stream(s) is between 40 and 50 m, and between the cultivation and springs is more than 50 m	60	50
	Neutral	The distance between cultivation and the stream(s) is between 30 and 40 m, and between the cultivation and springs is more than 50 m	40	0
Minimum		The distance between the cultivation and the stream(s) is less than 30 m, and between the cultivation and springs is less than 50 m	0	−100

After defining the value function, the reference impact levels were identified, which represent regions of "Good" and "Neutral" expectations—regions where the alternatives to be evaluated are neither very attractive nor very repulsive. Actions with an effect above the good level on the scale generate scores higher than 100 (one hundred), while actions below the neutral level generate negative scores. The transformation of the value function is performed using the linear Equation (3) of positive type.

The properties of this transformation are as follows:

$$v'(a) = v(a) \cdot \alpha + \beta \quad (3)$$

where $v'(a)$ is the score transformed for action a; $v(a)$ is the original score of action a; and α and β are linear constants of the scale, where $\alpha > 0$. The new scale should be obtained by solving a 1st-degree equation system with two unknowns. Further information is presented by [35]. Finally, all resulting weights were standardized into values defined between zero and one. Thus, once the descriptors and their respective value functions were in place, it was possible to measure the performance of alternatives intra-criteria.

To obtain a global evaluation of alternatives, in which all criteria and sub-criteria are simultaneously considered, it was necessary to weight the weights by means of compensation rates and sum-weighted coefficients. To achieve this, the balance weighting method was employed, which is based on attributing utility weights to each criterion via the linear scale ϵ [0,1], whereby utility value 1 refers to the best alternative and 0 to the worst [35,43]. The additive model used to aggregate the value functions into a global assessment is represented by Equation (2).

The model was tested in 14 agricultural establishments to measure and compare the pressures exerted by the four main agricultural production systems identified in the watershed (Figure 2). An agricultural production system is defined by combinations of crops, livestock, and technical factors available to a farm unit, such as labor, technical knowledge, agricultural area, equipment, and capital [20]. The results of the global assessment and the criteria impact profile are the result of the average score of the establishments belonging to each of the four production systems. The agricultural establishments were chosen by employing purposive sampling directed [37] with the support of the actors interviewed in the previous stage, especially by indication of rural extensionists and agents linked to the agrarian or environmental area of municipal governments.

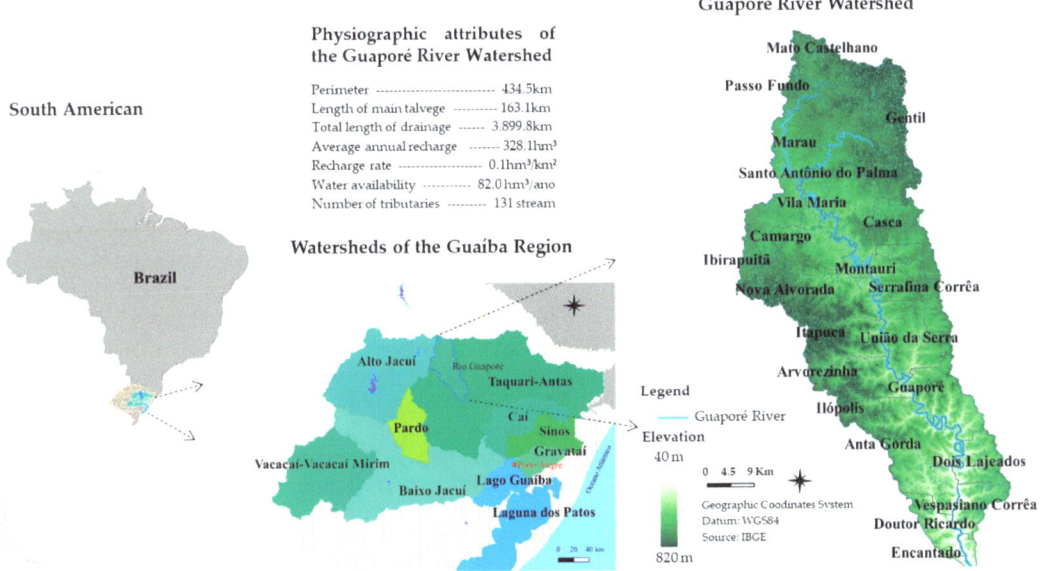

Figure 2. Guaporé River Watershed—RS, Brazil.

The techniques used were interviews and in loco passive observation to complement the information acquired via interviews [44]. The in loco observation technique helped to characterize the environment and the soil, including (a) classifying the slope of the cultivated land, (b) the degree of soil compaction, (c) the stability of soil aggregates, (d) verifying the presence of signs of erosion and gauging the percentage of soil coverage by residues The soil texture was determined based on the clay content presented in the soil analyses carried out by the farmers themselves.

2.2. Study Area

The Guaporé River Watershed (GRW) is located in the northeast region of the state of Rio Grande do Sul, Brazil (420.900–366.400 mE and 6.874.286–6.772.536 mS, zone 22S). With 2400 km^2 of drained area, it encompasses 25 municipalities, 5 of which are fully inserted in the HB (Figure 2). The GRW presents a wide and diverse drainage network with a predominantly dendritic shape. The average monthly flow rate was 31.3 m^3 s^{-1} in 2012 and 2013 [45]. The climate type of the region is subtropical and superhumid mesothermal, without a defined dry season [46], and with an average annual temperature of 17.9 °C (Cfa Köppen system). The average annual rainfall varies between 1550 and 1700 mm, with 1861 and 1434 mm in 2018 and 2019, respectively [47].

The topography varies from gently to moderately undulating in the northern portion of the GRW and strongly undulating to steep relief classes in the southern region. Approximately one-third of the GRW area has slopes between 15 and 30%. The Prevailing vegetation is composed of Seasonal Deciduous Forest, Mixed Ombrophylous Forest, and areas of Grassy Steppe. The geological formation is characterized by volcanic lava flows of the Serra Geral formation, typified by Caxias, Gramado, and Paranapanema, covering, respectively, 72.2, 26.1, and 1.7% of the area. The soil classes found in the catchment are as follows: Ferralsols (31.2%), Luvisols (24.2%), Nitosols (21.4%), Acrisols (16.6%), and Leptosols (6.6%) [16].

Annual crops cover 54.6% of the watershed area, being more expressive in the north, while forests cover 36.8% of the area, especially in the southern portion. Grassland covers 4.8% of the area; forestry, 2.6%; urban areas correspond to 0.7%, and the water body class to 0.5% of the watershed. According to the Brazilian Institute of Geography and Statistics (IBGE, as in Portuguese abbreviation), the average surface of agricultural establishments in the GRW is 34 hectares. The establishments with more than 100 hectares are equivalent to 5.3% of the total area, contrasting with more than 86% of the establishments with less than 50 ha, and more than 57% of them with less than 20 ha [48].

Soybeans (*Glycine max* L. Merr) and corn (*Zea mays*) under no-till systems are predominant in the northern part of the watershed in the spring/summer. At the same time, in the fall/winter, oat (*Avena sativa*) and wheat (*Triticum aestivum*) are usually grown in these areas. Land use in the southern region is more diversified. However, the conventional system, with intense soil disturbance predominates, especially in the tobacco (*Nicotiana tabacum* L.) production areas. In contrast, in sloping and even mountainous areas, yerba mate (*Ilex paraguariensis* A. St. Hil.) cultivation predominates, and the soil surface remains constantly protected. Although pig, poultry, and dairy farming in the intensive system is recurrent in the GRW, it is more notorious in the south–central region (Appendix B).

In summary, the watershed landscape is representative of family and farmer agriculture in South America: agriculture, animal husbandry, agro-industries, and small urban settlements imbricated in the rural landscape [49]. The agricultural production systems are typically based on family, and most of them occupy ecologically fragile environments and sloping areas, with the presence of springs and aquifer recharge [50]. Additionally, the GRW provides drinking water to more than half a million local inhabitants, and the Guaporé River is one of the main tributaries of the Taquari River, a tributary of the Guaíba River Watershed, which supplies a large part of the more than four million inhabitants of the metropolitan region of Porto Alegre [51].

3. Results and Discussion

There are two main results. One refers to the structuring of the model for assessing agricultural pressures on aquatic ecosystems, with criteria and weights assigned to them by the stakeholders who participated in the research. The other result is the performance assessment of the different production systems practiced by farmers living in the Guaporé River Watershed.

3.1. Structure of the Model

The model is designed to assess two broad clusters of criteria: (i) land cover and management, and (ii) agricultural residue production and disposal. Land cover, land management, and landscape characteristics are the main drivers of material transfer, deposition, and redistribution over time in a watershed. The impact of the pressure exerted by anthropic activities, especially by agriculture, on aquatic ecosystems depends on the natural ecosystem's characteristics, land occupation, and soil management. Our model, therefore, is conceptualized using three phenomenological criteria (three basic axes) in the first cluster: (a) land use—arrangement of crops in the agricultural space, (b) landscape features and soil properties—buffering capacity of the impacts of anthropic pressure, and (c) soil management. These axes were broken down into eleven less complex levels (sub-criteria) to detail the assessment.

A second criteria cluster—agricultural residues and discards—was included in the model, since Guaporé River is one of the watersheds with the highest concentration of pig, poultry, and dairy production in Latin America. These axes were also broken down into four criteria and fourteen less complex levels (sub-criteria) to detail the assessment. The amounts, application forms, and care with the use of (d) industrialized fertilizers, (e) animal wastes, and (f) pesticides by the farmers were monitored and included in the model. Finally, we introduced some data to the model regarding (g) disposal of products of industrial origin and products generated on the farms themselves. Specifically, we evaluated the disposal of agrochemical and medicine packaging, old tires, machinery and equipment parts, lubricating oil, grease, batteries, hardware, plastic pipes, and glass, among other industrial materials used in agriculture, as well as the disposal of dead animals and non-hazardous agricultural waste.

The result of structuring the multicriteria model to evaluate pressures exerted by agricultural production systems on aquatic ecosystems was configured in a tree arrangement of hierarchical ramifications of seven criteria (with different weights, whose sum is 100) and twenty-five sub-criteria. The sum of the weights of sub-criteria in the same level should be 100%, as pictured in Table 3.

Table 3. Hierarchical structure of the model: clusters, criteria, and sub-criteria with their respective weights.

Cluster	Criteria	Weights (%)		Sub-Criteria
(i) Land occupation and soil management	C1. Land use	28.2	41.6	C1.1. Ratio of area cultivated to the establishment's area
			29.2	C1.2. Distribution of crops and livestock in the landscape
			25.0	C1.3. Distance from the crop to the stream(s)
			4.2	C1.4. Access to water for animal desedentation
	C2. Landscape features and soil characteristics	18.3	45.5	C2.1. Degree of slope of the cultivated land
			31.8	C2.2. Potential erodibility of the cultivated soil
			13.6	C2.3. Average depth of cultivated soil
			9.1	C2.4. Texture of cultivated soil
	C3. Soil management	25.4	41.7	C3.1. Soil tillage
			33.3	C3.2. Soil cover
			25.0	C3.3. Physical barriers to water containment

Table 3. Cont.

Cluster	Criteria	Weights (%)		Sub-Criteria
(ii) Agricultural waste and discards	C4. Mineral fertilizers	4.2	62.5 31.2 6.3	C4.1. Fertilizer rates C4.2. Technique used to apply fertilizer C4.3. Climatic condition during applying fertilizer
	C5. Animal wastes	8.4	47.6 33.3 14.3 4.8	C5.1. Waste rates C5.2. Technique used to apply manure C5.3. Climatic condition during applying manure C5.4. Storage system
	C6. Pesticides	14.1	43.5 34.8 21.7	C6.1. Pesticides rates C6.2. Adoption of official recommendations C6.3. Weather condition during applying pesticides
	C7. Discards	1.4	51.3 35.9 7.7 5.1	C7.1. Pesticides packages C7.2. Dead animals C7.3. Reverse logistic products C7.4. Agrosilvopastoral waste non-hazardous

Coupling the remarks of farmers, technicians, and public managers with the scientific support of renowned agrarian science researchers, we attributed more than two-thirds of the total weight of our model of evaluation (71.9% of 100%) to the first cluster: land occupation and soil management (Figure 3). This result is because the transformation of natural biomes into agro-ecosystems has inexorably caused changes in natural biogeochemical cycles [52], among them, the water cycle is completely modified [7]. Both the quality and flow of water in the atmosphere–soil–vegetation are less favorable for biota (including plants), favoring its rapid transfer to the oceans. Changing natural land use (forest, savannah, native grasslands, caatinga, and other biomes) to agricultural production areas (reforestation, pasture, and especially annual crops) puts significant pressure on aquatic ecosystems [7].

Land use (28.2%) was represented mainly by the ratio between crop areas and natural vegetation (41.6% of 28.2%); soil management (25.4%) with emphasis on soil tillage (41.7% of 25.4%); and landscape features and soil properties (18.3%) were especially represented by the slope of cultivated areas (45.5% of 18.3%), constituting a starting point for planning the management of water resources considering agricultural activities carried out in the GRW. The weights assigned to soil characteristics, land use, and management are in accordance with soil erosion models, especially the Revised Universal Soil Loss Equation (RUSLE) model [53].

For the second cluster, agricultural residue production and disposal with a weight of 28.1% (of 100%) was assigned. Since the inputs used to meet the demands of farming systems and increase their production efficiencies have generated different types of waste [54], off-farm agricultural inputs and organic waste produced on the farm were parameterized and included in the model. Pesticides were given the highest scores (14.1%), organic wastes (manure animal) participated with 8.4%, industrialized fertilizers made up 4.2%, and other discards 1.4% (pesticides packages, dead animals, reverse logistic, products, agrosilvopastoral waste) (Table 3).

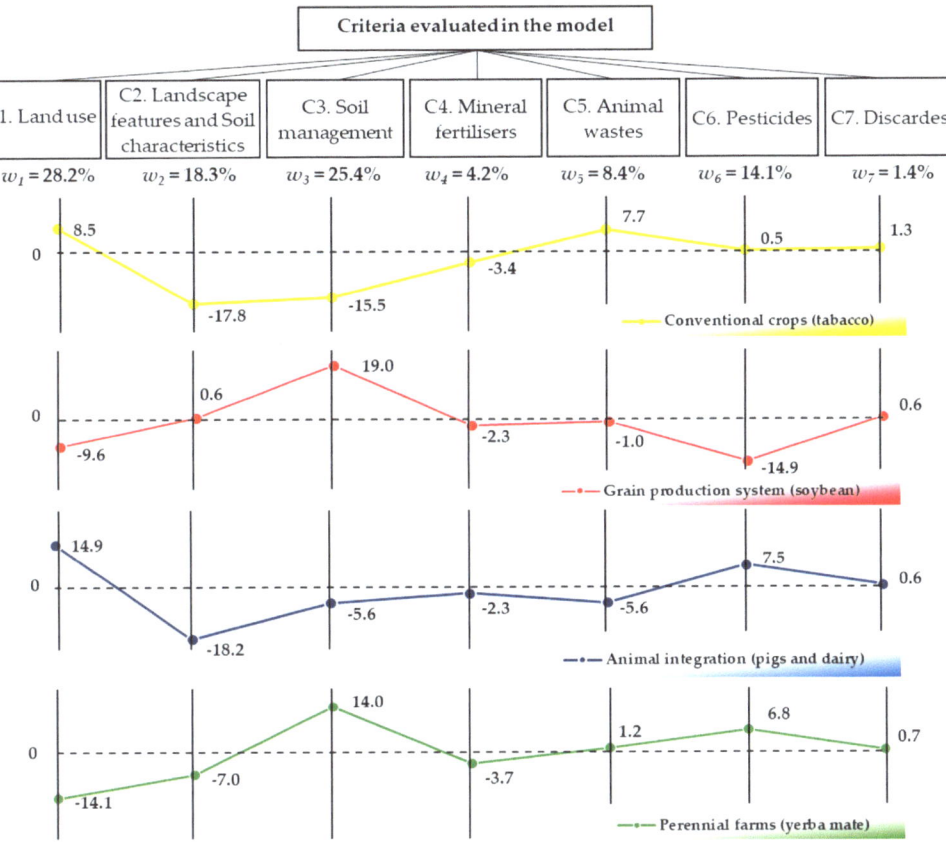

Figure 3. Level of pressure in each criterion in the four production systems analyzed (alternatives). $w_1, w_2, \ldots, w_i$ = weight of the different criteria in the model.

3.2. Performance of the Alternatives (Production Systems)

The Guaporé River Watershed can be divided into two large agricultural regions [12,16]. The Northern region has a smooth to moderately wavy relief, where deep soils predominate, cultivated with genetically modified soybeans and corn in the spring–summer, and cereals and forage crops in the autumn–winter, under no-till farming systems. There are farmers specialized in the production of milk, pigs, and poultry in the integration system with regional agribusinesses.

The second region, which occupies two-thirds of the GRW's surface area, is agroecologically very fragile: hilly terrain and shallow soils that make agricultural mechanization difficult. The land use is more diversified, with small parcels of land intermingled with natural or re-vegetated forest intensively exploited under the conventional tillage system [12,16]. In addition to the family's subsistence crops, there are many tobacco-growing farmers integrated in partnership with large tobacco corporations. In addition, this is one of the regions in Latin America with the highest density of pigs, poultry, and dairy cattle. Finally, this is the region where yerba mate was historically grown and has recently registered a significant increase in the area with this crop. Inclusively, in the last two decades, the conversion of small agricultural establishments producing yerba mate into agro–industrial complexes in the sector has been ongoing [55].

Therefore, we identified and organized crop and livestock systems into four alternatives: (1) grain no-tillage systems, represented by soybean fields; (2) production systems under conventional tillage, and areas cultivated with tobacco; (3) perennial systems, grown with yerba mate; and (4) animal integration of pigs and dairy cattle farming systems. To evaluate and compare the pressures these different agricultural production systems cause on aquatic ecosystems, the evaluation model presented above was adopted.

Through the local evaluation of the criteria (Figure 3) and sub-criteria (Figure 4), it was possible to identify the virtues and vulnerabilities of each of the selected production systems. Therefore, the environmental conflicts of the production systems conventional tillage (tobacco)—Alternative 2 of our model—can be explained mainly for three criteria The first of these is the technique used to apply fertilizers. The results are in line with the studies [56,57], identifying applied doses higher than those needed by the crop. In addition, the technique used to apply the mineral fertilizer is regularly mistaken, as it has been applied on the soil surface instead of applied in the furrow. This condition is aggravated since fertilization and transplanting of the tobacco seedlings takes place in September and October, when the most significant rainfall is recorded in the region.

The second criterion that contributes negatively to the evaluation of tobacco production is soil management. In the GRW, there are two groups of tobacco farmers in terms of soil management: those who use minimum tillage (they only turn the soil over to prepare the ridge), and those who use a conventional tillage system to later make the ridge. In the conventional tillage system, the soil surface remains uncovered much of the year, which favors the erosion process. The study in [58] has shown that soil management practiced by tobacco farmers leads to rapid, intense degradation of some natural soil properties, especially those related to the dynamics of soil organic matter, compared with more conservationist uses. Complementarily, the soils cultivated with tobacco in general are fragile, shallow, and of medium texture, and the environment is sloping, generally with a gradient above 16 degrees [56].

Thus, the landscape features and soil properties were the criterion that weighed most negatively in the evaluation of the tobacco farming system. These characteristics indicate a low water-storage capacity and the high susceptibility of the soil to erosion. Testing soil management systems for tobacco cropped using animal traction on shallow soil on steep lands showed that the total soil loss was 15 Mg ha^{-1} for conventional tillage, and was reduced about five times for minimal- and no-tillage systems [58]. This same trend was observed for total losses of phosphorus and potassium, where no-tillage systems reduced about 97- and 57-times the losses of these nutrients compared to conventional tillage.

Although tobacco cultivation is labeled for consuming high amounts of agrochemicals, our monitoring has shown that consumption is lower than in the crop fields under no-tillage for grain production. Herbicides are mostly used to control unwanted weeds and anti-sprouting agents that inhibit the growth of axial buds. Herbicides occasionally include 2,4-D, which is extremely toxic [12].

Tobacco cultivation systems have some peculiarities, such as high added value and the possibility of being cultivated in small areas on rocky slopes without large technological investments in equipment and facilities. Much of the work is performed manually, which allows for the inclusion of less capitalized farmers. Moreover, the tobacco production chain is very well structured in the GRW, especially regarding the availability of inputs, technical assistance to farmers, logistics, and demand (leaf tobacco).

Studies with perennial crops have shown that the degree of soil degradation and the contamination of surface watercourses are relatively lower than when used with annual crops [59]. For example, monitoring two-paired catchments (eucalyptus and degraded grassland) showed that twofold smaller surface runoff and sediment yield occurred in the eucalyptus catchment [60]. It was also found that the reconstitution of natural forests of degraded soils by conventional tillage recovered the soil carbon stock quickly. Forestlands have negligible soil losses in comparison with the other vegetation covers [61]. In contrast, the intense soil tillage and mechanical weed control for tea plantations in central China lead to high erosion rates, especially at slopes higher than 30°.

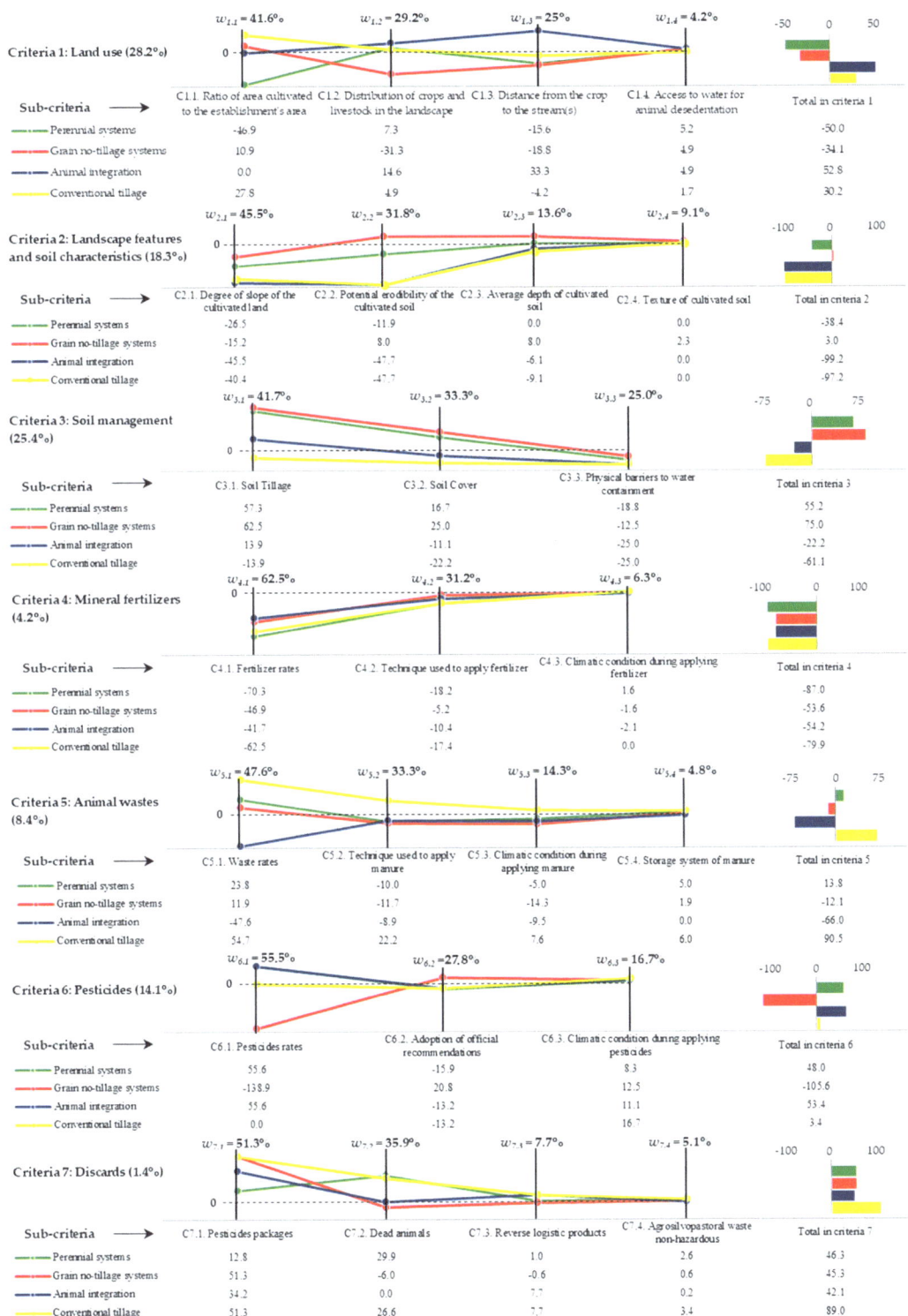

Figure 4. Level of pressure in each sub-criterion for the four production systems analyzed. $w_{1.1}, w_{1.2}, \ldots, w_{1.i}$ = weight of sub-criteria in each criterion.

In our study, although tobacco and yerba mate systems occupy similarly fragile environments—shallow soils and hilly terrain located on the slopes—in yerba mate cultivation there is no constant soil disturbance and, in most cases, the areas remain covered most of the year. The yerba mate production system, regarding the distribution of plots in the landscape, transplanting and establishment of the orchard, and cultural management, resembles Argentine organic or agroforestry systems [62]. Yerba mate has become an attractive production system for the region's farmers, consolidating it as the largest production hub in Brazil. In the last two decades, many farmers have seen this system as an interesting option in terms of income generation and low labor demand in an increasingly sparse and aged rural population. The greatest environmental adversity assessed is related to land occupation, since yerba mate is native to the region and farmers make the most of this comparative advantage by cultivating it in permanent preservation areas, provided for in Law 12.651 of 2012 [63]. With minor adjustments in land occupation, this production system can present even less harmful impacts on the environment.

The performance of Alternative 1—grain no-tillage systems (soybean)—presents intermediate values between the two crops presented above. The relief favors mechanization and the adoption of no-tillage systems. The grain crops practically occupy the entire surface of the establishments, including areas that legislation does not allow for agricultural use, as is the case of marginal strips along water courses. Furthermore, most establishments do not comply with what is determined by Law No. 12,651 of 2012 [63] regarding maintaining that 20% of the establishment's area be covered by native vegetation. A second aspect that disfavors assessment of the grain production system is the volume of pesticides applied to the crops. Soybean production, par example, may include up to ten applications of pesticides in a crop cycle. In general, 2,4-D is applied before sowing—to control glyphosate-resistant plants; after sowing (between 30 and 50 days), a second application is made to control undesirable plants, this time with glyphosate. During the soybeans cycle, between three and four applications of fungicides and three or four applications of insecticides are made. Accordingly, it is not surprising to find that several pesticides and their metabolites are present in the water of the dense drainage network of the GRW [12]. Of even greater concern is the fact biofilms are already impregnated with pesticides [64], including glyphosate and AMPA. Furthermore, the doses used are commonly higher than those officially recommended by the National Health Surveillance Agency (ANVISA, as in Portuguese abbreviation). According to data from the Brazilian Institute of Environment and Renewable Natural Resources, after glyphosate and 2,4-D, atrazine and simazine are the most commercialized pesticides in the southern region of Brazil [65]. The farmers of the GRW seem to follow this pattern of pesticide use.

Improvements in landscape and soil properties are positive impacts (buffering anthropic impacts on environmental degradation) of the grain production system. The soil is deep and well-structured, with high infiltration capacity, and practically 100% of the surface is managed with no-tillage. Studies have shown that no-till farming systems increase the stability of aggregates, the infiltration and availability of water, the cycling of nutrients by microbial action, the content of organic matter, and the capacity of the soil to retain nutrients [66]. However, we have found that farmers have removed the terraces to speed up sowing, cultivation, and harvesting operations. Also, we did not find any other type of physical barrier to runoff. The study in [8] has demonstrated that the absence of terraces makes no-tillage ineffective in controlling runoff and soil erosion. Over a long period of monitoring, they found that the presence of the terraces reduced peak flow rates by 79%, sediment yield from 0.44 to 0.16 Mg ha^{-1}, and the total surface runoff from 1622 to 363 m^3 ha^{-1} (reduced 77%).

The performance of Alternative 4—animal integration (pig and dairy cattle farming systems)—reflects the unfavorable characteristics of the environment and the soil in which most of the animal-raising systems in the GRW are found. The soils are fragile and located on an extremely rugged terrain. Generally, the rearing system is accompanied by corn crops, where the waste from animal rearing is distributed. These crops are predominantly managed with no-till farming systems, in which the lack of disturbance of the fragile soils favors water infiltration and regulates the flow into watercourses. However, the continuous use of waste in the same area, without eventual incorporation, may cause an imbalance in the soil's physical, chemical, and biological properties. High concentrations of nutrients in the topsoil increase the propensity of transfers to water bodies, mainly N and P, which can trigger the eutrophication process [67].

Although the environment in which the animal husbandry systems are located is considered to be ecologically fragile, it should be noted that the model positively evaluated land use (Figure 4) because the ratio between the establishment area and the cultivated area is significantly high compared with the other productive systems in the region. Pig and dairy cattle systems (also the poultry breeding systems) of the GRW occupy agricultural production units that have important reserves with natural forest areas. The cultivated areas are distributed in the natural landscape, which plays an important role in mitigating the negative effects of both agricultural practices and the inputs used. Similarly, pollutants need to travel long distances occupied by natural forests that separate crops from springs and streams, which hinders the transfer of agricultural residues from crops to water bodies.

Another less-impacting criterion of the animal husbandry system, verified using the model, is the fact that it uses a relatively low volume of agrochemicals. Normally, one or two applications a year are used to control undesirable plants in the crop areas that complement the production system of the establishments. Furthermore, most of the products used are classified as low or moderately toxic according to their toxicity class.

Based on the multicriteria model (Table 3), Alternative 3—perennial system (yerba mate)—is the one with the least negative interference for the maintenance of the natural characteristics of the water resources of the GRW. In contrast, conventional tillage (tobacco) manifested the worst conditions for water resources (Figure 5).

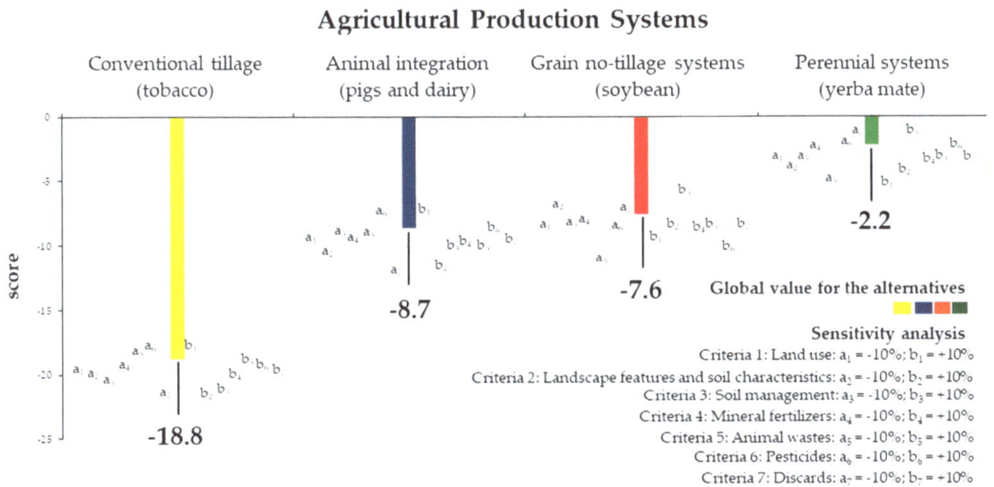

Figure 5. Global assessment of alternatives (production systems).

The representation provided by the multicriteria model derives from scaled measurements in mathematics, psychology, and philosophy [68]. Thus, it is recommended to produce small variations in the raw values of the compensation rates assigned to the criteria to verify model sensitivity. In this case, we proceeded with changes of 10% up and down. As can be seen in Figure 5, the model does not present significant changes in performance due to modifications in the compensation rates; therefore, it can be considered robust concerning the parameters evaluated.

Figure 6 shows the equations that represent the overall evaluation of the production systems as a function of the compensation rate between criteria. Straight lines represent the global evaluation of the alternatives as a function of the variation in the substitution rate of one of the model's criterion in graphical form [32].

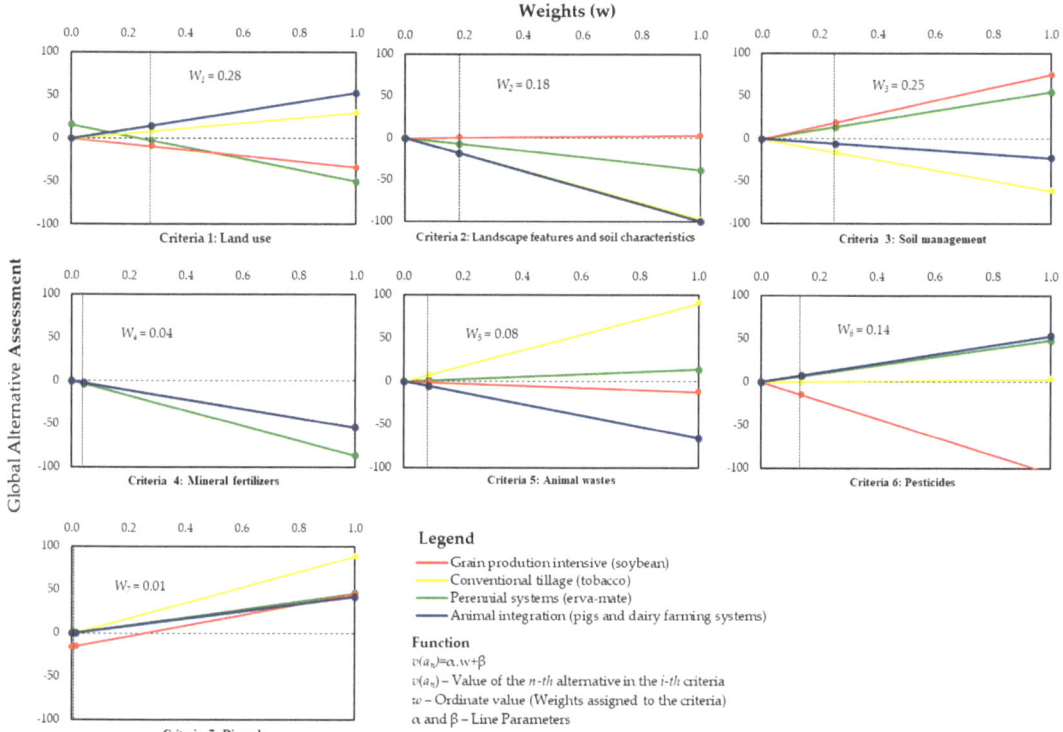

Figure 6. Representation of the evaluation as a function of variation in compensation rates.

Finally, it is necessary to emphasize that the scores attributed to the production systems in the multicriteria model are adimensional and, therefore, a physical degree is not applied to the estimated scores. The values of -2.2, -7.6, -8.7, and -18.8 (Figure 5), although satisfactory to represent the differences in pressure performance of the alternatives, do not necessarily represent the absolute polluting potential at the watershed scale. Therefore, it is necessary to relativize them. In this case, two conditioning factors were sought to understand the behavior of the model's pressures: cultivated surface and the number of establishments corresponding to each production system in the watershed.

The first test to relativize the pressures of the production systems in the GRW can be carried out using the proportional area occupied by the different cultivation systems in the watershed. Soybean, yerba mate, and tobacco crops represent 29.8%, 3.5%, and 1.4%, respectively, of the total watershed area. By multiplying the scores associated with each production system by the area occupied, it becomes evident that grain production systems (soybeans) have the greatest impact on water resources (-2.3) in comparison with the other systems assessed. Yerba mate and tobacco crops have much lower impacts than grain crops (-0.1 and -0.3, respectively).

The same procedure can be used to scale the score of production and rearing systems concerning the number of establishments in the GRW. Soybean, yerba mate, tobacco, and pig and cattle farming are present in 41.8%, 23.6%, 10.3%, and 44.9% of establishments, respectively, with scaled scores of -3.2; -0.5; -1.9, and -3.9. Despite the overall pressure of the tobacco farming system being the highest among the models analyzed (−18.8), this production system occupies a small area and is present in a restricted number of establishments if compared to other systems in the region. For this reason, it is inferred that the pressure related to tobacco cultivation is much more intense locally than on a regional scale considering the drainage area of the watershed BH. In contrast, the soybean cultivation system and the animal husbandry system demonstrate a greater negative influence on water resources at the regional level since they have a greater presence in terms of area and agricultural establishments in the region under analysis.

Brazil is a continental country possessing one of the planet's largest reserves of available fresh water, and it is one of the world's leading agricultural producers. If, on the one hand, "modern" Brazilian agriculture—more technical, represented by capitalized farmers and with commercial relations unified with global economic cycles—has contributed to significantly increasing the regional production of soy, corn, poultry, pigs, and dairy cattle, etc., on the other hand, it has stimulated the exploitation of natural resources to a level beyond which the environment can support.

This seems to be the situation detected in the agricultural area comprising the Guaporé River Watershed, whose water resources are part of the drainage area that contributes to the water supply for a population of almost 4 million inhabitants of the Porto Alegre metropolitan region. Without being too rigorous, the studied watershed represents the dynamics of the main productive systems developed in Brazil, so that the local manifestations identified correspond satisfactorily to a large part of the regional- and global-scale problems of Brazilian agriculture.

The model was designed and developed with the participation of interested parties; thus, the systematized information is associated with their perception of the regional context in which the watershed was studied. Therefore, to extrapolate the results to other watersheds (other regions), it is necessary to adapt the criteria and weights to the local agricultural systems. Moreover, methodologically, the research indicates a starting point for those interested in constructing environmental indicators for agricultural activities. Among the limiting factors is the need to consult numerous times with stakeholders to reach a consensus on the organization of the criteria and weights assigned to them.

4. Conclusions

The hypothesis that models structured through participatory processes can identify and solve complex problems, such as the management of natural ecosystems, has been confirmed. The participatory approach is fully capable of providing objective and useful data for a model that aims to assess how and to what extent different production systems put pressure on (or even impact) aquatic ecosystems. Such a model, which combines seven criteria and twenty-five sub-criteria, has proved to be robust enough to assess and compare different agricultural pressures at the watershed scale.

To reduce the pressures arising from regional (and even national) agriculture, the following is suggested:

i. Land use should be adjusted to agricultural suitability, and conservationist practices of soil and water management should be incorporated, restoring permanent preservation areas, especially in springs and waterways functioning as buffer zones from agricultural pressures.

ii. Legislative and governance structures should encourage agricultural models with lower impacts on natural ecosystems through compensatory policy instruments and market instruments to take advantage of positive linkages between economic development and the environment. It is also necessary to create a political/institutional environment favorable to sustainability that works through negotiation and is dialogical among all actors involved in the process (farmers and state agents).

iii. Control mechanisms foreseen in the Brazilian environmental policy must be applied, particularly (a) licenses to authorize the installation and operation of potentially polluting agricultural projects and activities; (b) environmental zoning to regulate land use; and (c) monitoring and guidance on the parameters and targets set for the emission of pollutants into the environment.

Author Contributions: Conceptualization, A.T., M.C.G. and D.d.S.R.; Data curation, T.T.; Formal analysis, A.T. and D.d.S.R.; Acquisition of funding, D.d.S.R.; Research, A.T.; Methodology, A.T. and M.C.G.; Project administration, J.M.R.; Resources, D.d.S.R. and A.T.; Supervision, J.M.R.; Validation, T.T.; Writing—original draft, A.T.; Writing—revision and editing, T.T., M.B.P., D.d.S.R. and J.M.R. All authors have read and agreed to the published version of the manuscript.

Funding: This research was funded by Coordination for the Improvement of Higher Education Personnel (Coordenação de Aperfeiçoamento de Pessoal de Nível Superior—CAPES—finance code 001) and Conselho Nacional de Desenvolvimento Cientifico e Tecnológico (CNPq) (152604/2022-7).

Data Availability Statement: Data are contained within the article.

Acknowledgments: Department of Soils in Federal University of Santa Maria and the postgraduate program in Family Farming Production Systems in Federal University of Pelotas.

Conflicts of Interest: The authors declare no conflicts of interest and the funders had no role in the design of the study; in the collection, analyses, or interpretation of data; in the writing of the manuscript; or in the decision to publish the results.

Appendix A

Table A1. Structure of the Model: Criteria and Sub-Criteria with Their Respective Weights, and the Impact Levels of the Descriptors.

Criteria	C1. Land use					28.2%
Sub-criteria	C1.1. Ratio of area cultivated to the establishment's area					41.6%
Attributes	Forest areas x < 30	Forest areas 20 < x > 30	Forest areas x > 30%			
Weights	−150	0	100			
Impact levels		Neutral	Good			
Sub-criteria	C1.2. Distribution of crops and livestock in the landscape					29.2%
Attributes	One plot of cultivated land	Two plots of cultivated land	Three plots of cultivated land	Equal four plots of cultivated land	More than four plots of cultivated land	
Weights	−100	0	50	100	150	
Impact levels		Neutral		Good		
Sub-criteria	C1.3. Distance from the crop to the stream(s)					25.0%
Attributes	x < 30 m streams and x < 50 m water sources	30 < x > 40 m streams and x > 50 m water sources	40 < x > 50 m streams and x > 50 m water sources	50 < x > 60 m streams and x > 50 m water sources	x > 60 m streams and x > 50 m water sources	
Weights	−100	0	50	100	150	
Impact levels		Neutral		Good		
Sub-criteria	C1.4. Access to water for animal desedentation					4.2%
Attributes	Access water through streams with no defined corridor	Access water through streams in specific corridors	Access the water by means of swamps	Access water by means of weirs	Do not access water in the natural environment	
Weights	−125	−50	0	100	125	
Impact levels			Neutral	Good		

Table A1. Cont.

Criteria	C2. Landscape features and soil characteristics					18.3%
Sub-criteria	C2.1. Degree of slope of the cultivated land					45.5%
Attributes	Slope x > 25% and occupied by annual crops	Slope x > 25% and occupied by natural pastures or forestry	Slope 16% > x < 25% and occupied by annual crops	Slope 16% > x < 25% and occupied by natural pastures or forestry	Slope < 16% and occupied by annual crops	Slope < 16% and occupied by natural pasture or forestry
Weights	−200	−166	−66	0	100	133
Impact levels				Neutral	Good	
Sub-criteria	C2.2. Potential erodibility of the cultivated soil					31.8%
Attributes	Soil erodibility potential is strong	Soil erodibility potential is moderate	Soil erodibility potential is incipient			
Weights	−150	0	100			
Impact levels		Neutral	Good			
Sub-critéria	C2.3. Average depth of cultivated soil					13.6%
Attributes	Average soil depth < 50 cm	Average soil depth ranges around 50 > x < 100 cm	Average soil depth varies around 100 > x < 150 cm	Average soil depth x > 150 cm		
Weights	−66	0	66	100		
Impact levels		Neutral		Good		
Sub-criteria	C2.4. Texture of cultivated soil					9.1%
Attributes	Clay content x < 15%	Clay content 15 > x < 35%	Clay content x > 35%			
Weights	−100	0	100			
Impact levels		Neutral	Good			
Criteria	C3. Soil management					25.4%
Sub-criteria	C3.1. Soil Tillage					41.7%
Attributes	Conventional tillage	Minimal tillage or where there is little soil movement between rows for perennial farms	No-tillage system or where there is no soil movement in the case of perennial farms	No-tillage system		
Weights	−100	0	100	150		
Impact levels		Neutral	Good			
Sub-criteria	C3.2. Soil Cover					33.3%
Attributes	<20% of soil surface covered in the post-harvest period until sowing/transplanting	25 > x < 40% of soil surface covered in the post-harvest period until sowing/transplanting	40 > x < 60% of the soil surface covered in the post-harvest period until sowing/transplanting	60 > x < 80% of soil surface area covered in the post-harvest period until sowing/transplanting	x > 80% of the soil surface covered by straw in the post-harvest period until sowing/transplanting	
Weights	−200	−100	0	100	133	
Impact levels			Neutral	Good		

105

Table A1. *Cont.*

Sub-criteria	C3.3. Physical barriers to water containment				25.0%
Attributes	No physical barriers are used to contain runoff, nor is level planting	No barriers are used to contain runoff, however, planting is on the level	Barriers are used to contain runoff and planting is level		
Weights	−100	0	100		
Impact levels		Neutral	Good		
Criteria	C4. Mineral fertilizers				4.2%
Sub-criteria	C4.1. Fertilizer rates				62.%
Attributes	Above the recommended dose	At the recommended dose	Below the recommended dose		
Weights	−150	0	100		
Impact levels		Neutral	Good		
Sub-criteria	C4.2. Technique used to apply fertilizer				31.2%
Attributes	All applied in the sowing	Incorporated in the seeding and part applied to the haulm	Incorporated in sowing	Incorporated by correction and part in the sowing by replacement	
Weights	−66	−33	0	100	
Impact levels			Neutral	Good	
Sub-criteria	C4.3. Climatic condition during applying fertilizer				6.3%
Attributes	Does not observe weather conditions	Sometimes observes weather conditions	Always observe the climatic conditions		
Weights	−100	0	100		
Impact levels		Neutral	Good		
Criteria	C5. Animal wastes				8.4%
Sub-criteria	C5.1. Waste rates				47.6%
Attributes	Pig x > 80 m^3 ha^{-1}; cattle x > 200 m^3 ha^{-1}; poultry x > 8 T ha^{-1}	Pig 60 > x < 80 m^3 ha^{-1}; cattle 150 < x < 200 m^3 ha^{-1}; poultry 4 > x < 8 T ha^{-1}	Pig 40 > x < 60 m^3 ha^{-1}; beef 100 < x < 150 m^3 ha^{-1}; poultry 3 > x < 5 T ha^{-1}	Pig x < 40 m^3 ha^{-1}; cattle x < 100 m^3 ha^{-1}; poultry x < 3 T ha^{-1}	Manure is not applied to the farm
Weights	−100	−44	0	Good	122
Impact levels			Neutral		
Sub-criteria	C5.2. Technique used to apply manure				33.3%
Attributes	Surface applied at post-planting or transplanting	Always applied to the soil surface	Surface application and sporadically incorporated into the soil	Not applied	
Weights	−100	−40	0	100	
Impact levels			Neutral	Good	

Table A1. Cont.

Sub-criteria	C5.3. Climatic condition during applying manure					14.3%
Attributes	Does not observe weather conditions	Sometimes observes weather conditions	Always observe the climatic conditions	Not applied		
Weights	−100	−40	0	100		
Impact levels			Neutral	Good		
Sub-criteria	C5.4. Storage system					4.8%
Attributes	Storage is not covered, waterproofed, or has a drainage channels	Storage is not covered, waterproofed, and has drainage channels	Storage is covered, waterproofed, without drainage channels	Storage is covered, waterproofed, and has drainage channels	It does not have a breeding system	
Weights	−60	0	60	100	140	
Impact levels		Neutral		Good		
Criteria	C6. Pesticides					14.1%
Sub-criteria	C6.1. Pesticides rates					43.5%
Attributes	Volume x > 10 L ha year^{-1}	Volume 10 > x > 5 L ha year^{-1}	Volume 5 > x > 3 L ha year^{-1}	Volume x < 3 L ha year^{-1}	Do not use pesticide	
Weights	−250	−125	0	100	250	
Impact levels			Neutral	Good		
Sub-criteria	C6.2. Adoption of official recommendations					34.8%
Attributes	Never adopts official recommendations, does not read package leaflets, does not observe markings, stripes and drawings on packages	Sometimes adopts official recommendations and does not always read package leaflets, observe colours, stripes and designs on packages	Sometimes adopts official recommendations and always reads package leaflets, observes colours, stripes and designs on packaging	Official recommendations adopted, do not always read package leaflets, observe colours, stripes and designs on packaging	Official recommendations are adopted and package leaflets are always read, and the colours, stripes and designs on the packaging are observed	
Weights	−142	−42	0	100	142	
Impact levels			Neutral	Good		
Sub-criteria	C6.3. Climatic condition during applying pesticides					21.7%
Attributes	Does not observe weather conditions	Sometimes observes weather conditions	Always observe the climatic conditions			
Weights	−100	0	100			
Impact levels		Neutral	Good			

Table A1. Cont.

Criteria	C7. Discards					1.4%
Sub-criteria	C7.1. Pesticides packages					51.3%
Attributes	Packages are discarded in an inadequate place and without triple washing	The packages are discarded in a place that is considered adequate, without carrying out the triple rinse	The packages are discarded at a place that is considered adequate, after being triple rinsed	The packages are delivered to the collection points after being triple rinsed	Packages are delivered to collection points without undergoing the triple rinse	
Weights	−125	−25	0	100	125	
Impact levels			Neutral	Good		
Sub-criteria	C7.2. Dead animals					35.9%
Attributes	Dead animals are disposed of "in the open"	Dead animals are buried in mass graves	Dead animals are disposed of in the conventional compost bin	Dead animals are incinerated	It has no animal husbandry system	
Weights	−111	−22	0	100	111	
Impact levels			Neutral	Good		
Sub-criteria	C7.3. Reverse logistic products					7.7%
Attributes	Are discarded without being separated	Are separated and discarded in a place considered appropriate	Are delivered to the collection points without being separated	They are separated and delivered to the collection points		
Weights	−50	0	100	116		
Impact levels		Neutral	Good			
Sub-criteria	C7.4. Agrosilvopastoral waste non-hazardous					5.1%
Attributes	It is disposed of "in the open"	It is burned	It is buried	It is destined for recycling without being separated	Separated according to its constitution or composition and destined for recycling	
Weights	−26	0	40	100	106	
Impact levels		Neutral		Good		

Appendix B

Land occupation and use

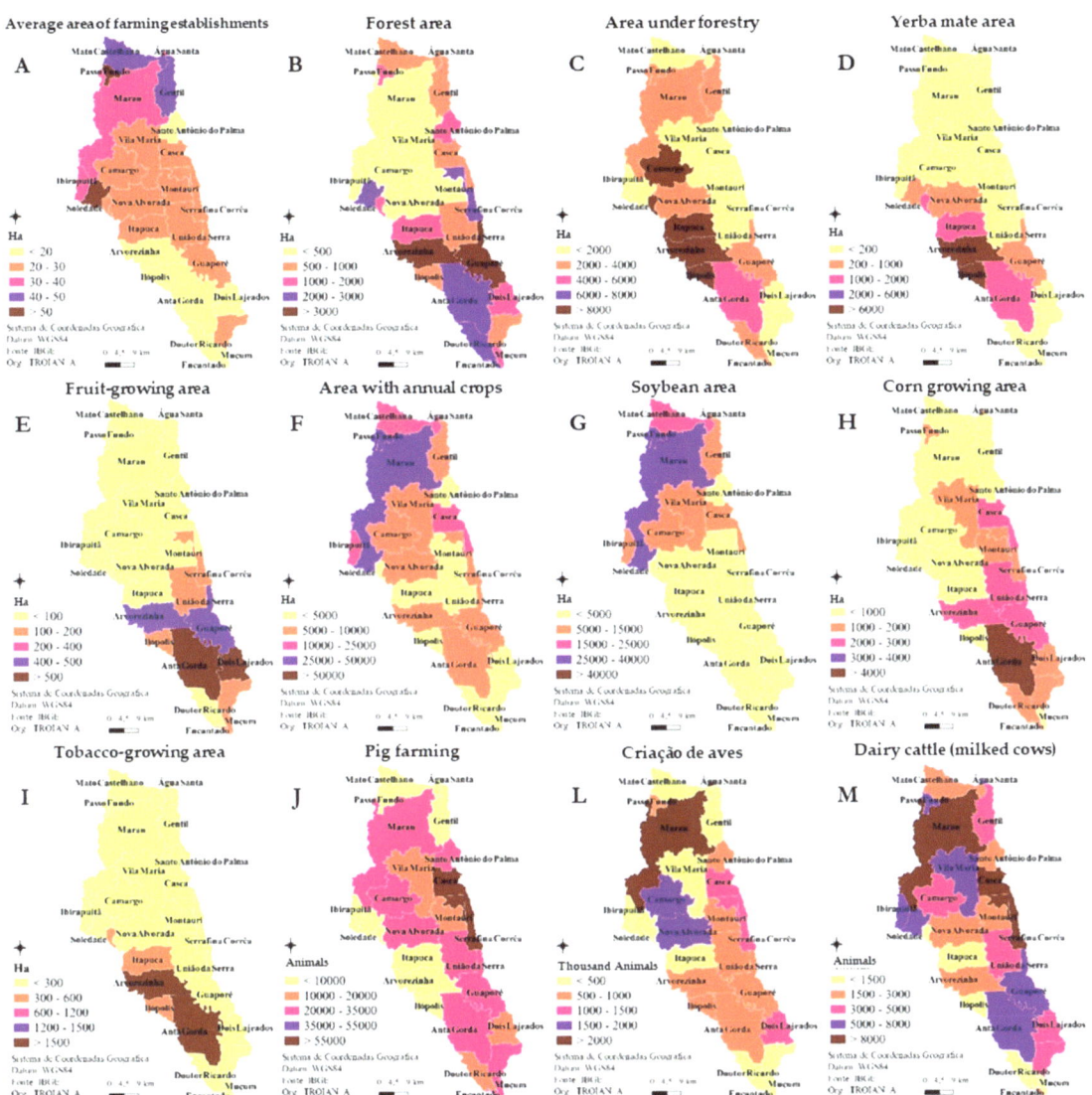

Figure A1. Land occupation and use.

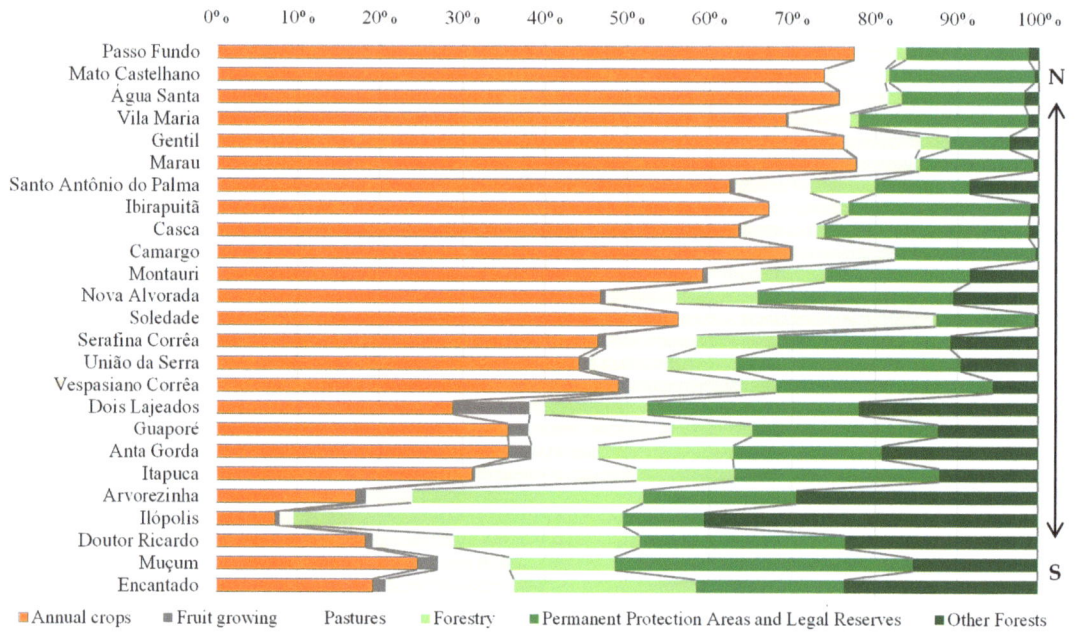

Figure A2. Land use and occupation.

References

1. Bayramoglu, B.; Chakir, R.; Lungarska, A. Impacts of land use and climate change on freshwater ecosystems in France. *Environ. Model. Assess.* **2020**, *25*, 147–172. [CrossRef]
2. IPCC (Intergovernmental Panel on Climate Change). Global Warming of 1.5 °C. An IPCC Special Report on the Impacts of Global Warming of 1.5 °C above Pre-Industrial Levels and Related Global Greenhouse Gas Emission Pathways, in the Context of Strengthening the Global Response to the Threat of Climate Change, Sustainable Development, and Efforts to Eradicate Poverty. 2018. Available online: https://www.ipcc.ch/sr15/download/#full (accessed on 24 August 2021).
3. Becker, A.G.; Moraes, B.S.; Menezes, C.C.; Loro, V.L.; Santos, D.R.; Reichert, J.M.; Baldisserotto, B. Pesticide contamination of water alters the metabolism of juvenile silver catfish, *Rhamdia quelen*. *Ecotoxicol. Environ. Saf.* **2009**, *72*, 1734–1739. [CrossRef]
4. Antrop, M. The Role of Cultural Values in Modern Landscapes. In *Landscape Interfaces*; Palang, H., Fry, G., Eds.; Landscape Series; Springer: Dordrecht, The Netherlands, 2003; pp. 91–108. [CrossRef]
5. Reichert, J.M.; Gubiani, P.I.; Rheinheimer dos Santos, D.; Reinert, D.J.; Aita, C.; Giacomini, S.J. Soil properties characterization for land-use planning and soil management in watersheds under family farming. *Soil Water Conserv. Res.* **2022**, *10*, 119–128. [CrossRef]
6. Troian, A.; Gomes, M.C.; Tiecher, T.; Berbel, J.; Gutiérrez-Martín, C. The drivers-pressures-state-impact-response model to structure cause-effect relationships between agriculture and aquatic ecosystems. *Sustainability* **2021**, *13*, 9365. [CrossRef]
7. Capel, P.D.; McCarthy, K.A.; Coupe, R.H.; Grey, K.M.; Amenumey, S.E.; Baker, N.T.; Johnson, R.L. *Agriculture—A River Runs through It—The Connections between Agriculture and Water Quality*; U.S. Geological Survey Circular; U.S. Geological Survey: Reston, Virginia, 2018; Volume 1433, 201p. [CrossRef]
8. Londero, A.L.; Minella, J.P.G.; Deuschle, D.; Schneider, F.J.A.; Boeni, M.; Merten, G.H. Impact of broad-based terraces on water and sediment losses in no-till (paired zero-order) catchments in southern Brazil. *J. Soils Sediments* **2017**, *18*, 1159–1175. [CrossRef]
9. Ebling, E.D.; Reichert, J.M.; Minella, J.P.G.; Holthusen, D.; Broetto, T.; Srinivasan, R. Rainfall event-based surface runoff and erosion in small watersheds under dairy and direct-seeding grain production. *Hydrol. Process.* **2022**, *36*, e14688. [CrossRef]
10. Silva, T.P.; Bressiani, D.; Ebling, E.D.; Deus Júnior, J.C.; Reichert, J.M. Evaluating hydrological and soil erosion processes in different time scales and land uses in southern Brazilian paired watersheds. *Hydrol. Sci. J.* **2023**, *68*, 1391–1408. [CrossRef]
11. Volf, G.; Atanasova, N.; Škerjanec, M.; Ožanić, N. Hybrid modeling approach for the northern Adriatic watershed management. *Sci. Total Environ.* **2017**, *635*, 353–363. [CrossRef]
12. Castro Lima, J.A.M.; Labanowski, J.; Bastos, M.C.; Zanella, R.; Prestes, O.; Damian, M.L.; Granado, E.; Tiecher, T.; Zafar, M.; Troian, A.; et al. "Modern agriculture" transfers many pesticides molecules to watercourses: A case study of a representative rural catchment of southern Brazil. *Environ. Sci. Pollut. Res.* **2020**, *27*, 10581–10598. [CrossRef]

13. UNDP (United Nations Development Programme). *A Água Para lá da Escassez: Poder, Pobreza e a Crise Mundial da Água*; Relatório do Desenvolvimento Humano: New York, NY, USA, 2006. Available online: http://hdr.undp.org/sites (accessed on 5 October 2021).
14. National Water Agency (ANA, as in Portuguese abbreviation). *Conjuntura dos Recursos Hídricos no Brasil 2019: Relatório Anual*; National Water Agency: Brasília, Brazil, 2019. Available online: https://www.ana.gov.br/ (accessed on 14 January 2020).
15. UNDP (United Nations Development Programme). *Perspectivas del Medio Ambiente Mundial, GEO 6: Planeta Sano, Personas Sanas*; Resumen Para Responsables de Formular Políticas: Nairobi, Kenya, 2019. Available online: https://www.unep.org/es/resources/perspectivas-del-medio-ambiente-mundial-6 (accessed on 5 October 2021).
16. Tiecher, T.; Minella, J.P.G.; Caner, L.; Zafar, M.; Capoane, V.; Evrard, O.; Le Gall, M.; Rheinheimer, D.S. Quantifying land use contributions to suspended sediment in a large cultivated catchment of Southern Brazil (Guaporé River, Rio Grande do Sul). *Agric. Ecosyst. Environ.* **2017**, *237*, 95–108. [CrossRef]
17. Barros, C.A.P.; Govers, G.; Minella, J.P.G.; Ramon, R. How water flow components affect sediment dynamics modeling in a Brazilian catchment. *J. Hydrol.* **2021**, *597*, 126411. [CrossRef]
18. Knott, J.; Mueller, M.; Pander, J.; Geist, J. Effectiveness of catchment erosion protection measures and scale-dependent response of stream biota. *Hydrobiologia* **2019**, *830*, 77–92. [CrossRef]
19. Bierschenk, A.M.; Mueller, M.; Pander, J.; Geist, J. Impact of catchment land use on fish community composition in the headwater areas of Elbe, Danube and Main. *Sci. Total Environ.* **2019**, *652*, 66–74. [CrossRef] [PubMed]
20. Mazoyer, M.; Roudart, L. *História das Agriculturas no Mundo: Do Neolítico à Crie Contemporânea*; Instituto Piaget: Lisboa, Portugal, 2001; 520p.
21. Dufumier, M. *Projetos de Desenvolvimento Agrícola: Manual Para Especialistas*, 2nd ed.; EDUFBA: Salvador, Brazil, 2010; 330p.
22. Stanners, D.; Bosch, P.; Dom, A.; Gabrielsen, P.; Gee, D.; Martin, J.; Weber, J.L. Frameworks for environmental assessment and indicators at the EEA. In *Sustainability Indicators: A Scientific Assessment*; HÁK, T., Moldan, B., Dahl, L.A., Eds.; Island Press: Covelo, CA, USA, 2007.
23. Westbury, D.B.; Park, J.; Mauchline, A.; Crane, R.; Mortimer, S. Assessing the environmental performance of English arable and livestock holdings using data from the Farm Accountancy Data Network (FADN). *J. Environ. Manag.* **2011**, *92*, 902–909. [CrossRef] [PubMed]
24. Ravier, C.; Prost, L.; Jeuffroy, M.; Wezel, A.; Paravano, L.; Reau, R. Multi-criteria and multi-stakeholder assessment of cropping systems for a result-oriented water quality preservation action programme. *Land Use Policy* **2015**, *42*, 131–140. [CrossRef]
25. Rousval, B. Aide Multicritère à L'évaluation de L'impact des Transports sur L'environnement. Modélisation et Simulation. 267 f. Ph.D. Thesis, University of Paris Dauphine, Paris, France, 2005. Available online: https://tel.archives-ouvertes.fr/tel-00543658v1 (accessed on 7 October 2021).
26. Funtowicz, S.; Ravetz, J. Ciência Pós-Normal e comunidades ampliadas dos pares face aos desafios ambientais. *Hist. Cienc. Saúde* **1997**, *4*, 219–230. [CrossRef]
27. Romero, C. *Teoría de la Decisión Multicriterio: Conceptos, Técnicas y Aplicaciones*; Alianza Editorial: Madrid, Spain, 1993; 98p.
28. Bana e Costa, C.A.; Pirlot, M. Thoughts on the Future of the Multicriteria Field: Basic Convictions and Outlines for a General Methodology. In *Multicriteria Analysis*; Clímaco, J., Ed.; Springer: Berlin/Heidelberg, Germany, 1997; pp. 562–568.
29. Romero, C. 1996. *Análisis de las Decisiones Multicriterios*; Conceptos, Técnicas y Aplicaciones; Gráficas Algorán: Madrid, Spain, 1996; 115p.
30. Doumpos, M.; Zopounidis, C. *Multicriteria Decision Aid Classification Methods*; Kluwer Academic Publishers: Dordrecht, The Netherlands, 2002; 271p.
31. Ehrgott, M.; Figueira, J.; Greco, S. *Trends in Multiple Criteria Decision Analysis*; Springer: New York, NY, USA, 2010; 429p. [CrossRef]
32. Ensslin, L.; Montibeller Neto, G.; Noronha, S.M. *Apoio à Decisão: Metodologia para Estruturação de Problemas e Avaliação Multicritério de Alternativas*; Insular: Florianópolis, Brazil, 2001; 293p.
33. Bouyssou, D. *Building Critcria: A Prerequisite for MCDA. In Readings in Mnltiple Criteria Decision Aid*; Bana e Costa, C.A., Ed.; Springer: Berlin/Heidelberg, Germany, 1990; pp. 91–151.
34. Roy, B. *Multicriteria Methodology for Decision Aiding*; Kluwer Academic Publishers: Dordrecht, The Netherlands, 1996; 303p.
35. Beinat, E. *Value Functions for Environmental Management*; Springer Science and Business Media Dordrecht: Dordrecht, The Netherlands, 1997; 249p.
36. Silva Neto, B. Análise-Diagnóstico de Sistemas Agrários: Uma interpretação baseada na Teoria da Complexidade e no Realismo Crítico. In *Desenvolvimento em Questão*; Unijuí: Ijuí, Brazil, 2007; pp. 33–58.
37. Patton, M.Q. *Qualitative Research and Evaluation Methods*, 3rd ed.; Sage Publications: London, UK, 2002; 179p.
38. Martins, R.C. *A Construção Social do Valor Econômico da Água: Estudo Sociológico Sobre Agricultura, Ruralidade e Valoração Ambiental no Estado de São Paulo*; Ph.D. Thesis, (Doctorte in Environmental Engineering Science). Escola de Engenharia de São Carlos, Universidade de São Paulo. 2004. Available online: https://www.teses.usp.br/?lang=pt-br (accessed on 5 September 2021).
39. Bana e Costa, C.A. *Processo de Apoio à Decisão: Problemáticas, Actores e Acções*; Apostila do Curso de Metodologias Multicritério em Apoio à Decisão; ENE, UFSC: Florianópolis, Brazil, 1995; 35p.
40. Eden, C.; Ackermann, F. Analysing and comparing idiographic causal maps. In *Managerial and Organizational Cognition*; Eden, C., Spender, J.C., Eds.; Sage: London, UK, 1998; 272p.
41. Keeney, R.L. *Value-Focused Thinking: A Path to Creative Decision Making*; Harvard University Press: Cambridge, MA, USA, 1992; 416p.

42. Xavier, J.H.V.; Gomes, M.C.; Sacco dos Anjos, F.; Scopel, E.; Macena, F.A.; Corbeels, M. Participatory multicriteria assessment of maize cropping systems in the context of family farmers in the Brazilian Cerrado. *Int J. Agric. Sustain.* **2020**, *18*, 410–426. [CrossRef]
43. Keeney, R.; Raiffa, H. *Decisions with Multiple Objectives: Preferences and Value Tradeoffs*; John Willey & Sons: Hoboken, NJ, USA, 1976; 565p.
44. Gil, A.C. *Métodos e Técnicas de Pesquisa Social*; Atlas: São Paulo, Brazil, 2008; 220p.
45. Scotto, M.A.L. *Fluxos de fósforo em uma Bacia Hidrográfica sob cultivo intensivo no sul do Brasil*. Master's dissertation (Master in Soil Science)–Programa de Pós-graduação em Ciência do Solo da Universidade Federal de Santa Maria, Santa Maria-RS/BR. 2014 Available online: https://repositorio.ufsm.br/ (accessed on 7 October 2021).
46. Alvares, C.A.; Stape, J.L.; Sentelhas, P.C.M.; Gonçalves, J.L.; Sparovek, G. Köppen's climate classification map for Brazil. *Meteorol. Z.* **2014**, *22*, 711–728. [CrossRef]
47. National Institute of Meteorology (INMET, as in Portuguese abbreviation). 2024 Estações Automáticas. Available online: https://portal.inmet.gov.br/paginas/catalogoaut (accessed on 21 February 2024).
48. Brazilian Institute of Geography and Statistics (IBGE, as in Portuguese abbreviation). Censo Agropecuário 2017a, Brasília, 2017a. Available online: https://censos.ibge.gov.br/agro/2017 (accessed on 14 October 2021).
49. Veiga, J.E. *Cidades Imaginárias. O Brasil é Menos Urbano do que se Calcula*; Editora Autores Associados: Campinas, Brazil, 2002; 304p.
50. Merten, G.H.; Minella, J.P. Qualidade da água em bacias hidrográficas rurais: Um desafio atual para a sobrevivência futura. *Agroecol. E Desenvolv. Rural. Sustentável* **2002**, *3*, 33–40.
51. Brazilian Institute of Geography and Statistics (IBGE, as in Portuguese abbreviation). *Estimativas de População*; Brasília, Brazil. 2018. Available online: https://sidra.ibge.gov.br/tabela/6579 (accessed on 14 October 2019).
52. Ellis, E.C.; Klein Goldewijk, K.; Siebert, S.; Lightman, D.; Ramankutty, N. Anthropogenic transformation of the biomes, 1700 to 2000. *Global Ecol. Biogeogr.* **2010**, *19*, 589–606. [CrossRef]
53. Thomas, J.; Joseph, S.; Thrivikramji, K.P. Assessment of soil erosion in a tropical mountain river basin of the southern Western Ghats, India using RUSLE and GIS. *Geosci. Front.* **2017**, *9*, 893–906. [CrossRef]
54. Rossol, C.D.; Filho, H.S.; Berté, L.N.; Jandrey, P.E.; Schwantes, D.; Gonçalves, A.C., Jr. Caracterização, classificação e destinação de resíduos da agricultura. *Sci. Agrar. Parana.* **2012**, *11*, 33–43. [CrossRef]
55. Brazilian Institute of Geography and Statistics (IBGE, as in Portuguese abbreviation). *Produção Agrícola Municipal*, Brasília, Brazil. 2017. Available online: https://sidra.ibge.gov.br/pesquisa/pam/tabelas (accessed on 7 October 2021).
56. Kaiser, D.R.; Sequinatto, L.; Reinert, D.J.; Reichert, J.M.; Rheinheimer, D.S.; Dalbianco, L. High nitrogen fertilization of tobacco crop in headwater watershed contaminates subsurface and well waters with nitrate. *J. Chem.* **2015**, *282500*, 283000. [CrossRef]
57. Bender, M.A.; Rheinheimer, D.S.; Tiecher, T.; Minella, J.P.G.; Barros, C.A.P.; Ramon, R. Phosphorus dynamics during storm events in a subtropical rural catchment in southern Brazil. *Agric. Ecosyst. Environ.* **2018**, *261*, 93–102. [CrossRef]
58. Reichert, J.M.; Pellegrini, A.; Rodrigues, M.F.; Tiecher, T.; Rheinheimer, D.S. Impact of tobacco management practices on soil, water and nutrients losses in steeplands with shallow soil. *Catena* **2019**, *183*, 104215. [CrossRef]
59. Hartemink, A.E. Soil erosion: Perennial crop plantations. In *Encyclopedia of Soil Science*; Board, A., Arnold, R.W., Finkl, C.W., Cortizas, A.M., Parkin, G., Semoka, J., Singer, A., Soon, Y.K., Spaargaren, O., Vázquez, F.M., Eds.; Taylor and Francis: New York, NY, USA, 2006; pp. 1613–1617.
60. Valente, M.L.; Reichert, J.M.; Legout, C.; Tiecher, T.; Cavalcante, R.B.L.; Evrard, O. Quantification of sediment source contributions in two paired catchments of the Brazilian Pampa using conventional and alternative fingerprinting approaches. *Hydrol. Process.* **2020**, *34*, 2965–2986. [CrossRef]
61. El Kateb, H.; Zhang, H.; Zhang, P.; Mosandl, R. Soil erosion and surface runoff on different vegetation covers and slope gradients: A field experiment in Southern Shaanxi Province, China. *Catena* **2013**, *105*, 1–10. [CrossRef]
62. Montagnini, F.; Eibl, B.I.; Barth, S.R. Organic yerba mate: An environmentally, socially and financially suitable agroforestry system. *Bois For. Trop.* **2011**, *308*, 59–74. [CrossRef]
63. Brazil. Law No. 12.651 of 25 May 2012. Provides for the protection of native vegetation; amends Laws Nos. 6.938, of 31 August 1981, 9.393, of 19 December 1996, and 11.428, of 22 December 2006; repeals Laws Nos. 4.771, of 15 September 1965, and 7.754, of 14 April 1989, and Provisional Measure No. 2.166-67, of 24 August 2001; and makes other provisions. *Official Gazette of the Federative Republic of Brazil*, Brasília, DF, 28 May 2012. Available online: https://www.planalto.gov.br/ccivil_03/_ato2011-2014/2012/lei/l12651.htm (accessed on 26 February 2024).
64. Rheinheimer, D.S.; Monteiro de Castro Lima, J.A.; Paranhos Rosa de Vargas, J.; Camotti Bastos, M.; Santanna dos Santos, M.A.; Mondamert, L.; Labanowski, J. Pesticide bioaccumulation in epilithic biofilms as a biomarker of agricultural activities in a representative watershed. *Environ. Monit. Assess.* **2020**, *192*. [CrossRef] [PubMed]
65. Brazilian Institute for the Environment and Renewable Natural Resources, (IBAMA, as in Portuguese abbreviation). *Boletins Anuais de Produção, Importação, Exportação e Vendas de Agrotóxicos no Brasil*; Brasília, Brazil. 2022. Available online: https://www.gov.br/ibama/pt-br (accessed on 25 March 2019).
66. Ambus, J.V.; Awe, G.O.; Faccio de Carvalho, P.C.; Reichert, J.M. Integrated crop-livestock systems in lowlands with rice cultivation improve root environment and maintain soil structure and functioning. *Soil Tillage Res.* **2023**, *227*, 105592. [CrossRef]

67. Ceretta, C.A.; Basso, C.J.; Vieira, F.C.B.; Herbes, M.G.; Moreira, I.C.L.; Berwanger, A.L. Dejeto líquido de suínos: I-perdas de nitrogênio e fósforo na solução escoada na superfície do solo, sob plantio direto. *Ciênc. Rural.* **2005**, *35*, 1296–1304. [CrossRef]
68. Saaty, T.L. How to Make a Decision: The Analytic Hierarchy Process. *Eur. J. Oper. Res.* **1990**, *48*, 9–26. [CrossRef]

Disclaimer/Publisher's Note: The statements, opinions and data contained in all publications are solely those of the individual author(s) and contributor(s) and not of MDPI and/or the editor(s). MDPI and/or the editor(s) disclaim responsibility for any injury to people or property resulting from any ideas, methods, instructions or products referred to in the content.

Article

A New Tool for Mapping Water Yield in Cold Alpine Regions

Linlin Zhao [1,2], Rensheng Chen [1,*], Yong Yang [1], Guohua Liu [3] and Xiqiang Wang [1]

[1] Qilian Alpine Ecology and Hydrology Research Station, Key Laboratory of Ecological Safety and Sustainable Development in Arid Lands, Northwest Institute of Eco-Environment and Resources, Chinese Academy of Sciences, Lanzhou 730000, China; zhaolinlin@nieer.ac.cn (L.Z.); yy177@lzb.zc.cn (Y.Y.); wangxq@lzb.ac.cn (X.W.)

[2] University of Chinese Academy of Sciences, Beijing 100049, China

[3] College of Geography and Tourism, Hengyang Normal University, Hengyang 421200, China; lgh1990@lzb.ac.cn

* Correspondence: crs2008@lzb.ac.cn

Citation: Zhao, L.; Chen, R.; Yang, Y.; Liu, G.; Wang, X. A New Tool for Mapping Water Yield in Cold Alpine Regions. *Water* **2023**, *15*, 2920. https://doi.org/10.3390/w15162920

Academic Editors: Xiaojie Li, Chenglong Zhang and Chenglong Zhang

Received: 6 July 2023
Revised: 4 August 2023
Accepted: 8 August 2023
Published: 13 August 2023

Copyright: © 2023 by the authors. Licensee MDPI, Basel, Switzerland. This article is an open access article distributed under the terms and conditions of the Creative Commons Attribution (CC BY) license (https://creativecommons.org/licenses/by/4.0/).

Abstract: Watershed management requires reliable information about hydrologic ecosystem services (HESs) to support decision-making. In cold alpine regions, the hydrology regime is largely affected by frozen ground and snow cover. However, existing special models of ecosystem services usually ignore cryosphere elements (such as frozen ground and snow cover) when mapping water yield, which limits their application and promotion in cold alpine regions. By considering the effects of frozen ground and snow cover on water yield, a new version of the Seasonal Water Yield model (SWY) in the Integrated Valuation of Ecosystem Services and Trade-offs (InVEST) was presented and applied in the Three-River Headwaters Region (TRHR) in southeastern Qinghai-Tibetan Plateau (QTP). Our study found that incorporating the effects of frozen ground and snow cover improved model performance. Frozen ground acts as a low permeable layer, reducing water infiltration, while snow cover affects water yield through processes of melting and sublimation. Both of these factors can significantly impact the distribution of spatial and temporal quickflow and baseflow. The annual average baseflow and water yield of the TRHR would be overestimated by 13 mm (47.58×10^8 m^3/yr) and 14 mm (51.24×10^8 m^3/yr), respectively, if the effect of snow cover on them is not considered. Furthermore, if the effect of frozen ground on water yield were not accounted for, there would be an average of 6 mm of quickflow misestimated as baseflow each year. Our study emphasizes that the effects of frozen ground and snow cover on water yield cannot be ignored, particularly over extended temporal horizons and in the context of climate change. It is crucial to consider their impacts on water resources in cold alpine regions when making water-related decisions. Our study widens the application of the SWY and contributes to water-related decision-making in cold alpine regions.

Keywords: water yield; InVEST; model revision; cold alpine regions

1. Introduction

Hydrologic ecosystem services (HESs) are the goods and services that ecosystems provide to people related to various uses of water, and include the provision of water for municipal, agricultural, and commercial purposes, mitigation of flow peaks to prevent inundations, reduction of sediment and nutrient loads in water, and the intrinsic value of natural hydrological systems for recreational activities [1,2]. HESs influence land management decisions through both regulations and investments aimed at safeguarding and enhancing water resources, and HES information is critical for water management and political decision-making [3,4]. Water yield is an important index for the assessment of water supply capacity and its assessment and improvement is often a priority for watershed management [5–7]. Decision-makers must have a clear understanding of how much water a parcel can yield and be able to identify regions that are vulnerable to floods or droughts. This will allow them to implement measures that benefit downstream residents [3]. In cold alpine regions, the vast expanses of frozen ground and snow cover have significant and far-reaching impacts on the ecological and hydrological processes within the basin [8–10]. As a result, there is a high demand for spatially explicit models to assess water yield in these regions.

The models used to evaluate water yield can be mainly classified into two categories: traditional hydrological models and special models of ecosystem services [3,4,11]. Traditional hydrological models are based on complex hydrological processes and often require specialized training for the operators, making it difficult for users to apply them accurately to watershed management, especially over a short period. Thus, they have been limited in their application in HES assessments. Although there are many hydrological models applied in the field of HES assessment, the methods are mostly the Soil and Water Assessment Tool (SWAT), while a few studies have also employed the Variable Infiltration Capacity model (VIC) model [12–17]. Similar to many process-based land surface models and hydrological models in cold alpine regions, although VIC can explain hydrological processes resulting from frozen ground and snow cover, it typically used for hydrology modeling and rarely for HESs due to its requirement for strong computing ability and specialized hydrological knowledge [18–23]. Even in cold alpine regions, many users without strong hydrological knowledge tend to ignore the impact of cryosphere elements and focus more on the impact of land use on HESs [24–26]. In contrast, models specifically designed for HESs are easy to use and contain fewer comprehensive algorithms [11]. However, changes in cryosphere elements will have an undeniable impact on hydrological ecosystems, including water yield [8–10]. Therefore, there is an urgent need to develop a model that is relatively easy to operate, requires relatively low specialized hydrological knowledge, and is more reliable for use in the assessment of HESs in cold alpine regions. Incorporating cryosphere elements into the special models of ecosystem services is a good choice. The Integrated Valuation of Ecosystem Services and Trade-offs tool (InVEST) was specifically developed for ecosystem services and has gained popularity in recent years [27–33]. The Annual Water Yield Model (AWY) of InVEST has been widely used to evaluate HESs [34,35]. Nonetheless, the influence of snow cover and frozen ground on water yield is highly seasonal, and inputting annual data makes it difficult to consider the effects of frozen ground and snow cover on hydrological processes by model revision. Although the AWY solely simulates the aggregate water yield, numerous applications necessitate an understanding of the seasonal water yield, which entails the differentiation between quickflow and baseflow [36].

Based on the Curve Number and water balance method, the Seasonal Water Yield model (SWY) of InVEST identifies differences between quickflow and baseflow [6]. It has been used in various geographical contexts, but its performance was found to be poorer in snow-dominated regions, based on evaluation [6,10,27,37]. Furthermore, incorporating the effects of the thickness and hydraulic conductivity of soil and topography makes it possible to take the effect of frozen ground into account according to temperature. Similarly, monthly inputs can be used to account for the effect of snow cover. Hamel et al. found that the model performance improved after incorporating snowmelt from the SWAT model into the baseflow output from the SWY [27]. However, this approach may not be feasible for all users if the SWAT snowmelt data is not available. On the other hand, Scordo et al. coupled a snow accumulation and ablation model to the SWY and found that it significantly improved the accuracy of predictions [10]. It should be noted that Scordo et al.'s study specified the quickflow and did not consider the baseflow component [10]. One of the main challenges for the current work is to integrate the effects of frozen ground and snow cover on water yield into the SWY to make it applicable in cold alpine regions.

The Three-River Headwaters Region (TRHR), as a crucial part of the Qinghai-Tibetan Plateau (QTP), is a typical cold alpine region with large areas of frozen ground and snow cover [38,39]. As a part of 'the Asian water tower', this region is essential to the human livelihood and ecological security of countries around the QTP [40,41]. In the past few decades, frozen ground and snow cover were degrading under climate warming [42–44]. In recent years, with climate change and ecosystem degradation, water resource management has become the priority task for decision-makers in the TRHR. Most HESs researched in this region were based on the traditional AWY, which focused on the total water yield instead of baseflow and quickflow. Pan et al. showed that the water supply in the TRHR was decreasing from 1980 to 2005 based on the AWY model [45]. It was found that the average annual water conservation of the QTP and the TRHR decreased from southeast to northwest using AWY [46,47]. However, the research above ignored the effects of the two cryosphere elements on the hydrological process. SWAT, with a snowmelt module, was applied in the TRHR to estimate the annual average water yield during the period of 1961–2010 but it ignored the effect of the frozen ground and the spatiotemporal change in water yield [48]. Hence, it is imperative to modify and implement the SWY in the TRHR to investigate and assess its applicability in cold alpine areas.

Given the limitations of the existing studies, and taking the TRHR as a case study, we directed our efforts towards refining the SWY by considering the impact of frozen ground and snow cover on water yield. The analyses aim at (1) developing a new version of SWY suitable for cold alpine regions and (2) learning how the frozen ground and snow cover affect water yield. The descriptions of the SWY and model revision are presented in the following section. Then, the characteristics of the study area, data, and methods for running, calibrating, and evaluating the new version of SWY and for sensitivity analyses of parameters are introduced. In addition, the results section also shows the effects of frozen ground and snow cover on water yield. Finally, the model performance of the new version of SWY and limitations of our study are discussed. There are two points that need clarification: (1) this model is not a traditional hydrological model but rather a special model of ecosystem services. The estimation of cryosphere elements in this model is relatively rough compared to traditional cold region hydrological models. Its advantage lies in its relatively simplicity in application. It is expected to provide a useful tool for watershed management decision-makers and to help to enhance ecological protection and construction in cold alpine regions. (2) When discussing the effect of frozen ground on water yield, the term "frozen ground" mainly refers to seasonal frozen ground and the active layer, without considering the effects of permafrost and its influence on water yield.

2. Methodology

2.1. Seasonal Water Yield Model

The SWY was computed through InVEST (https://naturalcapitalproject.stanford.edu/software/invest) (accessed on 5 March 2020) [36]. The SWY generates several outputs, including monthly and annual quickflow and annual baseflow rasters for the watershed. Quickflow refers to direct runoff reaching streams during or shortly after rain events. Baseflow is water that reaches streams later, between precipitation/snowmelt events that can have residence times of months or even years.

Firstly, the model calculates monthly quickflow (QF) on each pixel based on a modification curve number (CN) approach [36]:

$$QF_{i,m} = n_m \times \left((a_{i,m} - S_i) \exp\left(-\frac{0.2 S_i}{a_{i,m}}\right) + \frac{S_i^2}{a_{i,m}} \exp\left(\frac{0.8 S_i}{a_{i,m}}\right) E1\left(\frac{S_i}{a_{i,m}}\right) \right) \times \left(25.4 \left[\frac{mm}{in}\right]\right) \quad (1)$$

$$a_{i,m} = \frac{P_{i,m}}{25.4 \times n_{i,m}} \quad (2)$$

$$S_i = \frac{1000}{CN_i} - 10 [in] \quad (3)$$

$$E_1(x) = \int_1^\infty \frac{e^{-xt}}{t} dt \quad (4)$$

where $a_{i,m}$ is the mean rain depth on a rainy day at pixel i on month m (in), $n_{i,m}$ is the number of events at pixel i in month m, and $P_{i,m}$ is the monthly precipitation for pixel I at month m (mm). S_i is the potential maximum soil moisture retention after runoff begins (in). CN_i is the curve number for pixel i, tabulated as a function of the local land use land cover (LULC) and soil type. The figure 25.4 is a conversion factor from inches (used by the equation) to millimeters.

Thus, the annual quickflow QF_i can be calculated from the sum of monthly $QF_{i,m}$ values:

$$QF_i = \sum_{m=1}^{12} QF_{i,m} \tag{5}$$

Next, the model partitions the monthly available water between local recharge and evapotranspiration. The local recharge of a pixel is computed from the local water balance:

$$L_i = P_i - QF_i - AET_i \tag{6}$$

$$AET_{im} = min(PET_{i,m}; P_{i,m} - QF_{i,m} + \alpha_m \beta_i L_{sum.avail,i}) \tag{7}$$

where P_i and $P_{i,m}$ are the annual and monthly precipitation, respectively. α_m is the fraction of the upgradient subsidy that is available in month m (with a default value of $1/12$). β_i is a spatial availability parameter (0–1) which represents the fraction of the upslope subsidy that is available for downslope evapotranspiration. $L_{sum.avail, i}$ is the sum of upgradient subsurface water that is potentially available at pixel i:

$$L_{sum.avail,i} = \sum_{j \in \{neighbor\ pixels\ draining\ to\ pixel\ i\}} p_{ij} \cdot \left(L_{avail,j} + L_{sum.avail,j}\right) \tag{8}$$

$$L_{avail,i} = min(\gamma L_i, L_i) \tag{9}$$

where $p_{ij} \in [0, 1]$ is the proportion of flow from cell i to j, and $L_{avail,i}$ is the available recharge to a pixel, which is L_i whenever L_i is negative, and a proportion γ of L_i when it is positive. γ is the fraction of pixel recharge that is available to downslope pixels.

Finally, the model computes the baseflow index. The baseflow index represents the contribution of a pixel to baseflow (i.e., water that reaches the stream during the dry season). If the local recharge is negative, B is set to zero. Otherwise, B is a function of the amount of flow leaving the pixel and of the relative contribution to recharge of this pixel.

For a pixel that is not adjacent to the stream channel, the cumulative baseflow, $B_{sum,i}$, is proportional to the cumulative baseflow leaving the adjacent downslope pixels minus the cumulative baseflow that was generated on that same downslope pixel:

$$B_{sum,i} = \begin{cases} L_{sum,i} \cdot \sum_{j \in \{cells\ to\ which\ cell\ i\ pours\}} p_{ij} \left(1 - \frac{L_{avail,j}}{L_{sum,j}}\right) \frac{B_{sum,j}}{L_{sum,j} - L_i} & \text{if } j \text{ is a nonstream pixel} \\ L_{sum,i} \cdot \sum_{j \in \{cells\ to\ which\ cell\ i\ pours\}} p_{ij} & \text{if } j \text{ is a stream pixel} \end{cases} \tag{10}$$

$L_{sum, i}$ is the cumulative upstream recharge defined by:

$$L_{sum,i} = L_i + \sum_{j, all\ pixels\ draining\ to\ pixel\ i} L_{sum,j} \cdot p_{ij} \tag{11}$$

The baseflow, B_i can be directly derived from the proportion of the cumulative baseflow leaving cell i, with respect to the available recharge to the upstream cumulative recharge:

$$B_i = max\left(B_{sum,i} \cdot \frac{L_i}{L_{sum,i}}, 0\right) \tag{12}$$

Details of the SWY assumption, equation, and workflow are described in the model documentation [36].

2.2. Model Revision

2.2.1. Considering the Effect of Frozen Ground on Water Yield

The impermeability of frozen ground is linked to the presence of unfrozen water and ground ice in the frozen ground [48]. Experiments conducted in the field showed that hydraulic conductivities decrease as temperature decreases [49]. When negative temperature occurs, the saturated hydraulic conductivity (SHC) is found to decrease by approximately 4 orders of magnitude near 0 °C, and then decreases slowly as the temperature decreases [50–52]. The soil water is liquid when the temperature of soil is positive, completely frozen when the soil temperature is lower than a critical temperature threshold (T_f), and partially frozen when the local temperature is between 0 °C and T_f. The saturated hydraulic conductivity (SHC) of soil is different in the three frozen states (Equation (13)).

$$k'_0 = \begin{cases} k_0 & T_s > 0 \\ k_0 \times 10^{-4} & T_f \leq T_s \leq 0 \\ 0 & T_s < T_f \end{cases} \quad (13)$$

where k'_0 is the SHC (cm/d) after correction by soil temperature, k_0 is the SHC (cm/d) before correction, T_s is the temperature of the soil, and T_f denotes the temperature threshold of soil freezing.

The first step was to revise the monthly soil SHC of the TRHR to account for frozen ground using Equation (13). Then, the average hydraulic conductivity value for 12 months was calculated since the soil group data used in the model were on an annual scale. The soil groups were then divided into four categories (A, B, C, and D) based on the range of soil SHC using the method in chapter 7 of the United States Department of Agriculture (USDA), as shown in Figure 1a [53].

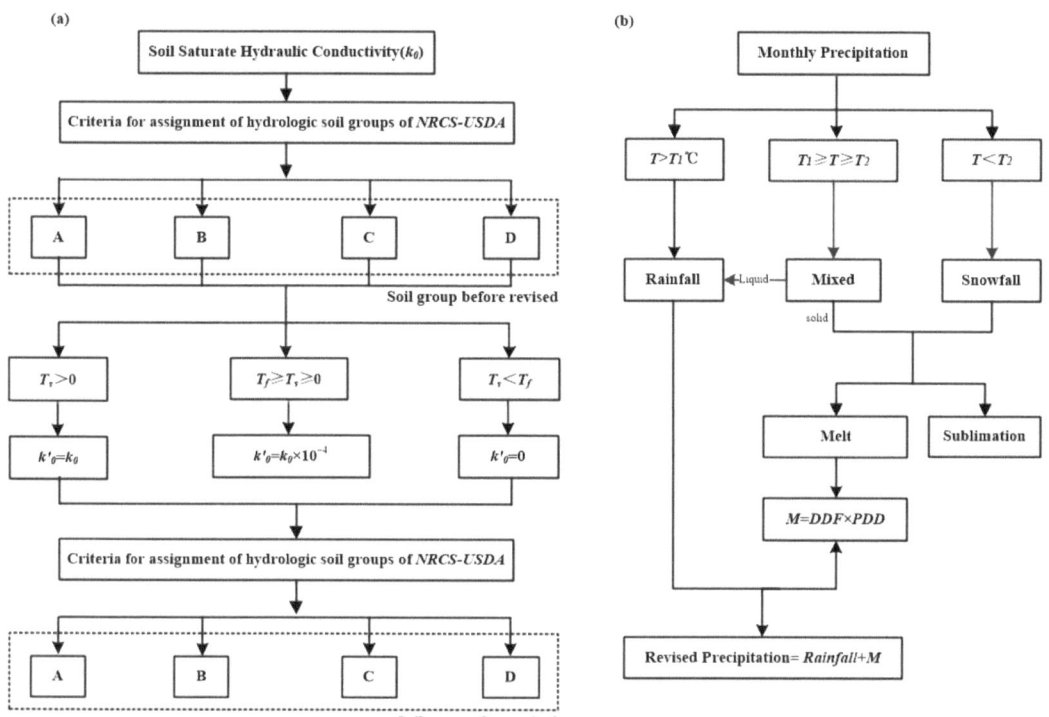

Figure 1. Methodological flowcharts considering the effect of frozen ground (**a**) and snow cover (**b**) on water yield (k'_0 is the saturated hydraulic conductivity (cm/d) after correction by soil temperature, k_0 is the saturated hydraulic conductivity (cm/d) before correction, T_s is the temperature of the soil, T_f denotes the temperature threshold of soil freezing, T_1 is the threshold temperature for differentiating rain and sleet, and T_2 is the threshold temperature for differentiating snow and sleet).

2.2.2. Considering the Effect of Snow Cover on Water Yield

Firstly, the precipitation was divided into rainfall, snowfall, and mixed by the temperature threshold method [54,55]. When the temperature is higher than the threshold temperature for differentiating rain and sleet (T_1), the precipitation form is rainfall; when the temperature is lower than the threshold temperature for differentiating snow and sleet (T_2), the precipitation form is snowfall; when the temperature is between the two threshold values, the precipitation is characterized sleet (mixed). Next, a positive degree-day factor method was used to calculate the snow melt amount [56] (Equation (14)). The estimation of snow sublimation is a rough approximation for the entire region and can be referenced from relevant literature on snow sublimation in the study area. Finally, the monthly liquid precipitation was obtained (Figure 1b).

$$M = DDF \cdot PDD \tag{14}$$

where M is the snow melt (mm); DDF is the degree-day factor; PDD is the sum of the mean cumulative positive temperature over a period named positive degree-day.

When precipitation reaches the threshold temperature, it transforms from a solid to a liquid state [54]. However, snow may not completely melt in the current month and instead accumulate and persist depending on the temperature. The sublimation of snow cover can lead to a decrease in precipitation amount on the underlying surface, and the melting of snow cover at different times can alter the annual distribution of precipitation [57,58].

2.2.3. Model Calibration, Validation, and Evaluation

There are two tasks that we need to complete in this section:

1. Determine if the model performance improves after the model revision.
2. Conduct sensitivity analysis (α, β, and γ) to identify the optimal parameter combination.

We established four different scenarios to compare their performance. The scenario that does not consider the impact of the cryosphere elements is referred to as SWY1. The scenario that considers the impact of frozen ground is referred to as SWY2. The scenario that considers the impact of snow cover is referred to as SWY3. Finally, the scenario that considers both the effects of frozen ground and snow cover on baseflow is referred to as SWY4. Considering the difficulty in obtaining observed quickflow and baseflow data, we validated the total water yield using observed annual streamflow data from the Yangtze River Source Region (YAR), Yellow River Source Region (YER), and Lantsang-Mekong River Source Region (LAR). To obtain the quickflow and baseflow data, we used the results of the Eckhardt filter method (Equations (15) and (16)) [59,60]. The Eckhardt filter is a widely used baseflow separation method that partitions streamflow into two components, quickflow and baseflow, and has been validated in many baseflow separation analyses [60–62].

$$y_k = f_k + b_k \tag{15}$$

where y is the total streamflow; f denotes the quickflow; b represents the baseflow; and k is the time step.

$$b_k = \frac{(1 - BFI_{max})\alpha b_{k-1} + (1 - \alpha)BFI_{max} y_k}{1 - \alpha BFI_{max}} \tag{16}$$

where α represents the recession constant, while BFI_{max} represents the maximum baseflow index (BFI). Three representative BFI_{max} values for different hydrological and hydrogeological conditions were introduced: 0.80 for perennial streams with porous aquifers; 0.50 for ephemeral streams with porous aquifers; and 0.25 for perennial streams with hard rock aquifers [59]. α is 0.98 for perennial streams, 0.963 for ephemeral streams with porous aquifers, and 0.955 for perennial streams with hard rock aquifers. Above all, we selected the suggested values of 0.98 for α and 0.80 for BFI_{max} [59].

To evaluate the model performance, we used three indicators: determination coefficient (R^2), Nash–Sutcliffe efficiency (NSE), and percent bias (PBIAS). R^2 determines the strength of correlation between the modeled and observed data (Equation (17)). NSE compares the residual variance of the modeled data to the measured data variance (Equation (18)). PBIAS compares the average tendency of the modeled data and observed data (Equation (19)) [63].

$$R^2 = \left\{ \frac{\sum_{i=1}^{n}(y_i - \overline{y})(f_i - \overline{f})}{\sqrt{\frac{1}{n}\sum_{i=1}^{n}(y_i - \overline{y})(f_i - \overline{f})^2}} \right\}^2 \tag{17}$$

$$NSE = 1 - \frac{\sum_{i=1}^{n}(f_i - y_i)^2}{\sum_{i=1}^{n}(y_i - \overline{y})^2} \tag{18}$$

$$PBIAS = \frac{\sum_{i=1}^{n}(y_i - f_i)}{\sum_{i=1}^{n} y_i} \tag{19}$$

where n denotes the number of time steps, y_i and f_i represent the observed/the results of the Eckhardt filter method and modeled results of the SWY, respectively, on the ith time step. $\overline{y}$ and $\overline{f}$ are the mean of observed data/the results of the Eckhardt filter method and modeled results of the SWY (y_i and f_i) across the n evaluation time steps.

The steps for model calibration, validation, and evaluation are shown in Figure 2 and are as follows:

1. Since α, β, and γ only affect baseflow and not quickflow, we compared the monthly quickflow values of SWY1, SWY2, SWY3, and SWY4 with those obtained from the Eckhardt filter method to select the best SWY for modeling quickflow at the preliminary stage (we calculated daily quickflow based on the Eckhardt filter method, then summed them into monthly).
2. We employed the method of Hamel et al. to analyze the sensitivity of parameters in YAR, YER, and LAR [27]. We started by using the default value for α (1/12) and γ (1) and changed β from 0 to 1 with the increments of 0.2 to analyze parameter sensitivity. Similarly, we set 1 for β and γ and adjusted α equal to 1/12, 1/6, and 1/3. We also repeated the previous analyses for γ by varying γ from 0 to 1 in increments of 0.2.
3. Using the various parameter combinations obtained from the best SWY for modeling quickflow, as determined in the first step, we compared the annual water yield values. First, we summed the 12 quickflow outputs of SWY to obtain the annual quickflow, and then added the annual quickflow and baseflow to calculate the annual water yield. We subsequently compared this to the observed streamflow data to determine the optimal parameter combination.
4. Using the optimal parameter combination obtained from the third step, we applied it to SWY1, SWY2, SWY3, and SWY4. We then compared the annual baseflow values obtained from these models with those from the Eckhardt filter method. To accomplish this, we initially computed the daily baseflow using the observed daily runoff data with the Eckhardt filter method. Subsequently, we summed the daily baseflow of the Eckhardt filter method to obtain the annual baseflow. This allowed us to investigate whether the best SWY for modeling baseflow was also the optimal choice for simulating quickflow.

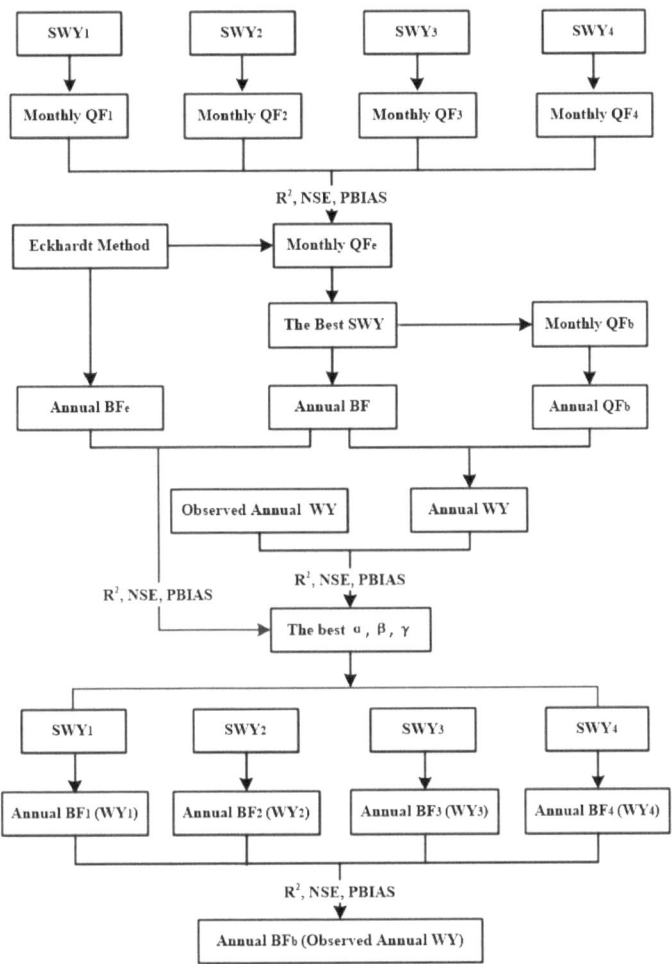

Figure 2. Flowcharts of model calibration, validation, and evaluation.

3. Case Study

In the model setup, SWY1, SWY2, SWY3, and SWY4 were run at 1×1 km^2 resolutions in the study area from 1981 to 2020. The output data generated by the models consisted of annual baseflow and monthly quickflow.

3.1. Study Area

The TRHR (31°39′–36°12′ N, 89°45′–102°23′ E) (the headwater region of the Yangtze River, the Yellow River, and the Lantsang-Mekong River), covers an area of 3.66×10^5 km^2 on the QTP, which is the largest National Natural Reserve in China (Figure 3a). It includes the YAR, YER, and LAR (Figure 3a). The study area is located within a region characterized by frozen ground, including areas of seasonal frozen ground and permafrost (Figure 3c). The mean maximum frozen depth of seasonal frozen ground in the TRHR from 1980 to 2014 was 132 cm (ranging from 48 cm to 213 cm), and the average of the active layer thickness in the permafrost zone from 2000 to 2015 was 215 cm (ranging from 134 cm to 410 cm) [64,65]. It should be clarified that the effect of permafrost change on water yield was not considered here. The area is characterized by a complex mountainous terrain with continuous and steep slopes, and an elevation range spanning from 2062 m to 6788 m, with an average elevation of 4400 m. The region experiences a typical continental plateau climate, characterized by intense radiation and cold, dry weather. Significantly, there is a discernible gradient of decreasing heat and water from the southeastern to the northwestern regions [39,66]. The annual average temperature in the study area is −3.94 °C, and the annual precipitation recorded between 1981 and 2020 is 463.4 mm. Grasslands account for 72% of the TRHR area, mainly including alpine grassland and alpine steppe [41] (Figure 3b) The study area is home to an abundance of rivers, lakes, and mountain snow cover, which serve as important sources of water for downstream regions (Figure 3c). The study area supplies a significant portion of the water for downstream regions, including 25% of the total water for the Yangtze River, 49% of the total water for the Yellow River, and 15% of the total water for the Lantsang-Mekong River [67]. Given its unique geographical location, abundant natural resources, and critical ecological functions, the study area serves as an important ecological barrier for the surrounding regions of the QTP.

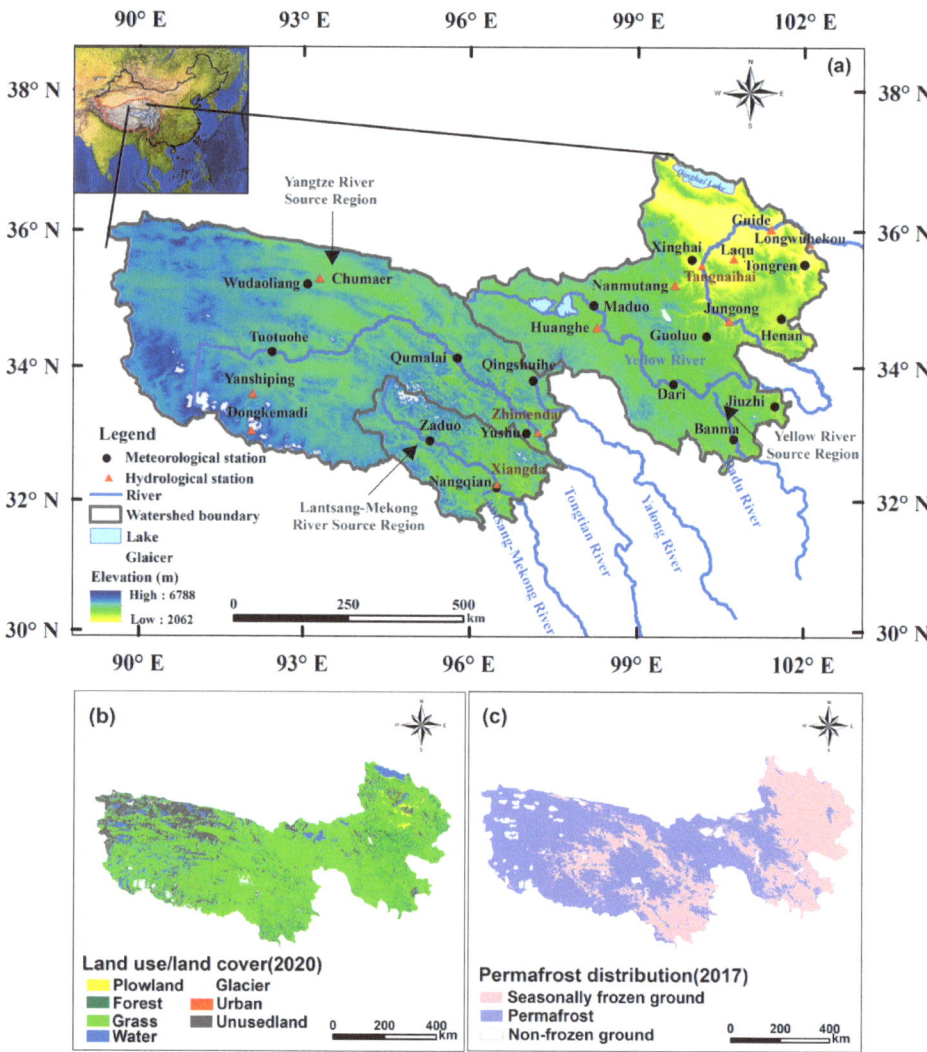

Figure 3. Location and topography (**a**), land use land cover (**b**), frozen ground distribution (**c**) of the TRHR.

3.2. Datasets

The data used in this study consisted of two main types: model input data (including revised model data) and validation data for the model results.

3.2.1. Meteorological Data

The meteorological data were the monthly average temperature, monthly precipitation, and pan evaporation of E601 during 1981–2020. In addition to the data from China Meteorological Data Service Center (http://data.cma.cn) (accessed on 1 January 2021), meteorological data from the hydrological station were also used. These station data were interpolated by the professional meteorological interpolation software ANUSPLIN version 4.4. Additionally, the monthly reference evapotranspiration (ET_0) was estimated by the method of Allen et al. [68].

3.2.2. Hydrological Data

The observed annual streamflow data of the YAR, YER, and LAR during 1981–2012, 1981–2015, and 1981–2010, respectively, were used to validate the annual water yield modeled by the SWY. The observed daily streamflow data of the YAR and YER were used for baseflow separation by the Eckhardt filter method to validate the monthly quickflow and annual baseflow modeled by the SWY. Notably, the daily data of the YAR cover the years 1981–1984, 1986, 2006–2009, and 2013–2014, and the daily data of the YER cover the years 1981–1997 and 2007–2014. Unfortunately, we have not obtained the daily streamflow data of the LAR.

3.2.3. Land Use/Land Cover (LULC) Data

The LULC dataset was downloaded from the Chinese Academy of Environmental Science data center (https://www.resdc.cn/) (accessed on 5 June 2021). It covered the years 1980, 1990, 1995, 2000, 2005, 2010, 2015, and 2020, with a spatial resolution of 1 km. It should be pointed out that according to the classification of this model (i.e., grass, forest, plowland, water, glacier, urban, and unused land), the LULC has not changed much in the past 40 years.

3.2.4. Soil Data

The InVEST User's Guide provides several soil parameter data and the ones used here are as follows:

(1) Hydrologic Soil Group raster (used as the soil group before revision) and Saturate Hydraulic Conductivity rasters (used to revise the soil group) from FutureWater (https://www.futurewater.eu/2015/07/soil-hydraulic-properties/) (accessed on 4 May 2021).
(2) China meteorological assimilation datasets for the SWAT model-soil temperature version 1.0 (http://data.tpdc.ac.cn/) (accessed on 5 November 2020) [69].
(3) Soil and runoff curve numbers (CN) of the Hydrologic Soil Groups [53,70]. The data inputs and source of the model are shown in Table 1.

Table 1. Data sources and format used as inputs of the SWY.

Data Inputs	Format	Source (before Processing into Model Inputs)
Monthly Precipitation (revised by the method shown in Figure 1b.)	Raster (1 km)	China Meteorological Data Service Center (http://data.cma.cn) (accessed on 1 January 2021).
Monthly Reference Evapotranspiration (ET_0)	Raster (1 km)	China Meteorological Data Service Center (http://data.cma.cn) (accessed on 1 January 2021).
Annual LULC Maps	Raster (1 km)	Chinese Academy of Environmental Science Data Center (https://www.resdc.cn/) (accessed on 5 June 2021).

Table 1. *Cont.*

Data Inputs	Format	Source (before Processing into Model Inputs)
Annual Soil Group (revised by the method shown in Figure 1a.)	Raster (1 km)	The soil temperature data was downloaded from the National Tibetan Plateau Data Center (http://data.tpdc.ac.cn/) (accessed on 5 November 2020) [69]. Hydrologic Soil Group raster (used as the soil group before revision) and Saturate Hydraulic Conductivity rasters (used to revise the soil group) from FutureWater (https://www.futurewater.eu/2015/07/soil-hydraulic-properties/) (accessed on 4 May 2021).
Biophysical Table (12 months in a CSV)	CSV	CN was downloaded from the United States Department of Agriculture [70]. Kc values were from FAO [68].
Rain Events (12 months in a CSV)	CSV	China Meteorological Data Service Center (http://data.cma.cn) (accessed on 1 January 2021).
DEM	Raster (1 km)	Geospatial Data Cloud http://www.gscloud.cn/ (accessed on 5 November 2020).
AOI (Area of Interest)	Vector	National Tibetan Plateau Data Center (http://data.tpdc.ac.cn/) (accessed on 5 November 2020) [71].
The Critical Temperature Threshold of Soil Freezing (T_f) [1]	-	−8 °C [72].
The Threshold Temperature For Differentiating Rain and Sleet (T_1) [1]	-	5 °C [54].
The Threshold Temperature For Differentiating Snow and Sleet (T_2) [1]	-	2 °C [55].
TFA (Threshold Flow Accumulation) [1]	-	3000
α; β; γ [1]	-	Default (1/12; 1; 1)

Note: [1] represents the parameters of the model, while the other ones in the table represent the model inputs.

3.3. Parameters

The critical temperature threshold of soil freezing (T_f) was set as −8 °C according to observation in northeast Tibetan Plateau [72]. The threshold temperature for differentiating rain and sleet (T_1) is 5 °C according to the study of precipitation type estimation and validation in China [54], and the threshold temperature for differentiating snow and sleet (T_2) is 2 °C according to research in the YER [55]. Simultaneously, the amount of snow sublimation was estimated according to observation in northeast Tibetan Plateau [73]. Threshold Flow Accumulation (TFA) was adjusted repeatedly by the user until the river in the TFA raster generated by the model was very close to reality, and the TFA was set as 3000. The details of the settings for α, β, and γ parameters can be found in Section 2.2.3.

4. Results

4.1. Sensitivity Analyses

Elevated values of α, β, and γ signify a heightened potential for the absorption of recharge from upslope pixels by evapotranspiration in downslope pixels, which may consequently result in a decline in baseflow. Our sensitivity analysis of α and β was consistent with the findings of Hamel et al., although they did not examine the sensitivity of baseflow to γ [27]. The results of our study showed that in both regions, baseflow decreased as α, β, and γ increased (Figure 4). In the TRHR region, we found that baseflow was most sensitive to changes in γ and least sensitive to changes in α. Varying α between 1/12 and 1/3 resulted in a reduction of baseflow in the YAR, YER, and LAR regions by 13.12% (7.32 mm), 6.90% (7.92 mm), and 2.25% (3.92 mm), respectively (Figure 4a). Varying β between 0 and one led to a reduction in baseflow in the YAR, YER, and LAR regions by 49.51% (54.74 mm), 36.93% (67.19 mm), and 35.67% (96.65 mm), respectively (Figure 4b). Varying γ between 0 and one resulted in a reduction in baseflow in the YAR, YER, and LAR regions by 84.80% (311.49 mm), 79.12% (434.85 mm), and 77.53% (601.54 mm), respectively

(Figure 4b). However, when γ was lower than 0.8, the baseflow values for all three regions were higher than the observed water yield values.

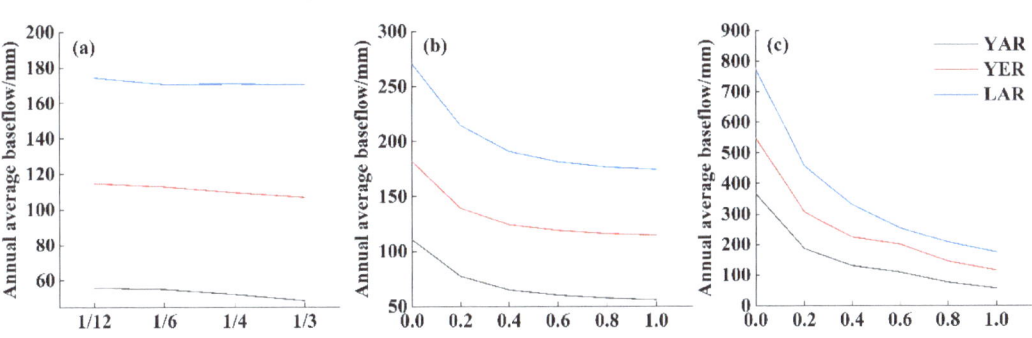

Figure 4. Sensitivity analyses of the baseflow modeled by SWY to α (**a**), β (**b**), and γ (**c**) parameters.

4.2. Model Calibration, Validation, and Evaluation

The value of β in the best parameter combinations was different in the three regions. For the YAR, the best parameter combination for modeling water yield was α = 1/12, β = 0.2, γ = 1, of which the R^2, NSE, and PBIAS were 0.47, 0.41, and −0.14, respectively. The best parameter combination for the YER and LAR was α = 1/12, β = 0.4, γ = 1, among which the R^2, NSE, and PBIAS of the YER were 0.76, 0.72, and 0.50, and 0.62, 0.32, and 1.02 in the LAR, respectively (Table 2). A value of 1/12 for α means there will be 1/12 of upslope annual available for local recharge each month. When γ is equal to one, nearly all the local recharge in a pixel is accessible to down-gradient pixels, which could occur in a steep watershed that rapidly releases water. A value of 0.2/0.4 for β indicates that 20%/40% of the up-gradient subsidy is available for down-gradient evapotranspiration.

Table 2. Model performance of the SWY with different parameter combinations.

Parameters	YAR			YER			LAR		
	R^2	NSE	PBIAS	R^2	NSE	PBIAS	R^2	NSE	PBIAS
α = 1/6, β = 1, γ = 1	0.40	−0.30	21.67	0.75	0.61	7.43	0.55	−0.49	9.80
α = 1/4, β = 1, γ = 1	0.40	0.07	14.69	0.74	0.68	3.77	0.55	−0.49	9.73
α = 1/3, β = 1, γ = 1	0.37	0.36	−1.56	0.73	0.69	−2.81	0.55	−0.46	9.54
α = 1/12, β = 0, γ = 1	0.43	−1.42	−32.66	0.69	−0.88	−33.03	0.56	−3.46	−28.96
α = 1/12, β = 0.2, γ = 1	0.47	0.41	−0.14	0.73	0.59	−7.98	0.57	0.12	−6.99
α = 1/12, β = 0.4, γ = 1	0.42	0.18	11.32	0.76	0.72	0.50	0.62	0.32	1.02
α = 1/12, β = 0.6, γ = 1	0.41	−0.02	16.57	0.75	0.68	3.98	0.59	0.03	5.53
α = 1/12, β = 0.8, γ = 1	0.41	−0.16	19.39	0.75	0.65	5.73	0.56	−0.20	7.66
α = 1/12, β = 1, γ = 0	0.45	−121.84	−286.37	0.71	−95.48	−260.84	0.60	−212.67	−224.99
α = 1/12, β = 1, γ = 0.2	0.51	−16.93	−107.33	0.77	−18.76	−113.03	0.61	−44.49	−101.77
α = 1/12, β = 1, γ = 0.4	0.34	−3.53	−47.22	0.77	−5.27	−62.10	0.60	−12.06	−50.84
α = 1/12, β = 1, γ = 0.6	0.32	−1.06	−23.75	0.73	−12.09	−48.07	0.60	−2.92	−22.50
α = 1/12, β = 1, γ = 0.8	0.42	0.21	0.67	0.70	0.32	−12.51	0.57	−0.49	−4.75
α = 1/12, β = 1, γ = 1	0.40	−0.27	21.16	0.74	−0.44	7.44	0.56	−0.33	8.62

Table 3 shows that the SWY4 model had the best performance in modeling quickflow, baseflow, and water yield in all three regions, while SWY2 and SWY3 performed better than SWY1 in modeling quickflow and baseflow. SWY1 and SWY2 had the same performance in modeling water yield, as did SWY3 and SWY4. The R^2 values between the modeled results of SWY4 and the observed data/the results of the Eckhardt filter passed the significance test ($p < 0.01$). For SWY4 in the YAR, the R^2, NSE, and PBIAS values for quickflow were 0.77, 0.76, and 1.25, respectively. The corresponding values for baseflow were 0.61, 0.55, and 3.73, and for water yield were 0.47, 0.41, and 0.54. For SWY4 in the YER, the R^2, NSE, and PBIAS values for quickflow were 0.72, 0.65, and 1.34, respectively. The corresponding values for baseflow were 0.82, 0.80, and −1.00, and for water yield were 0.76, 0.72, and 1.04. For the SWY4 model in the LAR, the R^2, NSE, and PBIAS values for water yield were 0.62, 0.32, and 1.02, respectively. Figure 5 illustrates the changing trends of the modeled results using SWY4 and the observed streamflow data/the results of the Eckhardt filter in the three regions.

Table 3. Model performance of quickflow (QF), baseflow (BF), and water yield (WY) of SWY1, SWY2, SWY3, and SWY4.

		YAR			YER			LAR		
		R^2	NSE	PBIAS	R^2	NSE	PBIAS	R^2	NSE	PBIAS
QF	SWY1	0.56	0.54	17.62	0.53	0.47	23.45	-	-	-
	SWY2	0.71	0.69	−5.91	0.65	0.55	−2.55	-	-	-
	SWY3	0.65	0.62	21.57	0.62	0.53	23.34	-	-	-
	SWY4	0.77	0.76	1.25	0.72	0.65	1.34	-	-	-
BF	SWY1	0.52	−0.02	−14.08	0.62	0.18	−17.39	-	-	-
	SWY2	0.56	0.26	−8.62	0.74	0.45	−12.41	-	-	-
	SWY3	0.58	0.50	−1.30	0.77	0.70	−5.66	-	-	-
	SWY4	0.61	0.55	3.73	0.82	0.80	−1.00	-	-	-
WY	SWY1	0.41	−0.02	−10.94	0.70	0.53	−8.80	0.57	−0.15	−8.30
	SWY2	0.41	−0.02	−10.94	0.70	0.53	−8.80	0.57	−0.15	−8.30
	SWY3	0.47	0.41	0.54	0.76	0.72	1.04	0.62	0.32	1.02
	SWY4	0.47	0.41	0.54	0.76	0.72	1.04	0.62	0.32	1.02

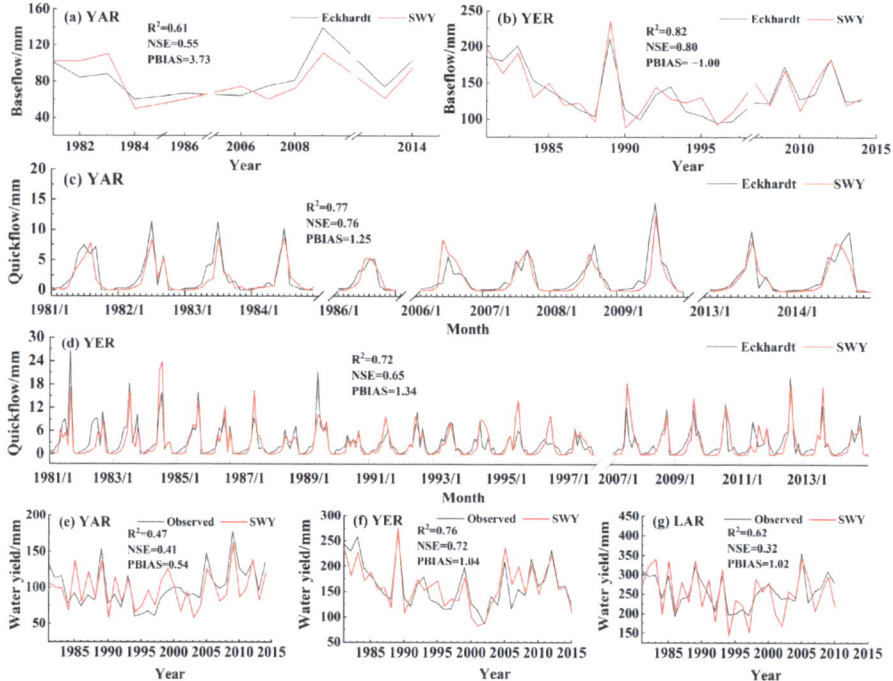

Figure 5. Comparisons of baseflow, quickflow, and water yield between modeled (SWY4) and observed data/the result of the Eckhardt filter method ((**a**,**b**) are the annual baseflow of the YAR of YER, (**c**,**d**) are the monthly quickflow of the YAR and YER, (**e**–**g**) are the annual water yield of the YAR, YER, and LAR, respectively).

The performance of the SWY scenarios improved as we moved from SWY1 to SWY4. Specifically, in the YAR and YER, the R^2 values for quickflow increased from 0.56 to 0.77 and from 0.53 to 0.72, respectively. The NSE values also improved from 0.54 to 0.76 and from 0.47 to 0.65, respectively, and the PBIAS values decreased from 17.62 to 1.25 and from 23.45 to 1.34, respectively. In the same regions, the R^2 values for baseflow increased from 0.52 to 0.61 and from 0.62 to 0.82, respectively. The NSE values increased from −0.02 to 0.55 and from 0.18 to 0.80, respectively, and the PBIAS values changed from −14.08 to 3.73 and from −17.39 to −1.00, respectively. Finally, for water yield in the YAR, YER, and LAR, the R^2 values increased from 0.41 to 0.47, from 0.70 to 0.76, and from 0.57 to 0.62, respectively. The NSE values increased from −0.02 to 0.41, from 0.63 to 0.72, and from 0.18 to 0.80, respectively, and the PBIAS values changed from −10.94 to 0.54, from 0.53 to 0.72, and from −8.30 to 1.02, respectively.

4.3. Effects of Frozen Ground and Snow Cover

Based on the Hydrologic Soil Groups of FutureWater, the soil group in the TRHR region was mainly composed of group C. When considering the effect of frozen ground, 47% of the area originally classified as group C was reclassified as group D (Figure 6a,b). The criteria for assigning hydrologic soil groups indicate that the SHC decreases from group A to group D, and low SHC can reduce infiltration. Frozen ground can act as a low permeable layer, which can limit the movement of water through the soil layers.

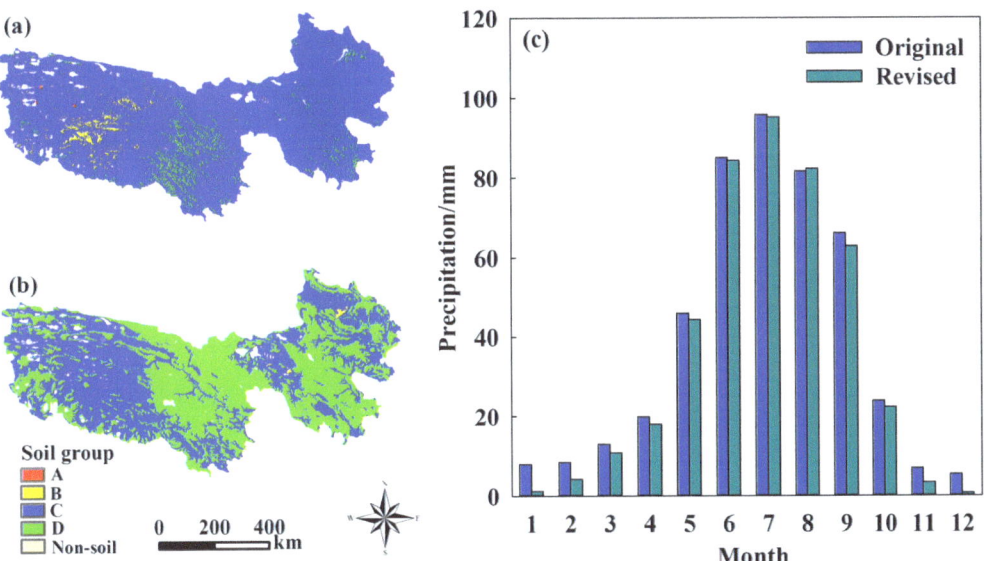

Figure 6. Soil groups of the TRHR before (**a**) and after (**b**) considering the effect of frozen ground and annual distribution of precipitation, and before and after considering the effect of snow cover (**c**).

Sublimation and melting of snow cover have opposite effects on liquid water. Sublimation reduces liquid water, while melting of snow cover increases it. After model revision, the amount of liquid water reaching the surface each month decreased (Figure 6c). Generally, the decrease in liquid water showed a downward trend from January to July and increased from September to December, except for August, when liquid water increased. Changes in the annual distribution of liquid water resulting from sublimation and melting of snow cover have a significant effect on the amount and spatiotemporal distribution of water yield.

From 1981 to 2020, the annual average quickflow of SWY1, SWY2, SWY3, and SWY4 was 26.04 mm (95.31 × 10^8 m^3/yr), 25.04 mm (91.65 × 10^8 m^3/yr), 32.64 mm

(119.46 × 10^8 m^3/yr), and 31.04 mm (113.61 × 10^8 m^3/yr), respectively. The annual average baseflow of SWY1, SWY2, SWY3, and SWY4 was 121.8 mm (445.79 × 10^8 m^3/yr), 108.8 mm (398.21 × 10^8 m^3/yr), 115.8 mm (423.83 × 10^8 m^3/yr), and 121.8 mm (445.79 × 10^8 m^3/yr), respectively. The annual average water yield of SWY1, SWY2, SWY3, and SWY4 was 133.8 mm (489.71 × 10^8 m^3/yr), 133.8 mm, 147.8 mm (540.95 × 10^8 m^3/yr), and 147.8 mm, respectively (Figure 7a). The differences between the results of SWY1, SWY2, SWY3, and SWY4 show that if the effect of frozen ground were not considered, the annual average quickflow would be underestimated by 6 mm (21.96 × 10^8 m^3/yr), baseflow would be overestimated by 6 mm (21.96 × 10^8 m^3/yr), and water yield would be unaffected. If the effect of snow cover were not considered, the annual average quickflow, baseflow, and water yield would be overestimated by 1 mm (3.66 × 10^8 m^3/yr), 13 mm (47.58 × 10^8 m^3/yr), and 14 mm (51.24 × 10^8 m^3/yr), respectively. If the effects of the frozen ground and snow cover were not considered, the three annual average flows would be underestimated by 5 mm (18.30 × 10^8 m^3/yr), overestimated by 18 mm (65.88 × 10^8 m^3/yr), and overestimated by 13 mm (47.58 × 10^8 m^3/yr).

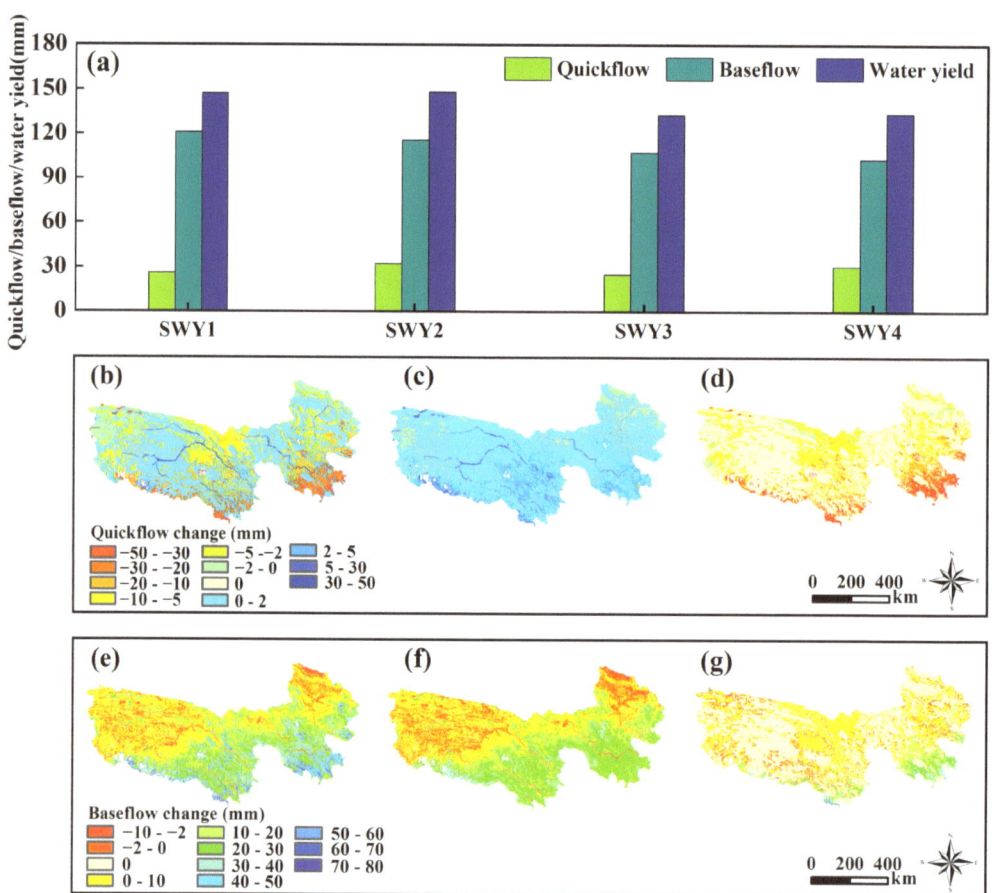

Figure 7. Annual average quickflow, baseflow, and water yield of SWY1, SWY2, SWY3, and SWY4 from 1981 to 2020 (**a**), and the spatial differences of quickflow and baseflow of SWY1, SWY2, and SWY3 compared with SWY4 ((**b–d**) are the spatial differences of quickflow of SWY1, SWY2, and SWY3 compared with SWY4; (**e–g**) are the spatial differences of baseflow of SWY1, SWY2, and SWY3 compared with SWY4).

Figure 7 shows the spatial differences between SWY1, SWY2, and SWY3, compared to SWY4. In terms of quickflow, the underestimation of SWY1 was mainly observed in areas where the soil type changed from type C to type D, with a range of 5–50 mm. The rest of the area was overestimated, with a range of 0–2 mm (Figure 7b). SWY2 was generally overestimated, with a relatively small range of 0–2 mm (Figure 7c). In the areas where the soil type remained unchanged after the model revision, the quickflow of SWY3 was not affected. In the rest of the area, it was underestimated with a range of 2–10 mm in the north and 10–50 mm in the south (Figure 7d). In terms of baseflow, SWY1 was generally overestimated, with the range increasing from the northwest (0–10 mm) to the southeast (70–80 mm). The underestimated area was mainly located in a small area of the northwest of the TRHR (Figure 7e). The difference between SWY2 and SWY4 was similar to that of SWY1 and SWY4, with the overestimation range increasing from the northwest (0–10 mm) to the southeast (30–40 mm), but with a slightly smaller range (Figure 7f). In the areas where the soil type remained unchanged after the model revision, the baseflow of SWY3 was not affected. However, in other areas, it was overestimated with the same range as the underestimated SWY3 (Figure 7g).

Based on the results presented above, it can be concluded that the effect of sublimation of snow in the study area was stronger than melting, and it mainly affected baseflow. Frozen ground acted as an aquitard, weakening the effect of infiltration. However, it did not affect the amount of total water yield.

5. Discussion

5.1. Incorporating Frozen Ground and Snow Cover Improved Model Performance

The revised SWY can be considered a useful tool for water resource management and decision-making in cold alpine regions. The effects of frozen ground and snow cover on water yield should be considered when using the model in these regions for accurate results. The improved model performance also highlights the importance of continuously revising and updating models to account for changing environmental conditions and to improve their accuracy and reliability. Further evaluation of the revised SWY showed that incorporating both the effects of frozen ground and snow cover led to optimal model performance. The improvements in model performance were lower when only one of these cryosphere elements was considered, but still higher than when neither were considered. Specifically, the performance improvements of SWY2 were higher than that of SWY3 for quickflow, while the opposite was observed for baseflow. The incorporation of snow cover improved the changing trend of quickflow, while the monthly average amount remained almost unchanged. However, both the changing trend and annual average amount of baseflow were improved after considering the effect of snow cover. The performance of SWY1 was similar to that of SWY2 in terms of water yield, while the performance of SWY3 was similar to that of SWY4. These results indicate that accounting for the effects of frozen ground and snow cover is crucial for accurately modeling water yield in cold alpine regions.

The results of the study can be explained from the perspective of how snow cover and frozen ground affect water yield, including quickflow and baseflow. Snow cover affects water yield by melting and sublimation. The melting of snow cover will increase water yield, while sublimation will cause snow loss, resulting in a decrease in water yield [74,75]. The effect of sublimation is greater than melting in the study area, resulting in a decrease in water yield. Quickflow is less affected by snow cover as it occurs during or shortly after rain events. The amount of annual sublimation almost offsets the amount of melting during the short period of quickflow generation. On the other hand, frozen ground acts as an aquitard, weakening water infiltration and affecting the proportion distribution of quickflow and baseflow. This results in a larger proportion of quickflow than that of the non-frozen ground underlying the surface, but it does not affect the total water yield. It is important to note that the phase of snow cover will not change immediately when the climate warms, indicating that appropriate temperature conditions must be met for changes to occur. However, the effects of snow cover and frozen ground on water yield should be taken seriously, especially in the long term and in the context of climate warming. The study found that the baseflow and water yield of the study area were overestimated by 13 mm (47.58×10^8 m^3/yr) and 14 mm (51.24×10^8 m^3/yr), respectively, without considering the effect of snow cover on them. Similarly, without considering the effect of frozen ground on water yield, there would be about 6 mm of quickflow misestimated as baseflow every year. The overestimation will be more obvious with the increase in precipitation (Figure 7e,f). Therefore, the effects of the two cryosphere elements on water yield should be considered for accurate results in water resource management and decision-making.

Scordo et al. showed that coupling the SNOW-17 model with SWY significantly improved the model performance by considering the effect of snow on water yield [10]. The principle of the SNOW-17 model is similar to the degree-day model used in this study, which calculates sublimation and melting based on temperature data to determine the proportion of snow in precipitation. However, the SNOW-17 model does not consider the effect of frozen ground and is more suitable for regions where snow has a greater impact. The advantage of this research is that it covers 749 watersheds in five climate zones of North America, providing a broad scope for study. Consequently, we plan to investigate the applicability of the revised SWY model in cold regions of China. This is also supported by the results of Hamel et al., in which the snowmelt data was derived from the SWAT, making it difficult for many users to obtain for their area of interest [27]. Our revised model can quantify the impact of snow cover on water yield using conventional temperature and precipitation data, which is more accessible for users. Hamel et al.'s research is more inclined to simulate and evaluate baseflow, which characterizes water supply in dry season and is particularly important in arid regions. In the next step, we can focus on the impact of the change in climate, cryosphere, and land use on regional baseflow and water yield to comprehensively evaluate the HESs in cold alpine regions.

5.2. Limitations of the Modeling Approach

Despite the improvement in model performance resulting from the incorporation of the effects of frozen ground and snow cover in the SWY when applied in the TRHR, a typical cold alpine region, there remain some limitations that require future improvement. Firstly, although spatially explicit models are generally appropriate for mapping ecosystem services, they have certain limitations in their application, as noted by [26]. Despite the SWY being based on a physically based approach, the equations employed are overly simplified, leading to a significant increase in the uncertainty of absolute values produced, particularly when using α, β, and γ to control flow routing. These parameters are not readily available for any given watershed, rendering them difficult to set accurately, thus further compounding uncertainties in the model outputs [3]. This study employed sensitivity analysis to set the values of α, β, and γ, which required constant parameter adjustments to arrive at the optimal combination ($\alpha = 1/12$, $\beta = 0.2$ for the YAR, and 0.4 for the YEA and LAR, $\gamma = 1$), resulting in a significant increase in our workload. A value of 1/12 for α indicates that each month of the year has a similar fraction of upslope annual available recharge, which is an ideal situation. Furthermore, variations in the optimal β across different regions within the TRHR suggest that it is necessary to set the optimal β for each sub-region when the study area is large. The value of γ in our optimal parameter combination is "1", which implies that all pixel recharge is available to downslope pixels, but there is no empirically based evidence to support its rationality. It is strongly recommended that the model developer provide a reference guide for setting the values of these three parameters, such as instructions for determining the optimal values based on terrain, soil properties, or other relevant factors using appropriate functions. Secondly, the lack of appropriate data poses a challenge to the validation of the tool. Although the new version of the SWY demonstrated an overall satisfactory performance following comprehensive evaluation, the validation of the model using observed data was limited to total water yield, without consideration of baseflow or quickflow. Furthermore, while the meteorological input of SWY was monthly, the output of baseflow was on an annual scale, which fails to reflect the variation characteristics of water yield during the year, including flood and dry periods. It is recommended that baseflow be calculated on a monthly scale instead, which is crucial for practical decision-making implications. Although some traditional hydrological models, such as SWAT and VIC, can overcome some of these limitations, they are difficult to operate and require more specialized hydrological knowledge and calculation ability, which increases the difficulty of using the models [3,4,11]. A more optimal approach could be for the developers of ecosystem service models to engage in deep collaboration with hydrologists to create HES models that are both simple to operate and consider critical hydrological physical processes and characteristics.

In addition to the inherent limitations of the model itself, there are also shortcomings in the improvement aspects of the SWY model. Although we have considered the effect of seasonal frozen ground and the active layer on water yield, we have overlooked the effect of permafrost and its changes on water yield. The method of revising soil saturated hydraulic conductivity based on soil temperature only reflects the effect of the low permeability of frozen ground on water yield. However, apart from being a low permeable layer, the storage capacity of the active layer also plays a significant role in water yield. Degradation of the permafrost leads to an increase in the thickness of the active layer, consequently increasing its storage capacity and resulting in a decrease in water yield [76]. Therefore, in future work, it is necessary to incorporate the effects of permafrost change on water yield into the SWY model to enhance its physical processes and improve its performance. One possible direction is to utilize satellite data, such as the Grace gravity satellite, to address the limitations.

6. Conclusions

To illustrate the application of the SWY in the TRHR, a new version was developed to account for the effects of frozen ground and snow cover on water yield, with the aim of supporting decision-making in cold alpine regions. In the first step, we set up four scenarios for the SWY, including a scenario that did not consider the effects of frozen ground and snow cover, a scenario that only considered the effect of frozen ground, a scenario that only considered the effect of snow cover, and a scenario that considered the effects of both frozen ground and snow cover. Subsequently, we evaluated and validated the four scenarios by comparing the results with observed streamflow data and those obtained using the Eckhardt filter method. Further analysis was conducted to elucidate how snow cover and frozen ground affect water yield. The results showed that:

(1) The performance of the SWY was best in the scenario that accounted for the effects of both frozen ground and snow cover, while the scenario that did not consider these effects had the worst performance. Furthermore, the model performance improved when considering the effects of frozen ground or snow cover on water yield.

(2) Snow cover affects water yield through processes of melting and sublimation, while frozen ground acts as an aquitard, reducing infiltration and thus affecting the distribution of both spatial and temporal quickflow and baseflow.

(3) Without considering the effect of snow cover on water yield, the annual average baseflow and water yield of the TRHR would be overestimated by 13 mm (47.58×10^8 m^3/yr) and 14 mm (51.24×10^8 m^3/yr), respectively. Similarly, if the effect of frozen ground on water yield were not considered, there would be an annual overestimation of about 6 mm of quickflow as baseflow.

A critical future research avenue in the field of water resource management is to evaluate and predict the impact of climate change on the spatiotemporal distribution of water yield in cold alpine regions. This research will be useful for decision-makers who are involved in the management of water-related ecosystem services programs in these regions. By understanding how climate change will affect water resources, policymakers can develop appropriate strategies to adapt to the changing conditions and ensure sustainable use of water resources in the future.

Author Contributions: Conceptualization, L.Z. and R.C.; methodology, L.Z.; software, G.L.; validation, L.Z.; data curation, Y.Y.; writing—original draft preparation, L.Z.; writing—review and editing Y.Y. and X.W.; funding acquisition, R.C. All authors have read and agreed to the published version of the manuscript.

Funding: This research was funded by the Joint Research Project of Three-River Headwaters National Park, Chinese Academy of Sciences and the People's Government of Qinghai Province (LHZX-2020-11), the National Natural Sciences Foundation of China (42171145), the Key Talent Program of Gansu Province, and the Gansu Provincial Science and Technology Program (22ZD6FA005).

Institutional Review Board Statement: Not applicable.

Informed Consent Statement: Not applicable.

Data Availability Statement: Not applicable.

Conflicts of Interest: The authors declare that they have no known competing financial interests or personal relationships that could have appeared to influence the work reported in this paper.

References

1. Brauman, K.A.; Daily, G.C.; Duarte, T.K.; Mooney, H.A. The Nature and Value of Ecosystem Services: An Overview Highlighting Hydrologic Services. *Annu. Rev. Environ. Resour.* **2007**, *32*, 67–98. [CrossRef]
2. Brauman, K.A. Hydrologic ecosystem services: Linking ecohydrologic processes to human well-being in water research and watershed management. *WIREs Water* **2015**, *2*, 345–358. [CrossRef]
3. Dennedy-Frank, P.J.; Muenich, R.L.; Chaubey, I.; Ziv, G. Comparing two tools for ecosystem service assessments regarding water resources decisions. *J. Environ. Manag.* **2016**, *177*, 331–340. [CrossRef] [PubMed]
4. Vigerstol, K.L.; Aukema, J.E. A comparison of tools for modeling freshwater ecosystem services. *J. Environ. Manag.* **2011**, *92*, 2403–2409. [CrossRef]
5. Fan, M.; Shibata, H. Spatial and Temporal Analysis of Hydrological Provision Ecosystem Services for Watershed Conservation Planning of Water Resources. *Water Resour. Manag.* **2014**, *28*, 3619–3636. [CrossRef]
6. Benra, F.; De Frutos, A.; Gaglio, M.; Álvarez-Garretón, C.; Felipe-Lucia, M.; Bonn, A. Mapping water ecosystem services: Evaluating InVEST model predictions in data scarce regions. *Environ. Model. Softw.* **2021**, *138*, 104982. [CrossRef]
7. Wang, Y.; Wang, H.; Liu, G.; Zhang, J.; Fang, Z. Factors driving water yield ecosystem services in the Yellow River Economic Belt, China: Spatial heterogeneity and spatial spillover perspectives. *J. Environ. Manag.* **2022**, *317*, 115477. [CrossRef]
8. Ward, R.C.; Robinson, M. *Principles of Hydrology*; McGraw Hill Book Company: London, UK, 1990; pp. 139–183.
9. Apurv, T.; Cai, X. Impact of Droughts on Water Supply in U.S. Watersheds: The Role of Renewable Surface and Groundwater Resources. *Earth's Futur.* **2020**, *8*, e2020EF001648. [CrossRef]
10. Scordo, F.; Lavender, T.M.; Seitz, C.; Perillo, V.L.; Rusak, J.A.; Piccolo, M.C.; Perillo, G.M.E. Modeling Water Yield: Assessing the Role of Site and Region-Specific Attributes in Determining Model Performance of the InVEST Seasonal Water Yield Model. *Water* **2018**, *10*, 1496. [CrossRef]
11. Cong, W.; Sun, X.; Guo, H.; Shan, R. Comparison of the SWAT and InVEST models to determine hydrological ecosystem service spatial patterns, priorities and trade-offs in a complex basin. *Ecol. Indic.* **2020**, *112*, 106089. [CrossRef]
12. Francesconi, W.; Srinivasan, R.; Pérez-Miñana, E.; Willcock, S.P.; Quintero, M. Using the Soil and Water Assessment Tool (SWAT) to model ecosystem services: A systematic review. *J. Hydrol.* **2016**, *535*, 625–636. [CrossRef]
13. Aznarez, C.; Jimeno-Sáez, P.; López-Ballesteros, A.; Pacheco, J.P.; Senent-Aparicio, J. Analysing the Impact of Climate Change on Hydrological Ecosystem Services in Laguna del Sauce (Uruguay) Using the SWAT Model and Remote Sensing Data. *Remote Sens.* **2021**, *13*, 2014. [CrossRef]
14. Uniyal, B.; Kosatica, E.; Koellner, T. Spatial and temporal variability of climate change impacts on ecosystem services in small agricultural catchments using the Soil and Water Assessment Tool (SWAT). *Sci. Total. Environ.* **2023**, *875*, 162520. [CrossRef] [PubMed]
15. Zhang, C.; Li, J.; Zhou, Z.; Sun, Y. Application of ecosystem service flows model in water security assessment: A case study in Weihe River Basin, China. *Ecol. Indic.* **2021**, *120*, 106974. [CrossRef]
16. Troin, M.; Caya, D. Evaluating the SWAT's snow hydrology over a Northern Quebec watershed. *Hydrol. Process.* **2014**, *28*, 1858–1873. [CrossRef]
17. Wu, Q.; Song, J.; Sun, H.; Huang, P.; Jing, K.; Xu, W.; Wang, H.; Liang, D. Spatiotemporal variations of water conservation function based on EOF analysis at multi time scales under different ecosystems of Heihe River Basin. *J. Environ. Manag.* **2023**, *325*, 116532. [CrossRef]
18. Agnihotri, J.; Behrangi, A.; Tavakoly, A.; Geheran, M.; Farmani, M.A.; Niu, G. Higher Frozen Soil Permeability Represented in a Hydrological Model Improves Spring Streamflow Prediction from River Basin to Continental Scales. *Water Resour. Res.* **2023**, *59*, e2022WR033075. [CrossRef]

19. Gädeke, A.; Krysanova, V.; Aryal, A.; Chang, J.; Grillakis, M.; Hanasaki, N.; Koutroulis, A.; Pokhrel, Y.; Satoh, Y.; Schaphoff, S.; et al. Performance evaluation of global hydrological models in six large Pan-Arctic watersheds. *Clim. Chang.* **2020**, *163*, 1329–1351. [CrossRef]
20. Sapriza-Azuri, G.; Gamazo, P.; Razavi, S.; Wheater, H.S. On the appropriate definition of soil profile configuration and initial conditions for land surface–hydrology models in cold regions. *Hydrol. Earth Syst. Sci.* **2018**, *22*, 3295–3309. [CrossRef]
21. Walvoord, M.A.; Kurylyk, B.L. Hydrologic Impacts of Thawing Permafrost—A Review. *Vadose Zone J.* **2016**, *15*, 1–20. [CrossRef]
22. Hiyama, T.; Park, H.; Kobayashi, K.; Lebedeva, L.; Gustafsson, D. Contribution of summer net precipitation to winter river discharge in permafrost zone of the Lena River basin. *J. Hydrol.* **2023**, *616*, 128797. [CrossRef]
23. Park, H.; Dibike, Y.; Su, F.; Shi, J.X. Cold Region Hydrologic Models and Applications. In *Arctic Hydrology, Permafrost and Ecosystems*; Yang, D., Kane, D.L., Eds.; Springer: Cham, Switzerland, 2021. [CrossRef]
24. Li, S.; Zhang, Y.; Wang, Z.; Li, L. Mapping human influence intensity in the Tibetan Plateau for conservation of ecological service functions. *Ecosyst. Serv.* **2018**, *30*, 276–286. [CrossRef]
25. Liu, J.; Qin, K.; Xie, G.; Xiao, Y.; Huang, M.; Gan, S. Is the 'water tower' reassuring? Viewing water security of Qinghai-Tibet Plateau from the perspective of ecosystem services 'supply-flow-demand'. *Environ. Res. Lett.* **2022**, *17*, 094043. [CrossRef]
26. Lü, D.; Lü, Y.; Gao, G.; Liu, S.; Wu, B.; Fu, B. Existent nature reserves not optimal for water service provision and conservation on the Qinghai-Tibet Plateau of China. *Glob. Ecol. Conserv.* **2021**, *32*, e01945. [CrossRef]
27. Hamel, P.; Valencia, J.; Schmitt, R.; Shrestha, T.; Piman, T.; Sharp, R.P.; Francesconi, W.; Guswa, A.J. Modeling seasonal water yield for landscape management: Applications in Peru and Myanmar. *J. Environ. Manag.* **2020**, *270*, 110792. [CrossRef]
28. Qin, H.; Chen, Y. Spatial non-stationarity of water conservation services and landscape patterns in Erhai Lake Basin, China. *Ecol. Indic.* **2023**, *146*, 109894. [CrossRef]
29. Posner, S.; Verutes, G.; Koh, I.; Denu, D.; Ricketts, T. Global use of ecosystem service models. *Ecosyst. Serv.* **2016**, *17*, 131–141. [CrossRef]
30. Dai, E.; Zhu, J.; Wang, X.; Xi, W. Multiple ecosystem services of monoculture and mixed plantations: A case study of the Huitong experimental forest of Southern China. *Land Use Policy* **2018**, *79*, 717–724. [CrossRef]
31. Bai, Y.; Wong, C.P.; Jiang, B.; Hughes, A.C.; Wang, M.; Wang, Q. Developing China's Ecological Redline Policy using ecosystem services assessments for land use planning. *Nat. Commun.* **2018**, *9*, 3034. [CrossRef]
32. Sun, S.; Shi, Q. Global Spatio-Temporal Assessment of Changes in Multiple Ecosystem Services Under Four IPCC SRES Land-use Scenarios. *Earth's Futur.* **2020**, *8*, e2020EF001668. [CrossRef]
33. Measho, S.; Chen, B.; Pellikka, P.; Trisurat, Y.; Guo, L.; Sun, S.; Zhang, H. Land Use/Land Cover Changes and Associated Impacts on Water Yield Availability and Variations in the Mereb-Gash River Basin in the Horn of Africa. *J. Geophys. Res. Biogeosci.* **2020**, *125*, e2020JG005632. [CrossRef]
34. Gao, J.; Jiang, Y.; Wang, H.; Zuo, L. Identification of Dominant Factors Affecting Soil Erosion and Water Yield within Ecological Red Line Areas. *Remote Sens.* **2020**, *12*, 399. [CrossRef]
35. Daneshi, A.; Brouwer, R.; Najafinejad, A.; Panahi, M.; Zarandian, A.; Maghsood, F.F. Modelling the impacts of climate and land use change on water security in a semi-arid forested watershed using InVEST. *J. Hydrol.* **2021**, *593*, 125621. [CrossRef]
36. Sharp, R.; Douglass, J.; Wolny, S.; Arkema, K.; Bernhardt, J.; Bierbower, W.; Chaumont, N.; Denu, D.; Fisher, D.; Glowinski, K.; et al. *InVEST 3.9.0. User's Guide*; The Natural Capital Project, Stanford University: Stanford, CA, USA, 2020.
37. Gaglio, M.; Aschonitis, V.; Pieretti, L.; Santos, L.; Gissi, E.; Castaldelli, G.; Fano, E. Modelling past, present and future Ecosystem Services supply in a protected floodplain under land use and climate changes. *Ecol. Model.* **2019**, *403*, 23–34. [CrossRef]
38. Immerzeel, W.W.; Lutz, A.F.; Andrade, M.; Bahl, A.; Biemans, H.; Bolch, T.; Hyde, S.; Brumby, S.; Davies, B.J.; Elmore, A.C.; et al. Importance and vulnerability of the world's water towers. *Nature* **2020**, *577*, 364–369. [CrossRef] [PubMed]
39. Wang, T.; Yang, D.; Yang, Y.; Piao, S.; Li, X.; Cheng, G.; Fu, B. Permafrost thawing puts the frozen carbon at risk over the Tibetan Plateau. *Sci. Adv.* **2020**, *6*, eaaz3513. [CrossRef]
40. Han, Z.; Song, W.; Deng, X.; Xu, X. Grassland ecosystem responses to climate change and human activities within the Three-River Headwaters region of China. *Sci. Rep.* **2018**, *8*, 9079. [CrossRef]
41. Bai, Y.; Guo, C.; Degen, A.A.; Ahmad, A.A.; Wang, W.; Zhang, T.; Li, W.; Ma, L.; Huang, M.; Zeng, H.; et al. Climate warming benefits alpine vegetation growth in Three-River Headwater Region, China. *Sci. Total. Environ.* **2020**, *742*, 140574. [CrossRef]
42. Gao, Z.; Lin, Z.; Niu, F.; Luo, J. Soil water dynamics in the active layers under different land-cover types in the permafrost regions of the Qinghai–Tibet Plateau, China. *Geoderma* **2020**, *364*, 114176. [CrossRef]
43. Thapa, S.; Zhang, F.; Zhang, H.; Zeng, C.; Wang, L.; Xu, C.-Y.; Thapa, A.; Nepal, S. Assessing the snow cover dynamics and its relationship with different hydro-climatic characteristics in Upper Ganges river basin and its sub-basins. *Sci. Total. Environ.* **2021**, *793*, 148648. [CrossRef]
44. Pan, T.; Wu, S.-H.; Dai, E.-F.; Liu, Y.-J. Spatiotemporal variation of water source supply service in Three Rivers Source Area of China based on InVEST model. *J. Appl. Ecol.* **2013**, *24*, 183–189. (In Chinese)
45. Wang, Y.; Ye, A.; Peng, D.; Miao, C.; Di, Z.; Gong, W. Spatiotemporal variations in water conservation function of the Tibetan Plateau under climate change based on InVEST model. *J. Hydrol. Reg. Stud.* **2022**, *41*, 101064. [CrossRef]
46. Xue, J.; Li, Z.; Feng, Q.; Gui, J.; Zhang, B. Spatiotemporal variations of water conservation and its influencing factors in ecological barrier region, Qinghai-Tibet Plateau. *J. Hydrol. Reg. Stud.* **2022**, *42*, 101164. [CrossRef]
47. Qiao, F.; Fu, G.; Xu, X.; An, L.; Lei, K.; Zhao, J.; Hao, C. Assessment of water conservation Function in the Three-River Headwaters Region. *Res. Environ. Sci.* **2018**, *31*, 1010–1018. (In Chinese)

48. Woo, M. *Permafrost Hydrology*; Springer Science & Business Media: Berlin/Heidelberg, Germany, 2012; pp. 16–17.
49. Luo, D.; Jin, H.; Bense, V.F.; Jin, X.; Li, X. Hydrothermal processes of near-surface warm permafrost in response to strong precipitation events in the Headwater Area of the Yellow River, Tibetan Plateau. *Geoderma* **2020**, *376*, 114531. [CrossRef]
50. Burt, T.P.; Williams, P.J. Hydraulic conductivity in frozen soils. *Earth Surf. Process. Landforms* **1976**, *1*, 349–360. [CrossRef]
51. Walvoord, M.A.; Voss, C.I.; Wellman, T.P. Influence of permafrost distribution on groundwater flow in the context of climate-driven permafrost thaw: Example from Yukon Flats Basin, Alaska, United States. *Water Resour. Res.* **2012**, *48*, 1–17. [CrossRef]
52. Ming, F.; Chen, L.; Li, D.; Wei, X. Estimation of hydraulic conductivity of saturated frozen soil from the soil freezing characteristic curve. *Sci. Total. Environ.* **2020**, *698*, 134132. [CrossRef] [PubMed]
53. NRCS-USDA. Chapter 7: Hydrologic Soil Groups. In *National Engineering Handbook*; USDA: Washington, DC, USA, 2009.
54. Chen, R.-S.; Liu, J.-F.; Song, Y.-X. Precipitation type estimation and validation in China. *J. Mt. Sci.* **2014**, *11*, 917–925. [CrossRef]
55. Shiyin, L.; Yong, Z.; Yingsong, Z.; Yongjian, D. Estimation of glacier runoff and future trends in the Yangtze River source region, China. *J. Glaciol.* **2009**, *55*, 353–362. [CrossRef]
56. Chen, R.; Wang, G.; Yang, Y.; Liu, J.; Han, C.; Song, Y.; Liu, Z.; Kang, E. Effects of Cryospheric Change on Alpine Hydrology: Combining a Model with Observations in the Upper Reaches of the Hei River, China. *J. Geophys. Res. Atmos.* **2018**, *123*, 3414–3442. [CrossRef]
57. Hancock, S.; Baxter, R.; Evans, J.; Huntley, B. Evaluating global snow water equivalent products for testing land surface models. *Remote Sens. Environ.* **2013**, *128*, 107–117. [CrossRef]
58. Winstral, A.; Marks, D. Simulating wind fields and snow redistribution using terrain-based parameters to model snow accumulation and melt over a semi-arid mountain catchment. *Hydrol. Process.* **2002**, *16*, 3585–3603. [CrossRef]
59. Eckhardt, K. How to construct recursive digital filters for baseflow separation. *Hydrol. Process.* **2005**, *19*, 507–515. [CrossRef]
60. Eckhardt, K. A comparison of baseflow indices, which were calculated with seven different baseflow separation methods. *J. Hydrol.* **2008**, *352*, 168–173. [CrossRef]
61. Ahiablame, L.; Chaubey, I.; Engel, B.; Cherkauer, K.; Merwade, V. Estimation of annual baseflow at ungauged sites in Indiana USA. *J. Hydrol.* **2013**, *476*, 13–27. [CrossRef]
62. Ahiablame, L.; Sheshukov, A.Y.; Rahmani, V.; Moriasi, D. Annual baseflow variations as influenced by climate variability and agricultural land use change in the Missouri River Basin. *J. Hydrol.* **2017**, *551*, 188–202. [CrossRef]
63. Ngo, T.S.; Nguyen, D.B.; Rajendra, P.S. Effect of land use change on runoff and sediment yield in Da River Basin of Hoa Binh province, Northwest Vietnam. *J. Mt. Sci.* **2015**, *12*, 1051–1064. [CrossRef]
64. Luo, S.; Fang, X.; Lyu, S.; Ma, D.; Chang, Y.; Song, M.; Chen, H. Frozen ground temperature trends associated with climate change in the Tibetan Plateau Three River Source Region from 1980 to 2014. *Clim. Res.* **2016**, *67*, 241–255. [CrossRef]
65. Ni, J.; Wu, T.; Zhu, X.; Hu, G.; Zou, D.; Wu, X.; Li, R.; Xie, C.; Qiao, Y.; Pang, Q.; et al. Simulation of the Present and Future Projection of Permafrost on the Qinghai-Tibet Plateau with Statistical and Machine Learning Models. *J. Geophys. Res. Atmos.* **2021**, *126*, e2020JD033402. [CrossRef]
66. Shen, X.; An, R.; Feng, L.; Ye, N.; Zhu, L.; Li, M. Vegetation changes in the Three-River Headwaters Region of the Tibetan Plateau of China. *Ecol. Indic.* **2018**, *93*, 804–812. [CrossRef]
67. Zhang, Y.; Zhang, S.; Zhai, X.; Xia, J. Runoff variation and its response to climate change in the Three Rivers Source Region. *J. Geogr. Sci.* **2012**, *22*, 781–794. [CrossRef]
68. Allen, R.G.; Pereira, L.S.; Raes, D.; Smith, M. Crop evapotranspiration-Guidelines for computing crop water requirements-FAO Irrigation and drainage paper 56. *FAO Rome* **1998**, *300*, D05109.
69. Meng, X.; Wang, H. *China Meteorological Assimilation Datasets for the SWAT Model-Soil Temperature Version 1.0 (2009–2013)*; National Tibetan Plateau Data Center: Beijing, China, 2018. [CrossRef]
70. NRCS-USDA. Chapter 9: Hydrologic Soil-Cover Complexes. In *National Engineering Handbook*; USDA: Washington, DC, USA, 2009.
71. Wei, Y. *The Boundaries of the Source Regions in Sanjiangyuan Region*; National Tibetan Plateau Data Center: Beijing, China, 2018. [CrossRef]
72. Chen, R.; Ding, Y.; Kang, E. Some knowledge on and parameters of China's alpine hydrology. *Adv. Water Sci.* **2014**, *25*, 307–317. (In Chinese)
73. Guo, S.; Chen, R.; Li, H. Surface Sublimation/Evaporation and Condensation/Deposition and Their Links to Westerlies during 2020 on the August-One Glacier, the Semi-Arid Qilian Mountains of Northeast Tibetan Plateau. *J. Geophys. Res. Atmos.* **2022**, *127*, e2022JD036494. [CrossRef]
74. Suzuki, K.; Liston, G.E.; Matsuo, K. Estimation of Continental-Basin-Scale Sublimation in the Lena River Basin, Siberia. *Adv. Meteorol.* **2015**, *2015*, 286206. [CrossRef]
75. Gascoin, S. Snowmelt and Snow Sublimation in the Indus Basin. *Water* **2021**, *13*, 2621. [CrossRef]
76. Suzuki, K.; Park, H.; Makarieva, O.; Kanamori, H.; Hori, M.; Matsuo, K.; Matsumura, S.; Nesterova, N.; Hiyama, T. Effect of Permafrost Thawing on Discharge of the Kolyma River, Northeastern Siberia. *Remote Sens.* **2021**, *13*, 4389. [CrossRef]

Disclaimer/Publisher's Note: The statements, opinions and data contained in all publications are solely those of the individual author(s) and contributor(s) and not of MDPI and/or the editor(s). MDPI and/or the editor(s) disclaim responsibility for any injury to people or property resulting from any ideas, methods, instructions or products referred to in the content.

Article

Landscape-Scale Mining and Water Management in a Hyper-Arid Catchment: The Cuajone Mine, Moquegua, Southern Peru

Morag Hunter [1,†], D. H. Nimalika Perera [2,†], Eustace P. G. Barnes [2], Hugo V. Lepage [2,*], Elias Escobedo-Pacheco [3], Noorhayati Idros [2], David Arvidsson-Shukur [2], Peter J. Newton [2], Luis de los Santos Valladares [2], Patrick A. Byrne [4] and Crispin H. W. Barnes [2]

[1] Earth Science Department, University of Cambridge, Downing Site, Cambridge CB2 3EQ, UK; mah1003@cam.ac.uk

[2] Cavendish Laboratory, University of Cambridge, JJ Thomson Avenue, Cambridge CB3 0HE, UK; nimalikap@yahoo.com (D.H.N.P.); eb651@cam.ac.uk (E.P.G.B.); idrosnoorhayati@gmail.com (N.I.); drma2@cam.ac.uk (D.A.-S.); pjn32@cam.ac.uk (P.J.N.); ld301@cam.ac.uk (L.d.l.S.V.); chwb101@cam.ac.uk (C.H.W.B.)

[3] Escuela Profesional de Ingeniería Agroindustrial, Universidad Nacional de Moquegua, Calle Ancash s/n, Ciudad Universitaria, Moquegua 18001, Peru; eescobedop@unam.edu.pe

[4] Faculty of Science, Liverpool John Moores University, Liverpool L3 3AF, UK; p.a.byrne@ljmu.ac.uk

* Correspondence: hl407@cam.ac.uk

† These authors contributed equally to this work.

Abstract: The expansion of copper mining on the hyper-arid pacific slope of Southern Peru has precipitated growing concern for scarce water resources in the region. Located in the headwaters of the Torata river, in the department of Moquegua, the Cuajone mine, owned by Southern Copper, provides a unique opportunity in a little-studied region to examine the relative impact of the landscape-scale mining on water resources in the region. Principal component and cluster analyses of the water chemistry data from 16 sites, collected over three seasons during 2017 and 2018, show distinct statistical groupings indicating that, above the settlement of Torata, water geochemistry is a function of chemical weathering processes acting upon underlying geological units, and confirming that the Cuajone mine does not significantly affect water quality in the Torata river. Impact mitigation strategies that firstly divert channel flow around the mine and secondly divert mine waste to the Toquepala river and tailings dam at Quebrada Honda remove the direct effects on the water quality in the Torata river for the foreseeable future. In the study area, our results further suggest that water quality has been more significantly impacted by urban effluents and agricultural runoff than the Cuajone mine. The increase in total dissolved solids in the waters of the lower catchment reflects the cumulative addition of dissolved ions through chemical weathering of the underlying geological units, supplemented by rapid recharge of surface waters contaminated by residues associated with agricultural and urban runoff through the porous alluvial aquifer. Concentrations in some of the major ions exceeded internationally recommended maxima for agricultural use, especially in the coastal region. Occasionally, arsenic and manganese contamination also reached unsafe levels for domestic consumption. In the lower catchment, below the Cuajone mine, data and multivariate analyses point to urban effluents and agricultural runoff rather than weathering of exposed rock units, natural or otherwise, as the main cause of contamination.

Keywords: landscape-scale mining; hyper-aridity; anthropogenic contamination; eco-toxicity; multivariate analysis

1. Introduction

Developmental processes that impact water quality and availability in hyper-arid regions are of particular concern when undertaking environmental impact assessments

for resource exploitation proposals [1]. A region may be defined as hyper-arid when the ratio between precipitation and maximum evapotranspiration (P/PE) is below 0.05 [2,3]. Determining the character of water chemistry and identifying the relative importance of both natural and anthropogenic factors in the contamination of water resources in such regions is a global priority, especially with predicted climate change impacts [4–6]. This is particularly important on the arid and mineral-rich western flank of the Andes, where two thirds of Peru's population share less than 2% of the country's water and where the world's largest copper deposits reside [7–10]. Increasing pressure from mineral exploitation on water resources poses a serious challenge to the ecology of these regions and regularly produces social conflict [11,12]. Stakeholder engagement is essential for both mining corporations and statutory bodies to ensure that adverse social and environmental impacts are minimized and developmental benefits are maximized for local communities as well as the national economy [13]. A copper mine producing 50 kt/d requires 30,000 m^3 per day of fresh water for processing. The Cuajone mine processes approximately 90 kt/d [14]. The projected nine-fold expansion of copper production in Southern Peru, already the second largest producer after Chile, can only accentuate these challenges. Since trace metals are a major factor in the toxicity of contaminated water [15,16], mapping their presence in potentially affected river systems and understanding their baseline geochemistry is an important preparatory challenge. The direct impacts of increased mining can be partly mitigated by re-routing rivers using canals and diverting waste water to tailings dams, but the increase in mining inevitably increases the human population in remote mineral-rich regions, with a concomitant increase in urban and agricultural development, both of which are often associated with contamination from trace metal and persistent residues [11,17–19].

In hyper-arid regions, superficial weathering regimes are dominated by geophysical processes, while subterranean weathering processes are primarily chemical and often play a dominant role in determining the geochemistry of water bodies [20,21]. Trace metal contamination in river systems is, in part, the product of natural acid rock drainage (ARD), in which the oxidation of sulfide-bearing minerals increases the acidity of runoff waters and enhances the further chemical breakdown of the underlying lithology. Hydrological systems dominated by acidic waters show increased trace metal mobilization and metal hydroxide precipitation [22]. The downstream eco-toxicity of these contaminants depends on their bio-availability and consequent accumulation in the food chain [23,24]. Polymetallic mining is a well-documented cause of increased ARD, where it is more commonly known as acid mine drainage (AMD). Sulfide minerals and ores, in which the dominant metal ion can be iron, zinc, copper or nickel, are commonly associated with AMD. Chalcopyrite (a copper-iron sulfide) is the most commonly mined copper-bearing ore, and it is frequently found alongside other sulfides, making copper mines particularly prone to AMD [11,18,25–27]. In geologically active regions, such as the Andes, ARD as a consequence of acidic epithermal waters entering the river catchment systems, produces elevated chemical weathering rates, increasing the addition of ions to river waters, the exact chemistry of which is dependent upon the geology of the source region [28]. The combination of geogenic and anthropogenic processes increases weathering processes at mine sites, especially in hyper-arid regions, such as those that exist in Southern Peru, making water resources less suitable for domestic consumption and agricultural irrigation. As such, water quality needs to be carefully monitored where mining is taking place [5,21].

Pesticide and fertilizer effluents and industrial and urban waste waters are also well-documented sources of trace metal contamination. Arsenic is used in herbicides and pesticides; cadmium in batteries and plastics; chromium in dyes and tanning processes; lead in batteries, wiring and cabling; mercury and manganese in pesticides and batteries; and zinc is found in pharmaceuticals, dyes and batteries. All of these elements can leach into the wider environment from agricultural and urban runoff. The uptake of trace metals depends upon soil characteristics, plant species and the metal concerned [18,29]. The trophic transfer of trace metals, even at low concentrations, is known to have mutagenic effects [30].

In addition to the more global anthropogenic and natural geogenic factors that contribute to the determination of water quality, there are factors specific to areas in which mining and catchment transfer projects have been developed [25,31]. The Torata river sub-catchment and lower reaches of the Moquegua river drainage basin represent a region in which both a landscape-scale mine and a regional-scale catchment transfer project could reasonably be thought to impact catchment dynamics, including water resource quality and availability (Figure 1). The exploitation of mineral resources increasingly necessitates environmental impact statements (EISs) with concomitant impact mitigation strategies which may become advisory or statutory obligations. Significant work has been undertaken to understand the impact of mining in this region, with a focus on community engagement and the development of integrated management strategies [12,32]. The development of landscape-scale mines increasingly requires such mitigation strategies to avoid the otherwise inevitable environmental consequences and political issues, although these may not be properly identified in many impact assessment processes [13,33–36]. The development of the Cuajone, Quellaveco and Toquepala mines in Southern Peru have all involved the implementation of significant impact mitigation, with both river channel and waste material diversion with a complex network of sealed and open channels, tunnels and existing river channels [14,32]. Waste materials from the Cuajone and Quellaveco mines do not enter the greater Moquegua river drainage basin but are transferred to the tailings dams at Quebrada Honda (17 28 19.52 S, 70 41 51.85 W) at 1135 m asl and Cortaderas 2 (17 12 02.49 S, 70 41 51.85 W) at 3086 m asl, all of which can be readily seen on satellite imagery. Additionally, a tunnel diverting Torata river channel flow to the north of the principal mine pit at Cuajone was constructed to ensure that the water quality of the headwaters is not affected by mine activity. Water scarcity in the region, combined with population growth and agricultural development, has increased the need for water resources and driven the development of the catchment transfer Pasto Grande project. This project transfers water from the headwaters of the Tambo river into the greater Moquegua river catchment in 262 km of open channels, tunnels and existing river channels, that include the Sajena, Torata and Moquegua rivers. The long-term impacts of this transfer on catchment ecology, hydrological dynamics and water chemistry have not been studied. Very few studies have been carried out on the impacts of catchment transfer projects, but they are widely considered to have drastic impacts on hydrological regimes, water chemistry, and aquatic biota [31]. The effect of the Pasto Grande project is the subject of further study.

As such, the Torata river was selected for several significant reasons: (i) the presence of the landscape-scale Cuajone copper mine, owned by Southern Copper, located in the headwaters, and the proposed expansion of mining in the region give the opportunity to look at the efficiency of water management measures associated with large-scale mining infrastructure; (ii) there is a global need for detailed studies of the controls on water chemistry in hyper-arid regions to guide resource management; (iii) the nature of the topography provides an ideal context in which to identify the sequential impacts of trace metal dispersion and concentration; (iv) the presence of two well-spaced urban settlements at 2200 m asl (Torata) and 1400 m asl (Moquegua), allowing the identification of distinct chemical signals; and (v) the presence of two distinct and well-separated agricultural areas associated with the above settlements (Figure 1).

The general aim of this paper is to establish the geogenic and anthropogenic determinants of water chemistry in the Torata river sub-catchment and lower Moquegua river, lying in a hyper-arid region in Southern Peru. The catchment is periodically affected by El Niño–Southern Oscillation (ENSO) events with Pacific-derived moisture and regional atmospheric instability producing heavy precipitation and catastrophic flooding in the Moquegua region as in 1998 [37] and more recently in 2016 [38].

Topographic conditions in the region provide an ideal transect from the high Andes to the Pacific coast with the Torata river and the lower Moquegua river crossing four distinct geological units, passing the landscape-scale Cuajone mine, the urban settlements of Torata and Moquegua, and two agricultural areas (Figure 1). A sample-sites network was established throughout the Torata river sub-catchment, above and below the Cuajone copper mine, and below the village of Torata and associated agricultural areas. Additionally, sites were adopted above and below the city of Moquegua and along the lower Moquegua river to the Pacific ocean. We describe the spatial and temporal variation in the chemistry of the river and its main tributaries, and use this to investigate the various contributors to the chemistry. Spatial patterns can be linked to processes specific to the upper, middle and lower catchments, particularly the underlying geology, any influence from mining and related acid–rock interactions, agriculture, and hydro-fluvial processes such as ground flow, surface runoff, distance from the ocean and elevation. Secondly, we discuss temporal patterns, linked to dry/wet seasons and El Niño and how these affect the baseline. Where the data allow, we have indicated qualitative source apportionment based on concentration data. In addition to the overall aim, we have, where the data allow, indicated qualitative source apportionment based on concentration data. In this, we have attempted to achieve four specific objectives: (1) to identify the impact of the Cuajone mine on the geochemistry of the Torata river; (2) to identify and describe the hydro-chemical impact of the rural settlement of Torata and associated agricultural lands on the Torata river; (3) to identify and describe the hydro-chemical impact of the urban settlement of Moquegua on the Moquegua river; and (4) to identify the the hydro-chemical impacts of the agricultural area lying along the lower Moquegua river valley.

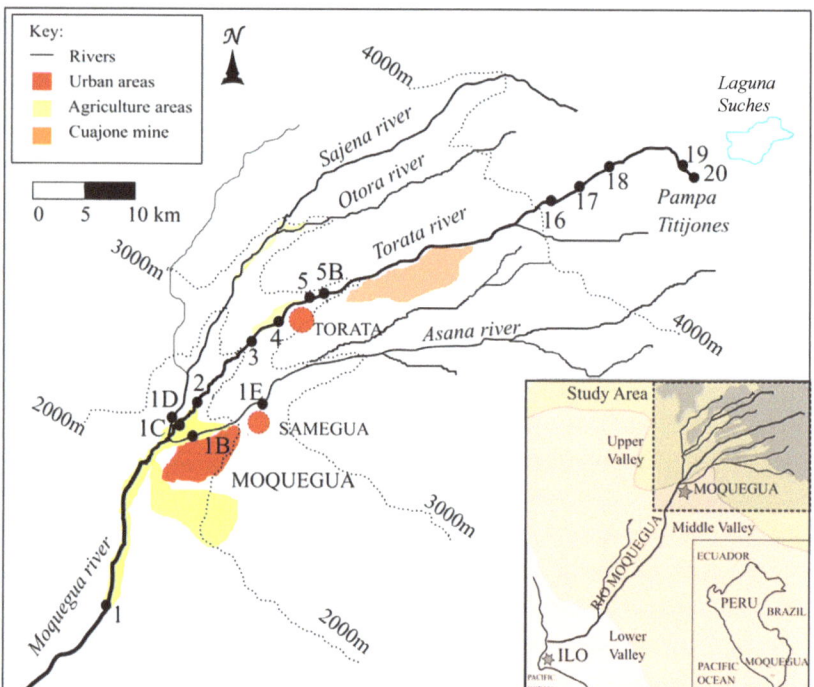

Figure 1. Map of selected study site locations in the foothill and headwaters (site 0A, 40 km downstream from site 1 near Ilo, is not shown). Inset shows underlying geological units adapted from Decou et al. [39]. Geological lithologies coded by color: pink—Coastal Batholith (intrusive 145–155 Ma); buff—Moquegua Group (sedimentary 50–54 Ma); green—Cretaceous volcanics and Eocene intrusives; grey—Miocene to recent pyroclastic deposits. Study area shown in dashed box.

2. Materials and Methods

2.1. Study Area

In the Peruvian Coastal Desert, the development of hyper-aridity evolved with the Andean orogenesis and consequent Pacific-slope rainshadow, together with the development of the Humboldt current and episodic El Niño phenomena [2]. As a consequence, the core Peruvian coastal desert region between 15° S and 30° S and from sea level to 3500 m asl is considered hyper-arid. It receives a mean annual precipitation of 3.5–4.5 mm. Today, the Pacific slope of the Andes supports a stable xeric environment, supporting little vegetation but which is affected by episodic catastrophic flooding [28,37,40,41]. The world's largest supergene copper deposits are also located in this hyper-arid zone, stretching from Southern Peru to Northern Chile [42,43]. Currently, there are three landscape-scale copper mines in the region; the Cuajone mine, the newly developed Quellaveco mine, and the Toquepala mine. The study area focuses on the Torata river in the Moquegua drainage. This river rises in the headwater bofedales at 4590 m asl at Pampa Titijones and flows to the Cuajone copper mine at 3500 m asl before descending through the settlement of Torata, approximately 11 km below the mine at 2200 m asl. The channel then passes through a narrow canyon to join the Sajena river and, subsequently, the Moquegua river at 1290 m asl below the city of Moquegua. Thereafter, the river flows through a narrow, intensively farmed floodplain before reaching the Pacific Ocean at Ilo [5].

2.2. Study Sites

The sampling network was designed to cover the principal tributaries in the Torata subcatchment and lower Moquegua river catchment. Data discussed here are from 16 sites within the catchment, 11 along the Torata river, 4 lying at or below the confluence of the Torata river with the Moquegua and Sajena rivers, and 1 site close to the ocean (Figure 1). Precise GPS locations and elevations of the sites are given in Table 1. Site 0A is in the arid tropical zone, near Ilo on the coast. Site 1 is located in the upper tropics, 20 km below both the city of Moquegua and an extensive area of irrigated agriculture. Sites 1B, 1C and 1D are located at the confluence of the Sajena, Torata and Moquegua rivers. Site 1E lies above all urban development and records the quality of the upper Asana River. Site 2 lies above all agricultural influence from the Moquegua city district, emerging from a canyon that starts in the Torata district. Sites 3 and 4 lie below the small rural town of Torata, and sites 5 and 5B above Torata, with 5B located immediately beneath the Cuajone mine. Directly above the Cuajone mine are sites 16, 17 and 18, and finally, sites 19 and 20 represent the source bofedal (peat wetland) and spring, respectively.

2.3. Sample Seasons

Three field investigations were undertaken: January 2017 (denoted 17-R), July 2017 (denoted 17-D), and January 2018 (denoted 18-R). The 17-R sampling campaign followed directly after an El Niño episode in which there was considerable flooding in the lower catchment. Despite this, it did not rain during the 17-R visit. During 17-D, the weather was dry and had been dry since the 17-R visit. Visit 18-R was nominally during the wet season but data collection preceded the onset of rains in the headwaters, and as such, many of the rivers retained dry season characteristics. Field visits were arranged to collect water quality data in situ and water samples for later analysis. Prior to sample and data collection, all field equipment was calibrated at the Universidad Nacional Autonoma de Moquegua (UNAM) laboratories. Owing to weather conditions and logistical constraints, it was not possible to collect samples at all sites on all field visits.

2.3.1. Water Quality Parameters and River Data Collection

At each site, the following data were collected: GPS coordinates, atmospheric and water temperature, altitude, dissolved oxygen (DO mg/L and %), pH, conductivity, total dissolved solids (TDS), and oxygen reduction potential (ORP). A simple assessment of nearby land usage was noted.

Table 1. Selected study sites including, approximate GPS coordinates, elevation above sea level, and distance from site 1.

Site	Site Name	GPS Coordinates		Altitude (m asl)	Distance from Site 1 (km)
Sites on Moquegua river (Osmore)					
0A	Lower Moquegua valley—El Algarrobal	17 37.705' S	71 17.579' W	68	45.8
1	Middle Moquegua valley—El Conde	17 20.089' S	70 59.889' W	955	0
Sites on Torata river					
1C	Above Sajena–Torata river confluence	17 10.720' S	70 57.037' W	1330	19.5
2	Puente Estuquina—beginning of agriculture area	17 09.015' S	70 55.01' W	1495	24.5
3	Puente Coplay—end of agriculture area	17 05.807' S	70 52.726' W	1941	32.6
4	Puente Canilay—downstream of Torata settlement	17 05.183' S	70 51.4303' W	2175	36.1
5	Fire station—beginning of agriculture area and upstream of Torata settlement	17 04.353' S	70 50.248' W	2225	37.7
5B	Middle Torata valley—the highest downstream of Cuajone mine	17 03.546' S	70 47.704' W	2700	45
16	Upper Torata valley—Lowest upstream of Cuajone mine	16 59.019' S	70 36.392' W	3939	66.2
17	Upper Torata valley—Middle	16 58.360' S	70 35.004' W	4100	69.3
18	Upper Torata valley—Highest	16 57.454' S	70 33.672' W	4225	72.2
19	Bofedale.at Torata river	16 57.120' S	70 29.698' W	4380	81.5
20	Torata river source	16 57.324' S	70 29.677' W	4400	82
Sites on Moquegua river (Asana river)					
1B	Below Moquegua city.	17 11.205' S	70 57.193' W	1318	18.7
1E	Above Moquegua and Samegua.	17 09.737' S	70 52.343' W	1662	28.6
Sites on Sajena/Otora river (Huaracane river)					
1D	Before Sajena–Torata confluence	17 10.690' S	70 57.052' W	1440	19.5

2.3.2. Water Sample Collection

At each site, 1-liter composite samples were collected in pre-washed polyethylene bottles. Subsamples for mercury (Hg) analysis were preserved with sulfuric acid and potassium dichromate to pH < 2, and samples for cyanide (CN) analysis were preserved with sodium hydroxide to pH > 12. Samples for total metal analysis were decanted into 50 mL tubes and preserved to pH < 2 with HNO_3 (99.9% trace metal basis) in the UNAM laboratory. All samples were packed in cool boxes for transport and stored at 4 °C until analysis.

2.4. Analysis of Water Samples

Chemical analyses for 44 components including major cations, major anions, trace metals, nutrients and biological components (see Table 2) were conducted by Northumbrian Water Scientific Services (Newcastle, UK), Greenwich University (Kent, UK) and the University of Cambridge (Cambridge, UK) laboratories. Samples for metal analysis were treated with acid digestion at 105 °C and analyzed by ICP-MS (7500ce, Agilent, London, UK). The anions were measured using ion chromatography (Dionex IC with Ionpac analytical column, Thermo Scientific, Oxford, UK) and nitrates were measured using discrete automated colorimetry (Aquakem 600, Thermo Scientific, UK). Hg was measured using atomic florescence (Millenium Merlin, PS Analytical, Kent, UK, LOD: 0.003 µg/L) and total cyanide (CN) was measured using a segmented flow analyzer (Skalar, Breda, The Netherlands, LOD: 0.02 mg/L). Total phosphorous (TP) was measured by ICP-OES (iCap 6500, Thermo Scientific, UK), and total nitrogen (TN) was measured using chemiluminescence (Shimadzu TOC-V, Milton Keynes, UK). Chemical oxygen demand (COD) was measured by digestion followed by colorimetry (Merck COD cell test kit, Darmstadt, Germany, COD 5–80 mg/L). *E. coli* and coliform were quantified using multiple-tube fermentation techniques. All analytical methods were carried out according to EPA and ISO protocols in quality-controlled labs. All samples were analyzed in duplicate for precision, and certified reference standards and internal calibration were used for accuracy measurements [44].

Table 2. Water parameters and their minimum and maximum values measured during the study: CN—cyanide; TP—total phosphorus; TN—total nitrogen; Cond—electrical conductivity; TDS—total dissolved solids; Temp—temperature; DO—dissolved oxygen; COD—chemical oxygen demand; Hard.—water hardness. Other symbols take their usual meanings.

Parameter	Min–Max	Parameter	Min–Max	Parameter	Min–Max
SO_4 (mg/L)	11–490	HCO_3^- (mg/L)	13.4–350	K (mg/L)	2.1–11
NO_2 (mg/L)	0–0.486	Hard. (mg/L $CaCO_3$)	17.3–764.3	Li (µg/L)	0–170
NO_3 (mg/L)	0–19.65	Al (mg/L)	0–2.0	Mg (mg/L)	1.8–36
PO_4 (mg/L)	0–0.57	As (µg/L)	2.1–21.0	Mn (mg/L)	0–0.65
CO_3 (mg/L)	0–0	Be (µg/L)	0–0.4	Mo (µg/L)	0.56–8.3
CN Total (mg/L)	0–0	Bi (mg/L)	0–0	Na (mg/L)	4.9–240
TP (mg/L as P)	0–0.41	Br (mg/L)	0–0.82	Ni (µg/L)	0–26
TN (mg/L)	0–8.1	Ca (mg/L)	4.5–251.8	Pb (µg/L)	0–3.9
pH	7.63–9.45	Cd (µg/L)	0–0.93	Se (µg/L)	0–4.8
Cond. (mS/cm)	95.4–2711.4	Cl (mg/L)	0–360.3	Si (mg/L)	2.5–25
TDS (mg/L)	0–1833	Cr (µg/L)	0–3.4	Te (mg/L)	0–0
Temp. (°C)	3.8–29.2	Co (µg/L)	0–4	Th (µg/L)	0–0.029
DO (mg/L)	0–12.4	Cu (µg/L)	0–53.0	U (µg/L)	0–13
COD	0–27.85	F (mg/L)	0–0.49	V (µg/L)	1.9–11
		Fe (mg/L)	0–1.5	Zn (µg/L)	0–55.0
		Hg (mg/L)	0–0		

2.5. Statistical Analysis

Cluster analysis (CA) and principal component analysis (PCA) were used to quantify the relative dominance of various physico-chemical components of the river water with the aim of discriminating the main controls over the river quality. These methods are widely used in the analysis of correlations in multivariate data [16,25]. Both CA and PCA were applied to the z-score normalized value of the measured parameters [45]. z-score normalization facilitates the discussion of correlations between parameters with vastly different magnitudes or amplitudes of fluctuations. CA provides a framework for creating subsets (clusters) of samples with high similarities. The internal structure of the data is ordered according to a dissimilarity measure, grouping similar samples. The analysis presented here uses Ward's hierarchical agglomerative CA method, together with a Euclidean dissimilarity measure [16,46,47]. This method finds the two closest samples (the data points with the smallest sum of squares of Euclidean distance), producing a secondary pseudo-sample. It then iterates this procedure until all the samples have been grouped, recording the dissimilarity distances $\{d_i\}$, where i indicates the ith link. In the CA dendrograms shown in this paper, the dissimilarity distance was reported as a fraction of the largest distance: $d_i/\max(\{d_i\})$. The cophenetic correlation coefficient (CCC) was used to measure how well the cluster dendrograms preserve the pairwise dissimilarity distances between original, un-clustered data [48]. A CCC of 100% corresponds to a dendrogram that perfectly preserves the original distances.

CA was employed in two ways in this study. Q-mode CA was used to find how the sample sites cluster with respect to the chemical parameters. This reveals spatial correlations of parameters over the set of sample sites. R-mode CA was used to find how the parameters cluster with respect to the sample sites. Clusters of similar parameters indicate that underlying factors affect them alike [49].

PCA is a methodology for the dimensional reduction in the number of parameters needed to describe a set of data [45,50]. The analysis begins with the calculation of the eigenvectors of the covariance matrix of the data, and the arrangement of them in descending order with respect to decreasing eigenvalues. If some eigenvalues are significantly larger (above some threshold) than the others, their corresponding eigenvectors are called "principal components". By picking out the principal components with the largest eigenvalues, it is possible to describe the covariance of the data to a level given by the ratio between the sum of their eigenvalues and the sum of all the eigenvalues. In this paper, PCA was employed on clusters with similarly behaving parameters to understand the intra-cluster correlation.

2.6. Contextual Materials and Analyses

In addition to the above methodologies applied to our sample site data, we also refer to data collected under our longer-term campaign, and data collated from published papers and technical reports on water management in the region. This provides a useful context in which to fully understand both the geogenic and anthropogenic determinants of water quality and availability, while underlining the need for further studies.

3. Results

Water chemistry data from the three field seasons (17-R, 17-D and 18-R) are presented in Figure 2. Measured chemical parameters with their minimum and maximum values are in Table 2. The following sections discuss the temporal and spatial variations in river water chemistry during the three sampling campaigns.

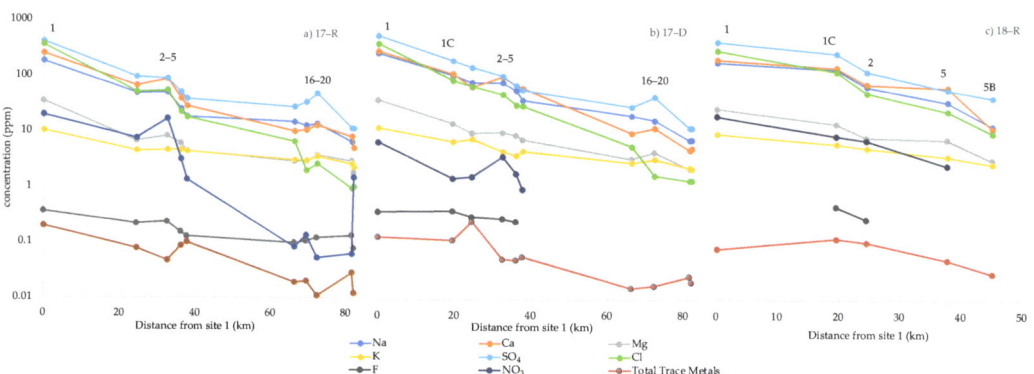

Figure 2. Spatial and temporal variation in cations (Na^+, Ca^{2+}, Mg^{2+}, K^+), anions (SO_4^{2-}, Cl^-, F^-, NO_3^-) and total trace metals (Cd, Cu, Co, Cr, Li, Pb, Mo, Ni, Se, U, V, and Zn) in water samples from Moquegua river sites plotted against distance upstream from site 1 in km. (**a**) January 2017 (17-R), (**b**) July 2017 (17-D) and (**c**) January 2018 (18-R). Site numbers are given in the labels.

3.1. Water Quality Parameters and Major Ion Composition

3.1.1. Headwater Sites above the Cuajone Mine (Sites 16, 17, 18, 19, 20)

These sites were neutral or slightly alkaline (pH 7.84 to 9.45) with most sites recording a pH suitable for aquatic life. Site 19 showed the most alkaline conditions with pH values up to 9.45 and water temperature readings ~5 °C higher than ambient temperature. This site is a bofedal (Andean peat wetland) just below the source spring, site 20, and water was collected at midday in 17-R but early morning in 17-D. Both dissolved oxygen (DO) levels and pH at this site were higher at midday (12.4 mg/L and pH 9.45 in 17-R) than in the early morning (6.62 mg/L and pH 8.52 in 17-D) and, when taken together, are consistent with photosynthetic activity consuming CO_2 [51,52] and adding free oxygen to the water during daylight hours. However, vegetative growth was more evident during the warmer wetter season (17-R) than the drier colder season (17-D), as to be expected, and therefore increased photosynthesis in season 17-R may also have contributed to the higher pH and DO levels. Similarly, the high quantity of decomposing organic matter related to the rich floral composition of the bofedal explains why the highest COD measurements in the Torata river (17 mg/L) system were observed at this site.

The headwater sites have the lowest concentrations of major cations ($Na^+ \gtrsim Ca^{2+} \gg Mg^{2+}$, K^+) and anions ($SO_4^{2-} \gg Cl^- > F^-$, NO_3^-), with total concentrations in all ions increasing downstream (Figure 2) as expected. Na/Cl equivalent molar ratios at the headwater sites, Figure 3a, are significantly higher (10.6-3.4, 17-R, 14.4-5.3 17-D) when compared to sites downstream of the Cuajone mine (2.2-1.4 17-R, 2.7-1.9 17-D), with the highest Na/Cl ratios recorded at site 19 (the bofedal, 10.6 during 17-R) and site 18 (14.4, during 17-D). SO$_4$/Cl equivalent molar ratios are also significantly higher in the headwater sites, with site 18 recording the highest values (13.45—17-R, 18.02—17-D). The Na/Cl equivalent molar ratio can be used as an indicator of ground water contributions to water bodies [53]. A high Na/Cl ratio is an indication of a significant contribution from mineral-rich ground water sources rather than precipitation, which usually has Na/Cl ratios between 0.8 and 1.0 [53,54]. The high Na/Cl and SO$_4$/Cl ratios, when combined with the higher water temperature, are consistent with a geothermal contribution to source waters in these upper Torata sites (c.f. findings in [5].)

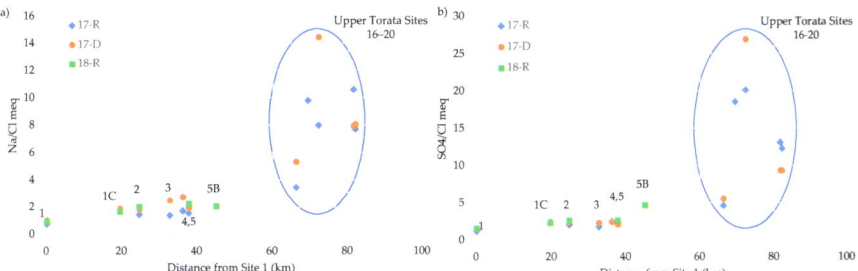

Figure 3. (a) Na/Cl equivalent molar ratio and (b) SO$_4$/Cl equivalent molar ratio for Moquegua river system.

Following the El Niño, the bofedal, site 19, had the highest Fe and Al concentrations of the headwater sites (1.5 mg/L and 0.09 mg/L in 17-R respectively, Figure 4). This Fe concentration exceeds the safe limit of 1 mg/L for aquatic life [55]. Both the high pH and the organic processes driving reducing conditions in the waters of the bofedal encourage dissolution of iron, and aluminum solubility also increases with higher pH and in the presence of sulfate in the water.

3.1.2. Foothill Sites below the Cuajone Mine (Sites 5B, 5, 4, 3, 2, 1B, 1C, 1D, 1E)

pH measurements and field observations at sites below the Cuajone mine did not show evidence of AMD. pH values at sites 5 and 5B, below the Cuajone mine, ranged from 7.98 to 8.56, little different from site 16 above the mine where the pH ranged from 7.84 to 7.97. A healthy riparian eco-system dominates the valley at this altitude. This highly sensitive environmental indicator suggests that the water quality in this area is good. Foothill sites 2, 5, and 5B recorded elevated Al, Fe and Mn concentrations compared to other sites, and levels were particularly high at these sites and site 4 during 17-R, after the El Niño episode. Cretaceous (91–65 Ma) volcanic and sedimentary rocks, with Eocene intrusions (55–45 Ma), underlie the area (Figure 1) and represent the oldest geological units in the sample area. Chemical weathering leads to formation of clay minerals and insoluble oxides, which are relatively high in Al and Fe. Changes in the runoff rate will have elevated the amount of very fine particulate matter in the rivers, and the observed levels of iron and aluminum probably reflect this. The Asana river (sites 1B and 1E) also shows relatively higher concentrations of these elements than equivalent sites on the Torata and Sajena rivers. Figure 4 shows the spatial distribution of Al, Fe, and Mn along the Torata and Moquegua rivers for the three seasons. Interestingly, although Al and Mn levels were not detectable and Fe was only found in moderate concentrations (0.18–0.24 mg/L) above the Cuajone mine (sites 16–18), all three metals were found in higher concentrations below the mine (site 5: Al, Fe, and Mn at 2.0, 1.5 and 0.33 mg/L, respectively). USEPA guidelines stipulate that recoverable Al exceeding 0.087 mg/L (pH 6.5–9.0) and Fe exceeding 1 mg/L is unsuitable for aquatic life, and the secondary drinking water standards specify that Al, Fe, and Mn exceeding, 0.2, 0.3 and 0.05 mg/L, respectively, is undesirable in drinking water. Mn concentrations exceed the safe limits for irrigation use (0.2 mg/L). Although these metals are associated with both AMD and neutral mine drainage (NMD) [23,24,56–58], the concentrations detected were lower than those normally found under AMD or NMD conditions [59–61]. Under alternating flooding and draining conditions, such as those produced during an El Niño event and following dry seasons, changes in the pH of the water will greatly affect Al, Mn, and Fe solubility via influencing either reductive dissolution or carbonate formation [62]. Therefore, it seems that seasonal variability controls the concentrations recorded in the data and suggests that these metal concentrations probably derive from surface runoff, soil erosion and throughflow during the recent El Niño event [63–66]. Observed concentrations were thus considered to be the result of differential weathering of the underlying bedrock and trace metal mobilization during the episodic rainfall during the recent El Niño event rather than mine leachate.

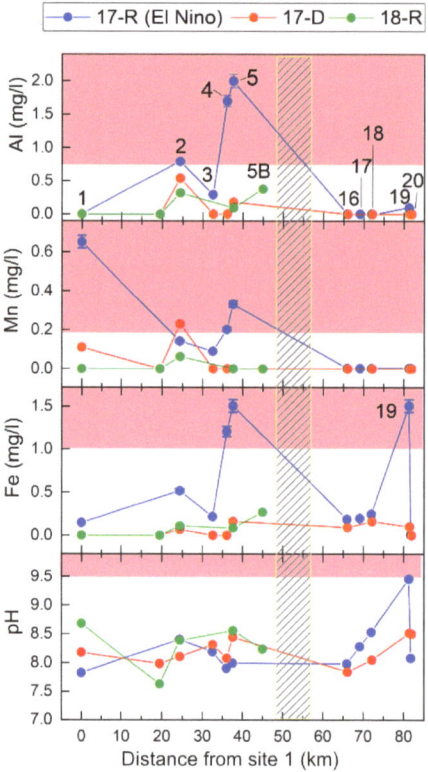

Figure 4. Al, Fe, and Mn concentrations and pH, showing elevated wet season metal concentrations in the Torata river below the Cuajone mine. Distance is measured upstream from site 1. Red shaded regions show the concentration levels above the safe limits for aquatic life. The Cuajone mine is located between site 5B and site 16 (vertical shaded region). Error bars represent standard deviation of measurement fluctuation. Labels represent site numbers.

The metal concentrations decrease downstream from site 5, and increase again at site 2 irrespective of the season (Figure 4). Such an increase in metal concentration could reflect evaporation as the river passes through the desert landscape; however, the concentrations of conservative ions such as Cl^- did not show a similar spike, pointing to other origins. Cation enrichment from atmospheric aerosols is a possibility. Regional high pressure and sinking air in the foothills concentrates atmospheric aerosols, which then settle out in the lower sections of the Moquegua, Sajena, and Torata rivers, possibly affecting the water chemistry. However, data from other studies [5] show that the river chemistry here is more significantly determined by contaminated groundwater. The aquifer is recharged by the filtration of surface waters through the porous alluvial deposits of the Moquegua valley floor. The chemistry of these surface waters south of the city provide numerous pollutants from agriculture and the city itself. Therefore, the elevated cations (and bicarbonate) in the river at this altitude are most likely due to recycled contaminated surface waters through the porous alluvial deposits.

The nitrate (NO_3^-) concentration in water can be used to assess chemical addition to the river from local agriculture. We observed raised NO_3^- concentrations, in the wet seasons 17-R and 18-R (0–16.64 mg/L and 0–7.9 mg/L), compared to the dry season 17-D (0–3.4 mg/L). The highest NO_3^- concentrations were recorded at site 3, located in the agricultural area downstream of Torata. The concentration drops at site 2 after the river has passed through 8 km of desert canyon. Interestingly, the NO_3^- concentration at site 1C in the farmland around Moquegua is similar to that at site 2 in both 17-D and 17R, possibly deriving from the greater emphasis on arable versus pastoral farming in this region. NO_3^- concentrations are controlled by increased surface runoff and leaching during the wet season, and relative concentration is linked to land use, being elevated in areas used for arable farming.

3.1.3. Sites Downstream of Moquegua City (Sites 1 and 0A)

The lower stretches of the Moquegua/Osmore river below Moquegua city record slightly alkaline (site 1, pH 7.81–8.68) conditions during all seasons. Water composition at site 1 (20 km outside Moquegua and below an extensive area of more intensively farmed land) was dominated by Ca^{2+} (180–250 mg/L), Na^+ (160–230 mg/L), SO_4^{2-} (370–470 mg/L) and Cl^- (260–360 mg/L). The concentrations of Na^+, SO_4^{2-} and Cl^- were above or close to unsafe levels for drinking water (EPA and WHO standards, 200 mg/L, 250 mg/L respectively) for all seasons. The tributaries (sites 1B, 1C, 1D, 1E) feeding into the river above the confluence have lower concentrations and slightly more alkaline pH values (e.g., site 1B—9.09, site 1E—8.10). At sites 1C (Torata river) and 1D (Sajena river), above the confluence with the Sajena and Torata rivers, metal concentrations were below EPA and WHO safe limits set for drinking water, although relative concentrations are seasonally affected, and Moquegua sites 1B and 1E (Asana river) recorded lower cation and anion concentrations in all seasons. Mn concentrations at site 1, during both wet and dry seasons (0.65 mg/L and 0.11 mg/L, respectively), are also higher than those recorded from the tributaries (see Figure 4)—site 1B (0.064 mg/L), site 1D (<0.05 mg/L), and site 1C (<0.05 mg/L)—with the exception of data from 18-R, when Mn concentration was 0.12 mg/L at site 1D, showing a significantly elevated contribution of this metal from the Sajena river. These results suggest that water from the Torata, Sajena and Asana/Moquegua tributaries does not significantly contribute to the elevated cation and anion concentrations at sites 1 and 0A, and the data are consistent with an extraneous source for the elevated chemical levels measured at site 1.

Cl^- concentration was above that considered safe for aquatic life over long periods of time during all seasons (chronic, >230 mg/L) and for the irrigation of sensitive crops such as avocados [67], a common crop in the Moquegua region. The highest K^+ concentration (8.7–11.0 mg/L) for all study sites was also found at site 1. As a result of these high concentrations of cations and anions, specific electrical conductivity (EC, 2040–2711 µS/cm) and total dissolved solids (TDS, 1020–1833 ppm) at site 1 were close to or above the maximum limits set by the Peruvian water authority for agricultural use (category 3: 2500 µS/cm and 1800 ppm, respectively) [68]. EC and TDS values exceeded the safe limits for agriculture use in the dry season 17-D.

In general, cation and anion concentrations will gradually increase downstream along river systems due to accumulation from natural mineral weathering, evaporation and ground water discharge. However, the concentrations at site 1 were 2–3 times higher than those recorded at sites 1C and 2 in all seasons (Figure 2: line graphs), which suggests an additional process causing increased concentration. It is unlikely that evaporation alone (see Figure 5) could account for this dramatic increase, so it is likely to be a result of changes in natural mineral weathering or the composition of groundwater discharge. Softer, more friable sedimentary rocks of the Moquegua Group, which are more susceptible to physical and therefore chemical erosion, underlie the land between sites 1 and 2, in contrast to the more competent volcanic extrusive units found above Moquegua. Site 1 also lies downstream of the agricultural land below Moquegua. It is not possible to separate the relative influence of these two factors with the current data, but given the low precipitation rates (and therefore reduced weathering potential), such a dramatic increase in concentration probably reflects the influence of urban and agricultural activities in the lower valley (c.f. [69]).

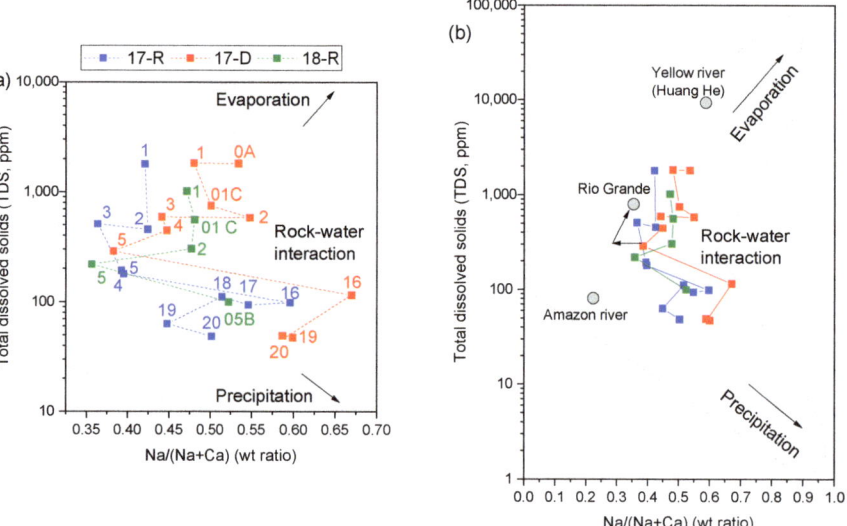

Figure 5. (**a**) Gibbs plot for the Moquegua river showing changes in water chemistry along the river and (**b**) compared with other major rivers, Hauang He river, China and Amazon river, rio Grande, AB, USA, and Amazon river, South America. Data from [70,71].

SO_4^{2-}, Cl^- and K^+ were among the main contaminants at site 1, and they are commonly found in agricultural fertilizers, in the form of potassium sulfate (K_2SO_4, 50% K and 17% S), potassium chloride (KCl, 60% K), and magnesium potassium sulfate (MgK(SO_4), 22% S) [72]. These anions may also derive from urban waste [73].

Site 1 also showed the highest NO_3^- concentrations in the study site network (19.65 mg/L 17-R, 17.7 mg/L 18-R, 6.1 mg/L 17-D). NO_3^- concentrations from the tributaries were much lower throughout the sampling campaign (site 1D, 2.2–2.4 mg/L; 1B and 1E 0-0.96–0-1.5 mg/L). NO_3^- is highly soluble and commonly associated with the use of agricultural fertilizers and domestic waste [16]. To analyze the effect of the latter, samples from site 1 were further analyzed for *E. coli* and thermal coliform, which are good indicators of sewage contamination. Both *E. coli* and thermal coliform were found to be below detection levels (<1.8 MPN) indicating little or no domestic sewage enters the river above site 1. Commonly used nitrogen-based fertilizers in the Moquegua district [72] are urea (CO(NH_2)$_2$, which has the highest nitrogen content (46% N), ammonium nitrate (NH_4NO_3, 33% N), and ammonium sulfate ((NH_4)$_2SO_4$, 21% N). Another nitrogen-phosphoric fertilizer that is commonly used is monoammonium phosphate ($NH_4H_2PO_4$, 10% N and 48% P), also a source of highly soluble phosphate [72]. Similar increases in total nitrogen (TN) and total phosphate (TP) were observed at site 1 ($r^2 = 0.99$, $p < 0.05$), where aquatic plant growth was prolific. Correspondingly, an increase in Mn concentration at this site could also arise from agriculture since Mn is less readily complexed by organic ligands, resulting in high mobility in agriculture areas [74].

These data and the discussion above are consistent with urban effluent and agricultural runoff and leaching, especially during the wet season, being a major source of raised cationic and anionic concentrations at sites downstream of both the city of Moquegua and the associated downstream industrial agriculture area (c.f. [69]). However, Figure 2 shows that there are no significant correlations between the trace metals and cation–anion concentrations, suggesting that the actual system is complex with multivariate determination.

3.1.4. Endmember Controls on Water Chemistry in the Moquegua River Catchment

Water chemistry in rivers is controlled by precipitation, absorption, evaporation, crystallization and weathering. The Gibbs plot of TDS against the weight ratio of $Na^+/(Na^+ + Ca^{2+})$ is a useful tool to identify these mechanisms in a river system [71,75,76]. The Gibbs plot for the Torata river sites is shown in Figure 5a. This shows that, irrespective of season, site 16 was the most Na-abundant (0.59–0.67 Na/Ca ratio), and site 5 was the most Ca-abundant (0.35–0.39 Na/Ca ratio), while the TDS remained between 98 and 115 mg/L at site 16 and 179–288 mg/L at site 5. This indicates a major change in water chemistry determinants between site 16 and site 5. The Gibbs plot confirms the dominance of precipitation and silicate weathering at headwater sites on water chemistry, and the change from Na abundance to Ca abundance reflects the change in the underlying geology between sites 16 and 5. TDS naturally increases downstream, and site 1 records concentration levels higher, by a factor of 1000, than the rest of the river system in all seasons. Evaporation becomes increasingly important in the lower parts of the river system, but the data indicate that both agricultural leachates and evaporation play significant roles in the chemistry of the lower catchment.

3.2. Trace Metals in the Moquegua River System

Total concentration of trace metals (Cd, Cu, Co, Cr, Pb, Li, Mo, Ni, Se, U, V, Zn) along the Torata and Moquegua rivers for each season are shown in Figure 2 (bar charts). Samples were analyzed for trace metals and other compounds associated with AMD and the natural geochemistry of the region (Table 2). Trace metals can have significant health impacts and are therefore of particular interest. Water analysis showed that Hg, Bi, Th, and Te concentrations in all water samples were below detection limits (<0.06, <2, <4 and <1 µg/L, respectively). The study also confirmed that cyanide (CN) remains below detection limits for all study sites (<20 µg/L). In all water samples, Be, Cd, Co, Cr, Cu, Li, Pb, Mo, Ni, Se, U, V and Zn concentrations were well below safe limits for drinking and irrigation purposes, established by USEPA, WHO, FAO and the Peruvian water authority [55,68,77,78]. Generally, the concentrations of trace metals increase downstream. In our study catchment Be, Co, Pb, Se, U and Zn were only detectable below the Cuajone mine.

For both wet and dry seasons, a notable increase in total trace metal (Cd, Cu, Co, Cr, Pb, Li, Mo, Ni, Se, U, V, and Zn) concentration was recorded below the Cuajone mine, with a 5-fold increase in the wet season and a 3.5-fold increase in the dry season measured between site 16 and site 5. The highest total metal concentration at site 5 was recorded after the El Niño episode (96 µg/L, 17-R). At site 5, significant levels of Li were recorded in 17-D and 18-R (33, 20 µg/L), while Cu was dominant in 17-R (53 µg/L: maximum recorded level in this study at any site, but still within safe limits). Pb was detected only at site 5 and site 4 in 17-R (2.9–3.9 µg/L). These trace metals occur naturally in river waters arising from typical Andean geological units, but their concentration is generally increased by AMD. However, apart from a slight peak at sites 4 and 5 after the El Niño episode in 17-R, trace metal concentrations are not dramatically elevated above site 2 but follow a natural concentration profile concurrent with the cations and anions.

Spatially, total trace metal concentration peaked below Moquegua at site 1 during 17-R and at site-2 during 17-D. Trace metals are used in a wide variety of agricultural chemicals and domestic goods and, as such, these peaks may derive from increased runoff and domestic waste in the lower valley, or increased weathering processes and the dominance of evaporation in the hyper-arid climes of the tropics. It is also true that trace metals can be easily transported by Fe, Mn and humic substances [64,79], which were found abundantly throughout the sample-site network. The trace metals data back up the conclusions from the major cations and anions above—that the dominant endmember in the lower catchment is agricultural and urban runoff.

It should be noted that the hardness of water, determined by Ca^{2+} and Mg^{2+} ion concentrations, has a major influence on the toxicity of trace metals [80,81]. Water samples from the Torata headwaters (sites 16–20) showed limited hardness (17–44 mg/L $CaCO_3$), while at lower elevations (sites 5 to 1C), water was permanently hard (75–378 mg/L $CaCO_3$). Irrespective of the season, below Moquegua, water became extremely hard (548–764 mg/L $CaCO_3$). Soft water (<17 mg/L $CaCO_3$) is generally associated with increased trace metal ecotoxicity [80–82]. High levels of water hardness in the region almost certainly reduce the bioavailability, and therefore ecotoxicity, of trace metals in the Moquegua region. The elevated hardness could be an artifact of hyper aridity.

3.3. Arsenic in the Torata and Lower Moquegua River System

Arsenic (As) concentrations are shown in Figure 6. The As levels along the Torata river were frequently above the recommended levels for human consumption and irrigated agriculture (10 µg As/L, WHO and USEPA standards). The highest As concentrations were recorded at upper Torata valley sites 17 and 18 (14–21 µg As/L), where aresenic containing sulfidic minerals deposited by epithermal and hydrothermal processes associated with the active arc are more prevalent. Site 1D along the Sajena river also recorded elevated As in 17-D (11 µg As/L). By contrast, the Asana river sites 1B and 1E showed consistently low concentrations of As (2.1–4.3 µg As/L) for both 17-D and 17-R seasons. The highest As concentration observed in the Moquegua river, below the confluence of the Sajena, Asana and Torata rivers, was at site 1 during 17-R (10 µg As/L), which followed extensive flooding, owing to the 2017 El Niño episode.

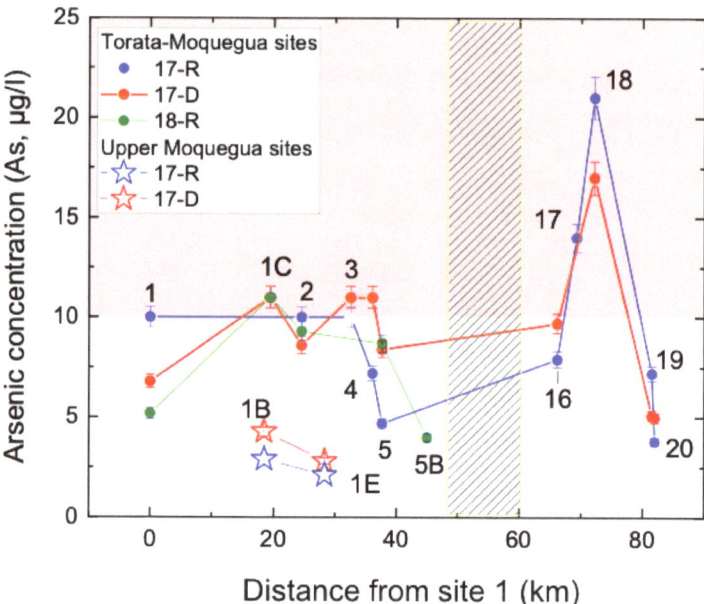

Figure 6. Arsenic concentrations in samples from Torata and Moquegua river site. Error bars show the standard deviations of measurement. Red-shaded regions show the concentration levels above the safe limits for potable use. The Cuajone mine is located between site 5B and site 16 (vertical shaded region).

Throughout the Andes of Southern Peru, river catchments are known to contain geogenic As [83]. The high Andean sites in the study area (16–20) are underlain by relatively young volcanoclastic deposits that are undergoing low temperature hydrothermal alteration and localized sulfidization. Previous studies have reported that As is predominantly mobilized from such rocks and their weathering products [83]. Therefore, the elevated As levels found in the upper Torata sites are most likely to be geogenic in origin.

The dominant minerals found at the Cuajone Copper Mine are sulfides, chiefly, chalcopyrite ($CuFeS_2$), chalcosine (Cu_2S), molybdenite (MoS_2), and pyrite (FeS_2), as well as arsenic-bearing minerals, including enargite (Cu_3AsS_4) and tennantite ($Cu_{12}As_4S_{13}$) [84]. These arsenic-bearing minerals, frequently associated with copper–porphyry deposits and generally marginal to the main orebody, develop during sulfide mineralization [84]. Mineralization is very localized, concentrated within a 3 km alternation halo, and does not extend significantly beyond this. It has been shown that the weathering of porphyry–Cu deposits and associated As-bearing minerals results in the dispersion of As into the groundwater and surface water bodies, depending on the redox state and pH of the system [85]. Site 5B sits just below the mine and does not show elevated levels of arsenic. However, sites further downstream show variably elevated arsenic levels depending on the season, consistent with remobilization via groundwater and periods of increased runoff and recharge through the shallow aquifer.

It is therefore likely that the pattern of arsenic concentrations observed, Figure 6, reflects both local geochemistry and groundwater dispersion.

3.4. Multivariate Analysis of Water Parameters

Multivariate analysis was used to give an unbiased assessment of statistical correlations in our data. Correlations between water quality parameters and correlations between sites were analyzed using both CA and PCA. Our analysis took into account the 33 water quality parameters listed in Table 2. Parameters NO_2, CN, Be, Bi, Te, Th, Hg, Pb were not included because their concentrations were below measurement accuracy. Parameters CO_3 and DO were omitted because the datasets were incomplete.

Figure 7a presents a dendrogram of an R-mode CA for all sites visited in 17-R, 17-D and 18-R. It shows how the parameters cluster according to sites. It was produced with a CCC of 82%. Parameters that have a similar distribution across the sites cluster. At the highest level, we see three clusters. The right cluster contains major ions and trace metals. These parameters arise from the natural water chemistry and dynamics in combination with the local geology. The left cluster contains nitrates, phosphates, trace metals, pH, and COD. These parameters are indicative of agricultural or biological impact. The central cluster contains Al, As, Fe, Si, Mn, and trace metals associated with the weathering of geological units with economic potential and/or urban effluent [86,87]. Peaks in these elements are also associated with periods of increased turbulence and greater suspended load (see above).

Figure 7b,c present a dendrogram from a Q-mode CA and corresponding Beck map (similar to a subway map with sites as stops) for the same 33 water quality parameters. They show how the sites cluster according to parameters. It was produced with a CCC of 84%. Sites with similar patterns of parameters cluster. At the highest level, we see three clusters; the left cluster contains the two lowland sites downstream of Moquegua city, 0A and 1, the central cluster contains foothill sites, and the right cluster mostly contains the highland sites, reinforcing the conclusion that the chemistry is controlled by geographical location and regional inputs.

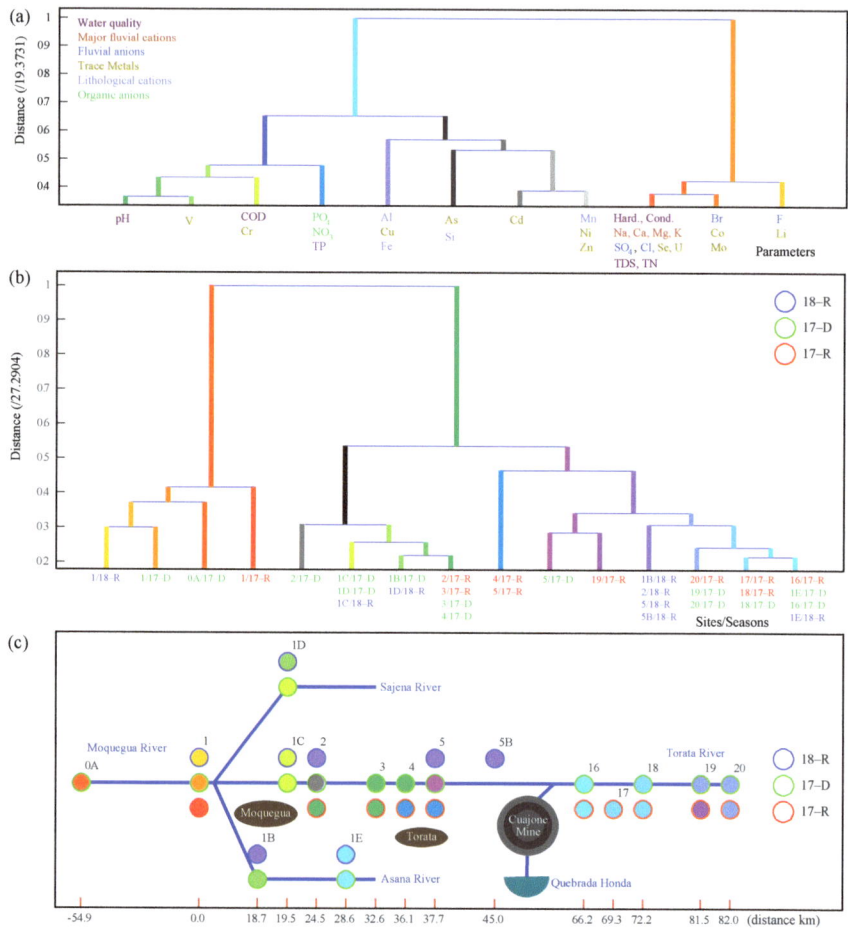

Figure 7. (**a**) Dendrogram showing the Euclidean dissimilarity between the 33 measured water parameters taking 17 Torata–Moquegua sample sites and seasons 17-R, 17-D, and 18-R into account. (**b**) Dendrogram showing the largest dissimilarity between the 17 Torata–Moquegua sample sites in different seasons (identified by the code [SITE]/[YEAR]–[SEASON]), taking 33 measured water parameters into account. (**c**) Beck map showing the spatial distribution of site clusters. The color map is the same as in (**b**). The color of the outer ring of each circle identifies the season using the key on the right.

PCA analysis carried out on the same data shows that 74% of the variance can be accounted for by the first four components. The first component PC1 accounts for 46% of the variance, PC2 accounts for 12%, PC3 accounts for 9%, and PC4 for 7%. Results are shown in Figure 8a—biplot of PC1 against PC2, and Figure 8b—PC1 against PC3. The parameter vectors in the biplots are colored to indicate which parameter cluster they belong to in Figure 7a. As can be seen in Figure 8, the right cluster (red, orange; those controlled by natural hydrological processes and underlying geology) aligns well with PC1, the central cluster (light blue, grey; those controlled by the suspended load and increased runoff) aligns with PC2, and the left cluster (green, blue; those controlled principally by anthropogenic activity) aligns with PC3. As and Si align with PC4. Colored circles are used in the biplot to indicate the site clusters from Figure 7b,c. The color of the outer rings indicates the season.

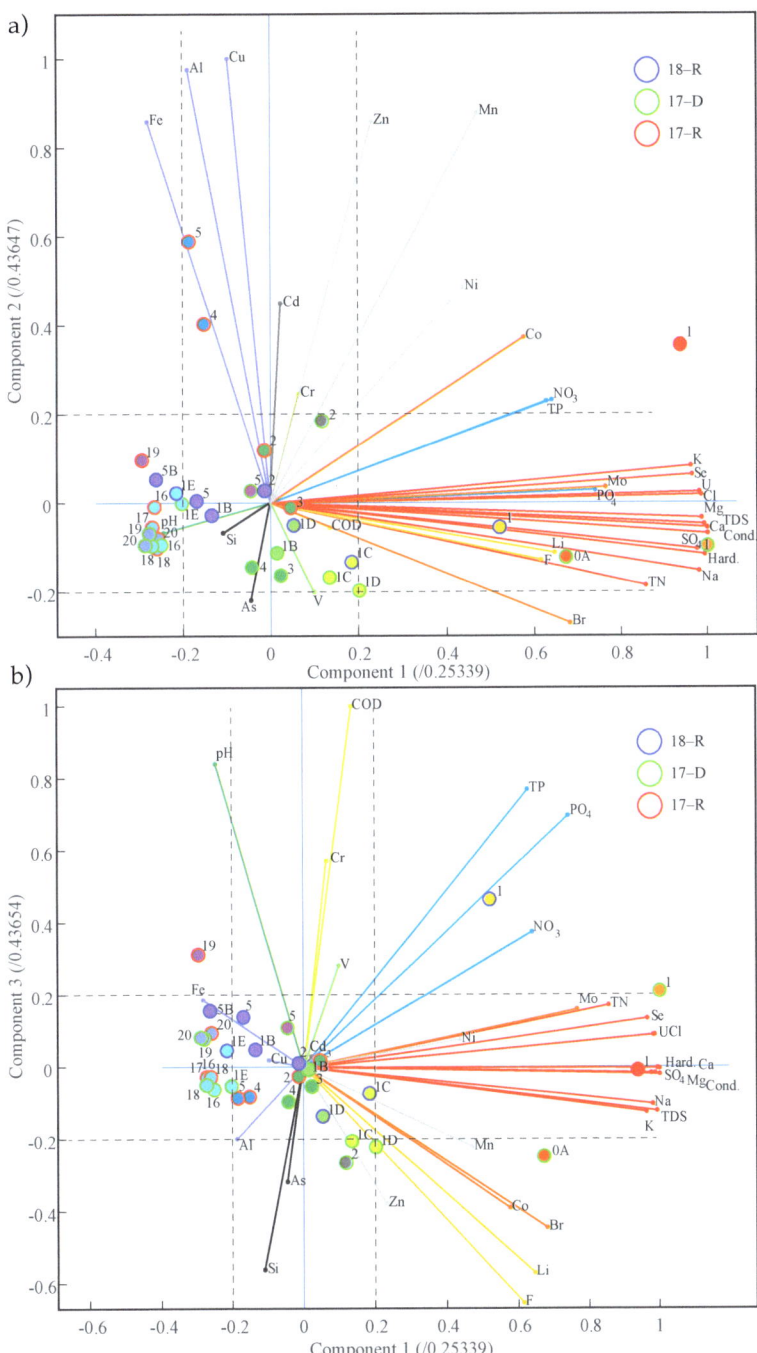

Figure 8. PCA biplots using 33 water parameters measured at 17 Torata–Moquegua sample sites in the three seasons 17-R, 17-D and 18-R. The colors of the parameter vectors are taken from Figure 7a. The colors for site points are taken from Figure 7b,c. (**a**) PC1 vs. PC2, (**b**) PC1 vs. PC3. Dashed lines at ±0.2 indicate the qualitative boundary used to indicate where components become significant at a site.

3.4.1. Headwater Sites

In general, the headwater sites 16–20 cluster with site 5 (and site 4 in 17-R) below the Cuajone mine. This precludes chemical addition to the river through mining activity. PC1 and PC2 are low for the headwater sites (in contrast to sites 4 and 5, see below), suggesting that the main controls on river chemistry up to this point are hydrological. Contribution from PC3 (chemicals associated with organics) is also low except for site 19, the bofedal, and this is interpreted as the naturally high levels of biogeochemical components in an Andean peatland. Data from the Asana and Sajena rivers were collected during 17-D, and site 1E on the Asana river also clusters with headwater sites, suggesting that this river has similar chemistry and dynamics to the Torata river to this point. In 18-R, headwater sites were inaccessible.

3.4.2. Foothill Sites

As mentioned above, sites 4 and 5 (5B) (between the mine and Torata) show strong alliance with the headwater sites but share some affinity with sites 2 and 3, below Torata All four sites showed an average contribution from PC1 as expected from their location in the middle of the river system. During 17-R, sites (4, 5) showed some contribution from PC2, whereas the headwater sites and sites (2, 3) showed average values. Elevated contribution from PC2 is consistent with increased runoff, creating turbulent flow following El Niño. All foothill sites showed an average value for PC3. Samples collected from the confluence of the Torata with the Osana and Sajena rivers (sites 1B, 1C, and 1D) during 17-D cluster with the other foothill sites (2, 3, 4). These sites showed average values for PC1, PC2 and PC3, consistent with the season being dry and the hydrographic location of these sites. The specific chemistry of the tributary sites in this area is seasonally regulated; sometimes, sites share characteristics with the foothill sites and other times share characteristics with headwater sites. This is consistent with variable runoff rates in different seasons changing the exact chemical elements added to the river system at various sites. In general though, hydrographic processes determine the chemistry of the foothill sites (see Figure 5), and the multivariate analysis for these sites shows that the Cuajone mine had little or no effect on the river system over the observation period. The fact that this period included an El Niño episode shows that the measures to isolate the mine from the Torata river are effective.

3.4.3. Lowland Sites

The lowland sites (sites 1, 0) formed an isolated cluster irrespective of season. PC1 was high in all seasons, consistent with these sites being towards the lower reaches of a river running through a hyper-arid zone with high levels of evaporation. In addition, elemental concentrations were elevated by the accumulation of dissolved ions through the chemical weathering of all three geological zones, from several agricultural regions and the two main urban centers, Torata and Moquegua. In 17-R, PC2 was high, indicating elevated wash caused by the El Niño episode. In 17-D and 18-R, PC3 was high, consistent with the agricultural activities along the river at that time.

3.5. Analysis of ANA and INGEMMET Data

In Figures 9 and 10, we present a collation of data for the Torata river system, including our surface water data, available survey data collected by ANA, the National Water Authority in Peru, from 2013 and 2014 [88–90] and groundwater analyzed by INGEMMET around the same time [5]. Figure 9 shows the downstream concentration of a selection of major ions (calcium, sodium, sulfate) plotted against altitude. The data show an increase in concentration from source to sink, consistent with the conservative behavior of the ions controlled predominantly by the cumulative addition of elements along the water course through weathering, and further concentration through evaporation. Points plotting to the right of the main cluster at just below 1000 m asl are surface waters affected by contaminated groundwater (c.f. conclusions drawn by INGEMET [5]). The Cuajone mine is marked by a grey box with diagonal lines. If the mine had a significant impact on the surface waters, we would expect to see a similar deviation of the surface waters to the right in the vicinity of the mine. In fact, the two grey data points at 1100 m asl are samples of runoff analyzed by ANA from the base of the tailing fan in 2013 and 2014. These data clearly show elevated concentrations of major ions. However, these elevated concentrations are not spatially distributed, and data gathered only a few hundred meters from the tailing fan show normal concentrations. The ions are quickly scavenged from the surface water through sorption and precipitation and do not travel far from the source.

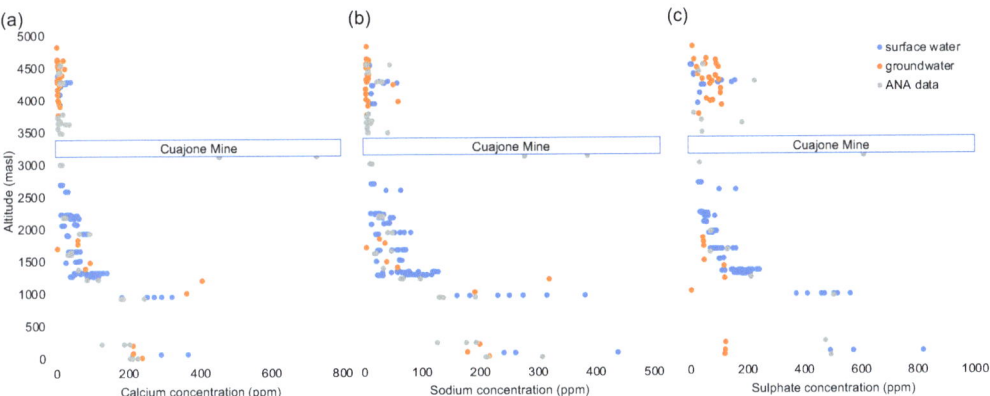

Figure 9. Combined data sources from field work, ANA data and INGEMMET groundwater reports [5,88–90]. Concentration of major ions (**a**) calcium, (**b**) sodium, and (**c**) sulphate in the Torata river system.

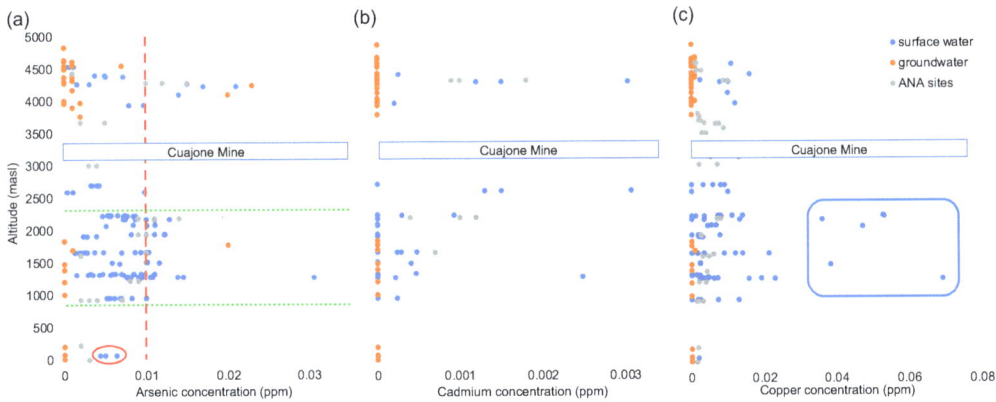

Figure 10. Combined data sources from field work, ANA data and INGEMMET groundwater report [5,88–90]. Concentration of trace elements in the Torata river system. The red dashed line in panel (**a**) indicates the safe limit for arsenic. The safety limits for cadmium and copper concentrations do not appear in panels (**b**,**c**), as they are above every data point. The green dotted lines in panel (**a**) delimit the area of intensive agricultural land ranging from above Torata to below Moquegua. The red circle in panel (**a**) identifies data taken at Site O. The blue box in panel (**c**) outlines data that were taken after a flash flood event.

Figure 10 shows the concentration of arsenic, cadmium and copper plotted against altitude. The concentration of these three trace elements in the water is a major concern for the people who live in the area. Data are not controlled by conservative processes as in the major ions but are more dispersed. Of note are the digressions to higher concentrations around 4250 m asl in all three elements. The higher altitude headwaters are underlain by young volcanic arc units, with very high permeability. There is extensive epithermal and hydrothermal activity associated with the active arc. This allows mobilization of these cations via natural acidic chemical weathering of the mantle derived volcanic material, and rapid recharge into the surface waters as a result of the highly porous nature of the lithology (c.f. findings by INGEMMET [5]). The riverine geochemistry at these high altitudes in Southern Peru and Northern Chile is composed of a mix of precipitation, surface water runoff and associated chemical weathering, and additional input from chemically enriched groundwaters via thermal springs. Typically, these waters increase not only the concentration of trace elements such as arsenic and boron but also the major anions such as chlorine and sulfate [91,92]. The Cuajone mine is once again represented by a grey box with dashed lines. There is no significant increase in concentration of any of the three trace metals at the altitude of the mine. In contrast, the green lines just below 2500 masl and 1000 m asl represent the start and end of the irrigated land between Torata and Moquegua. The data in all three plots are very dispersed in this region. Surface waters contaminated by the presence of the urban settlements and agricultural practices are rapidly mixed into the ground water through the porous alluvial valley fill at this altitude, and recharge leads to increased concentrations in surface waters downstream. The concentrations of cadmium and copper are below the guidelines for safe water throughout the river network, both for drinking and use in agriculture. The levels of arsenic are generally below the limits quoted by USEPA guidelines [77] except a few sites in the headwaters, where aggressive chemical weathering takes place, and also at times of elevated flow following heavy rains and/or flash floods in the agricultural plains.

4. Discussion

Hyper-arid regions rely largely on river systems as the principal source of water. Changes in both river water availability and chemistry therefore have a direct and immediate impact on dependent systems in such regions. This is particularly true in the Andean transect in this study, where the Moquegua river system passes through the hyper-arid Atacama Desert and is used for human consumption and agricultural and industrial use throughout its length.

The location of the Cuajone mine above Torata required that both channel and waste material diversion mitigation strategies are essential to avoid impacting water quality in heavily settled areas beneath the mine. To this end, a diversionary channel was constructed around the projected mine pit at the Cuajone mine. Additionally, the construction of tunnels and concrete-lined channels enables waste materials to be diverted to the tailings dams at Quebrada Honda and Cortaderas 2 (see Appendix A). The expansion of urban areas and the extension and modernization of agriculture in the region has necessitated the development of a regional irrigation project. The Pasto Grande catchment transfer project diverts water from the Tambo catchment, to the north, to sub-catchment tributaries in the greater Moquegua drainage basin. The catchment transfer project has significantly increased channel flow in the Sajena, Otora, Torata and Moquegua rivers and will undoubtedly add physico-chemical characteristics of the upper Tambo river to the Moquegua–Torata river system. Our data indicate a clear urban and agricultural signal in the lower parts of the Torata and Moquegua river sample sites, while data from Pasto Grande sample sites have not been analyzed to identify a clear signal from that source in the lower Moquegua drainage basin.

Water availability in the Moquegua river system arises from seasonal rains and geothermal sources. In the month prior to the 17-R investigation, an El Niño episode had led to catastrophic flooding and a temporary superabundance of turbid river flow. River courses were altered, farmland destroyed, and bridges and roads damaged. However, climatic conditions during the study period were generally dry, and the rivers had already returned to dry season flow rates during the sampling campaign. The same was observed for 17-D and 18-R. The cycles of flooding and drought in this region, which disrupt water availability and complicate water resource allocation, are irregular and therefore difficult to predict.

Irregular seasonality affecting water chemistry, through the erosion and concentration of trace metal containing silts, further complicates water resource allocation and treatment.

Elevated concentrations of Al in drinking water is often considered to increase the risk of Alzheimer's disease, with prolonged exposure leading to systematic toxicity and renal failure; increased Fe and Mn is associated with aesthetic problems such as the staining of laundry, and an unpleasant smell and taste [93,94]; and elevated As concentrations in drinking water is associated with skin, liver and kidney cancers [95,96]. Our data show that harmful hydro-chemistry, derived from elevated levels of Al, Fe, Mn, and As, is found to be highly variable across the sample site network and period of work. As such, the associated social and economic cost, while difficult to determine, will most likely be attributed to bio-accumulation over longer periods of time.

The study indicates multi-factoral determination of river chemistry at many sites, including, more significantly, seasonality. This has significant implications for the direct management of water resources in hyper-arid regions as well as pre-development stakeholder engagement to ensure effective social and environmental impact management [12]. These areas are particularly prone to episodic and localized precipitation patterns, as well as distinct seasonality, which may suffer increased climatic volatility in the future [97,98]. The data show increased mobilization of trace metals into the river system following episodes of more intense rainfall. These have a relatively short residence time and are scavenged from the water by sorption and precipitation, within a matter of months (17-R c.f. 17-D). However, the management of the purity of the water in the lower catchment requires consideration of agricultural practices and supervision of the disposal of anthropogenic waste, as these contaminated superficial waters are rapidly recycled into the river through the porous aquifer of the alluvial plain [5].

This study finds three distinct geochemical and anthropogenic determinants to the water chemistry of the Torata River system: natural hydrological factors, the addition of elements through channel flow volatility, and biogeochemical parameters associated with elevated organics, both natural and anthropogenic. Sites (16, 17, 18, 19, 20) fall within the arid temperate zone from 2800 to 4600 m asl and are all above the Cuajone mine. Water quality here is determined by geological conditions and weathering processes. Cation, anion, and trace metal levels were well below the safe limits for human consumption, with the exception of arsenic concentrations, which exceeded safe limits for human consumption at sites 17 and 18. This can be attributed to geochemical weathering processes, but the water chemistry also suggests that the river is fed by Na^+- and SO_4^{2-}-rich ground waters, which can also be a source of As contamination [83]. Sites 3, 4, 5, and 5B fall within the arid subtropics from 1400 to 2800 m asl below the Cuajone Mine, town of Torata, and its associated agricultural areas. These sites showed moderate water quality with significant seasonal variation. While agricultural leachates are still detectable, naturally weathered contaminant loading is more prevalent. Increased surface runoff following El Niño, led to elevated Al, Fe and Mn in the water, where concentrations were recorded above recommended limits for domestic consumption and agricultural use. An increase in total trace metals Li, Cu, and Pb concentrations was also found below the Cuajone mine, but no indication of AMD was seen. Cation–anion composition at these sites showed water chemistry undergoing a transition between the Cuajone mine and Torata village, in which the water composition changes from Na-dominant to Ca-dominant. This reflects a change in dominance from precipitative to evaporative processes in determining the hydrochemistry of the river.

Sites 0A, 1, 1B, 1C, 1D, 1E, and 2 fall within a hot, hyper-arid desert extending from sea level to 1400 m asl. Sites 1B, 1C, 1D, 1E, and 2 show more chemical affinity with the sites in the arid subtropical zone, with water chemistry controlled though natural hydrological processes and a minor concentration of some elements during periods of high runoff. Sites-1 and 0A, below Moquegua, exhibited the worst water quality, irrespective of season. The data presented here are consistent with elevated levels of these chemicals from urban waste and agricultural leachates rather than natural weathering or hydrological processes. Significant levels of cations (Ca^{2+}, Na^+, K^+) and anions (SO_4^{2-}, Cl^-) as well as high EC and TDS levels were recorded. Intermittently, trace metals and metalloids (Mn, As) exceeded the guideline values. The water quality in this section of the catchment did not meet strict criteria for safe domestic consumption and, on occasion, agricultural use.

The preceding discussion of chemical analyses and cluster analysis (CA) of site data show that the impact of the Cuajone mine is less significant than agricultural and urban activities in the contamination of this river system. Owing to the hyper-aridity in the region, the low weathering regime reduces the mobilization of soil minerals, and it is difficult to conclude that ARD occurs to any significant extent in this catchment, or that the surface water quality has been greatly affected. Direct contamination as a result of AMD from the mine itself is reduced by the bypass of the Torata river around the Cuajone mine through a system of canals and diversions, and an upstream dam that largely prevents flooding from affecting the lower catchment. However, although AMD from open pit mining may be controlled during operation periods, it can cause long-term environmental problems after closure [23], and it is therefore important to develop a strategy to maintain good control over the mine drainage and tailing disposal to protect this drainage basin long after mine closure.

The data show that the impact of agriculture and urban activities is more significant than mining in determining cation–anion concentrations in the hyper-arid river system in this study. These rivers, in combination with the local climate and alluvial soils, provide ideal conditions for the development of highly productive small- and medium-scale agriculture, which has been practiced for centuries [78]. The modernization of agriculture has led to the rapid increase in the use of pesticides and fertilizers, which in sensitive environments will have consequences for the quality of water resources and their value for domestic consumption and agricultural irrigation. Additionally, the use of river water and river channels to dispose of urban waste and industrial effluents is clearly deleterious to water quality and environmental integrity. The effective regulation of agrochemical use and tight control of urban and industrial effluent and sewage disposal in Peru will be key for the long-term environmental stability and management of a very scarce resource and, therefore, the long-term sustainability of development.

The desert micro-habitats in the region, although depauperate, have a high level of biological endemicity [99]. Hyper-arid environments are particularly sensitive to perturbation, and recuperation is slow. Thus, the impacts on biodiversity and ecological integrity from irregular water contamination are disproportionately high, making the monitoring of water quality and control of contamination all the more consequential in the region. The impact of elevated metal, metalloid, and anion concentrations in this river system on xeric ecologies is not yet known, and remains a research priority for the wider region. It is of note that the Peruvian National Water authority (ANA) monitors only 98 of the 159 river catchments identified in Peru. Only 60% of these monitored rivers meet nationally agreed environmental quality standards [68]. Given the scale of mining in the region and the volume of fresh water required for processing ores as described above, this study highlights the need for careful monitoring of water availability as well as quality, in a region of extreme water scarcity. As weathering processes under hyper-arid conditions are known to amplify the concentration of harmful geochemistry, monitoring temporal variability in water quality is also important. The effects of temporal variability and long-term volatility in channel flow on bioaccumulation and trophic transfer in the region's xeric ecosystems are as yet unquantified but are critical to understanding the full anthropogenic impact.

In summary, this investigation indicates that the geochemistry of the Torata river subcatchment is substantially determined by the underlying geological units and, in combination with weathering processes, characteristic of conditions found from the high Puna, at 4500 m asl, to the lower coastal deserts. There does appear to be an indication of seasonal variation, perhaps linked to the episodic El Niño phenomena, reflecting the dominance of geogenic factors.

Our data show that the Cuajone mine did not affect water quality parameters in the Torata river, nor significantly alter the geochemistry of the river during the study period. We do not have data prior to the onset of mining at the site, and so cannot constrain any quantitative changes in the geological contribution to the river chemistry as a result of the mine, but the data are consistent with chemical weathering being the dominant control on the fluvial geochemistry, both above and below the mine site during the period of investigation. It seems, at least in the time period concerned, that mine activity has had no adverse effect on water qualities necessary for aquatic life.

Data indicate a clear signal from the rural settlement of Torata and associated agricultural lands beneath the Cuajone mine. A clear geochemical signal is also discernible below the city of Moquegua in the Moquegua river, and also below the confluence of the Sajena and Torata with the Moquegua river. The data are consistent with contamination by urban and agro-chemical usage in the built-up and agricultural areas lying below the city of Moquegua. This is the result of the rapid recharge of contaminated surface waters from the urban and agricultural areas through the highly porous alluvial fill in the Moquegua valley. Additionally, significant increases in organic matter and sulfates were identified at El Conde, 18 km below the city.

5. Conclusions

This study presents time-limited findings of research undertaken from 2015 to 2020 in the Torata and greater Moquegua river systems in Southern Peru and offers an insight into the determinants of water quality and availability in hyper-arid catchments in the region and, in furtherance of our understanding of geochemical dynamics in hyper-arid river systems, represents a unique case study. Our findings indicate the need for further investigations into both mine impact mitigation strategies and water resource exploitation, for both resource management and environmental integrity. In Southern Peru, this would involve investigating the wider environmental implications associated with landscape-scale mining, including impact mitigation strategies employed prior to, during, and long after mining operations cease, as well as the direct impacts of the mine activity on groundwater flow and storage, as well as headwater ecology in this hyper-arid region.

Author Contributions: Conceptualization, D.H.N.P., E.P.G.B., L.d.l.S.V. and C.H.W.B.; methodology, M.H., D.H.N.P., E.P.G.B., E.E.-P., P.J.N. and C.H.W.B.; software, D.A.-S. and C.H.W.B.; validation, M.H., D.H.N.P., E.P.G.B., P.J.N., P.A.B. and C.H.W.B.; formal analysis, M.H., D.H.N.P., E.P.G.B., P.A.B. and C.H.W.B.; investigation, M.H., D.H.N.P., E.P.G.B. and E.E.-P.; resources, D.H.N.P., E.E.-P., D.A.-S. and C.H.W.B.; data curation, M.H., D.H.N.P., E.E.-P. and C.H.W.B.; writing—original draft preparation, M.H., D.H.N.P., E.P.G.B. and C.H.W.B.; writing—review and editing, M.H., E.P.G.B., H.V.L., D.H.N.P., P.J.N., P.A.B. and C.H.W.B.; visualization, M.H., D.H.N.P., E.P.G.B., H.V.L., N.I. and C.H.W.B.; supervision, D.H.N.P., E.P.G.B., H.V.L. and C.H.W.B.; project administration, E.P.G.B., E.E.-P. and C.H.W.B.; funding acquisition, E.P.G.B., L.d.l.S.V. and C.H.W.B. All authors have read and agreed to the published version of the manuscript.

Funding: This research is supported through a collaborative agreement between the National University of Moquegua (UNAM) and the University of Cambridge (grant RG85120).

Data Availability Statement: All data are provided in Excel files and can be downloaded from www.repository.cam.ac.uk (accessed on 20 January 2024).

Acknowledgments: The authors acknowledge the valuable support of A. Quispe and W. Zeballos at UNAM; Himantha Cooray for his assistance in developing the Mathematica program for data analysis; John Forrest for his contribution in arranging fieldwork; and Julia Porturas for her administrative support.

Conflicts of Interest: The authors declare no conflicts of interest.

Abbreviations

The following abbreviations are used in this manuscript:

DOAJ	Directory of Open Access Journals
ARD	Acid rock drainage
AMD	Acid mine drainage
EIS	Environmental impact statements
masl	Meters above sea level
17-R	January 2017
17-D	July 2017
18-R	January 2018
UNAM	Universidad Nacional Autonoma de Moquegua
GPS	Global Positioning System
DO	Dissolved oxygen
TDS	Total dissolved solids
ICP-MS	Inductively coupled plasma mass spectrometry
ICP-OES	Inductively coupled plasma - optical emission spectrometry
CN	Cyanide
TP	Total phosphorous
TN	Total nitrogen
COD	Chemical oxygen demand
ENSO	El Niño–Southern Oscillation
Ma	Millions of years
CA	Cluster analysis
PCA	Principal component analysis
PC	Principal component
EPA	Environmental Protection Agency
ISO	International Organization for Standardization
CCC	Cophenetic correlation coefficient
EC	Electrical conductivity
USEPA	United States Environmental Protection Agency
WHO	World Health Organization
FAO	Food and Agriculture Organization of the United Nations

Appendix A. Waste Canal from the Cuajone Mine

Figure A1 shows the study area with mining activities outlined. The canal identified in yellow collects waste products from the Cuajone mine and carries them out to the Quebrada Honda tailings dam. These channels are made up of a series of tunnels, concrete-lined open channels, and existing natural river channels.

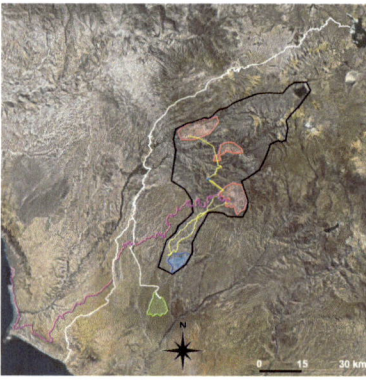

Figure A1. Black polygon: area of direct mining and mine related infrastructural impact. Red polygons: upper red polygon = Southern Copper Cuajone mine, middle red polygon = Anglo American

Quellaveco mine, lower red polygon = Southern Copper Toquepala mine. Yellow lines: mine waste channels. Blue polygons: Cortaderas 2 and Quebrada Honda tailings dams. Purple line: railway line to Ilo smelter. Pink polygon: Ilo smelter. White lines: Pasto Grande project irrigation canals (tunnels, concrete-lined channels, and existing river channels). Green polygon: AH Pampa Sitana irrigation project.

References

1. Ling, H.; Zhang, P.; Xu, H.; Zhang, G. Determining the ecological water allocation in a hyper-arid catchment with increasing competition for water resources. *Glob. Planet. Chang.* **2016**, *145*, 143–152. [CrossRef]
2. Houston, J.; Hartley, A.J. The central Andean west-slope rainshadow and its potential contribution to the origin of hyper-aridity in the Atacama Desert. *Int. J. Climatol. A J. R. Meteorol. Soc.* **2003**, *23*, 1453–1464. [CrossRef]
3. Villalobos-Puma, E.; Flores-Rojas, J.L.; Martinez-Castro, D.; Morales, A.; Lavado-Casimiro, W.; Mosquera-Vásquez, K.; Silva, Y. Summertime precipitation extremes and the influence of atmospheric flows on the western slopes of the southern Andes of Perú. *Int. J. Climatol.* **2022**, *42*, 9909–9930. [CrossRef]
4. Zarch, M.A.A.; Sivakumar, B.; Malekinezhad, H.; Sharma, A. Future aridity under conditions of global climate change. *J. Hydrol.* **2017**, *554*, 451–469. [CrossRef]
5. Ng Cutipa, W.L.; Peña Laureano, F.; Acosta Pereira, H. *Hidrogeología de la Cuenca del río Ilo-Moquegua (13172), Región Moquegua [Boletín–H 6]*; Repositorio Ingemmet: Lima, Peru, 2019.
6. Brêda, J.P.L.; de Paiva, R.C.D.; Siqueira, V.A.; Collischonn, W. Assessing climate change impact on flood discharge in South America and the influence of its main drivers. *J. Hydrol.* **2023**, *619*, 129284. [CrossRef]
7. Clark, A.H.; Farrar, E.; Kontak, D.J.; Langridge, R.J.; Arenas F, M.J.; France, L.J.; McBride, S.L.; Woodman, P.L.; Wasteneys, H.A.; Sandeman, H.A.; et al. Geologic and geochronologic constraints on the metallogenic evolution of the Andes of southeastern Peru. *Econ. Geol.* **1990**, *85*, 1520–1583. [CrossRef]
8. Fraser, B. Water wars come to the Andes. *Sci. Am.* **2009**, *19*, 1–3.
9. Reich, M.; Palacios, C.; Vargas, G.; Luo, S.; Cameron, E.M.; Leybourne, M.I.; Parada, M.A.; Zúñiga, A.; You, C.F. Supergene enrichment of copper deposits since the onset of modern hyperaridity in the Atacama Desert, Chile. *Miner. Depos.* **2009**, *44*, 497–504. [CrossRef]
10. Lepage, H.V.; Barnes, E.; Kor, E.; Hunter, M.; Barnes, C.H. Greening and Browning Trends on the Pacific Slope of Peru and Northern Chile. *Remote Sens.* **2023**, *15*, 3628. [CrossRef]
11. Bebbington, A.; Williams, M. Water and mining conflicts in Peru. *Mt. Res. Dev.* **2008**, *28*, 190–195. [CrossRef]
12. Saenz, C. Stakeholders, social and environmental impact management as key factors of the corporate social management in the mining industry: A Peruvian case study. *Corp. Soc. Responsib. Environ. Manag.* **2024**, *31*, 697–708. [CrossRef]
13. Saenz, C. Enhancing community development management and the management of social and environmental impacts to get social license to operate in the mining industry: A Peruvian case study. *Bus. Strategy Dev.* **2023**, *6*, 873–884. [CrossRef]
14. Wood Group USA Inc. Cuajone Operations Peru Technical Report Summary. 2023. Available online: https://minedocs.com/24/Cuajone-TR-12312022.pdf (accessed on 12 February 2024).
15. Ali, M.M.; Ali, M.L.; Islam, M.S.; Rahman, M.Z. Preliminary assessment of heavy metals in water and sediment of Karnaphuli River, Bangladesh. *Environ. Nanotechnol. Monit. Manag.* **2016**, *5*, 27–35. [CrossRef]
16. Varol, M.; Şen, B. Assessment of nutrient and heavy metal contamination in surface water and sediments of the upper Tigris River, Turkey. *Catena* **2012**, *92*, 1–10. [CrossRef]
17. Budds, J.; Hinojosa, L. Restructuring and rescaling water governance in mining contexts: The co-production of waterscapes in Peru. *Water Altern.* **2012**, *5*, 119.
18. Han, Y.S.; Youm, S.J.; Oh, C.; Cho, Y.C.; Ahn, J.S. Geochemical and eco-toxicological characteristics of stream water and its sediments affected by acid mine drainage. *Catena* **2017**, *148*, 52–59. [CrossRef]
19. Tume, P.; González, E.; Reyes, F.; Fuentes, J.P.; Roca, N.; Bech, J.; Medina, G. Sources analysis and health risk assessment of trace elements in urban soils of Hualpen, Chile. *Catena* **2019**, *175*, 304–316. [CrossRef]
20. Trudgill, S.T.; Goudie, A.S.; Viles, H.A. *Weathering Processes and Forms*; Geological Society: London, UK, 2022.
21. Visitación-Bustamante, K.; Ramos-Fernandez, L.; Visitación-Figueroa, L. Characterization of the hydrochemistry in a high Andean sub-basin in the region of Moquegua, Peru. *Tecnol. Y Cienc. Agua* **2023**, *14*, 257–289. [CrossRef]
22. Olías, M.; Nieto, J.M.; Pérez-López, R.; Cánovas, C.R.; Macías, F.; Sarmiento, A.M.; Galván, L. Controls on acid mine water composition from the Iberian Pyrite Belt (SW Spain). *Catena* **2016**, *137*, 12–23. [CrossRef]
23. Akcil, A.; Koldas, S. Acid Mine Drainage (AMD): Causes, treatment and case studies. *J. Clean. Prod.* **2006**, *14*, 1139–1145. [CrossRef]
24. Nordstrom, D.K.; Blowes, D.W.; Ptacek, C.J. Hydrogeochemistry and microbiology of mine drainage: An update. *Appl. Geochem.* **2015**, *57*, 3–16. [CrossRef]
25. Rey, J.; Martínez, J.; Hidalgo, M.; Rojas, D. Heavy metal pollution in the Quaternary Garza basin: A multidisciplinary study of the environmental risks posed by mining (Linares, southern Spain). *Catena* **2013**, *110*, 234–242. [CrossRef]
26. Martin, C.W. Trace metal concentrations along tributary streams of historically mined areas, Lower Lahn and Dill River basins, central Germany. *Catena* **2019**, *174*, 174–183. [CrossRef]

27. Punia, A. Role of temperature, wind, and precipitation in heavy metal contamination at copper mines: A review. *Environ. Sci. Pollut. Res.* **2021**, *28*, 4056–4072. [CrossRef]
28. Sun, T.; Bao, H.; Reich, M.; Hemming, S.R. More than ten million years of hyper-aridity recorded in the Atacama Gravels. *Geochim. Cosmochim. Acta* **2018**, *227*, 123–132. [CrossRef]
29. Colica, A.; Benvenuti, M.; Chiarantini, L.; Costagliola, P.; Lattanzi, P.; Rimondi, V.; Rinaldi, M. From point source to diffuse source of contaminants: The example of mercury dispersion in the Paglia River (Central Italy). *Catena* **2019**, *172*, 488–500. [CrossRef]
30. Shaban, N.S.; Abdou, K.A.; Hassan, N.E.H.Y. Impact of toxic heavy metals and pesticide residues in herbal products. *Beni-Suef Univ. J. Basic Appl. Sci.* **2016**, *5*, 102–106. [CrossRef]
31. Davies, B.R.; Thoms, M.; Meador, M. An assessment of the ecological impacts of inter-basin water transfers, and their threats to river basin integrity and conservation. *Aquat. Conserv. Mar. Freshw. Ecosyst.* **1992**, *2*, 325–349. [CrossRef]
32. Li, F. Documenting accountability: Environmental impact assessment in a Peruvian mining project. *PoLAR Political Leg. Anthropol. Rev.* **2009**, *32*, 218–236. [CrossRef]
33. Bebbington, A.J.; Bury, J.T. Institutional challenges for mining and sustainability in Peru. *Proc. Natl. Acad. Sci. USA* **2009**, *106*, 17296–17301. [CrossRef]
34. Samimi Namin, F.; Shahriar, K.; Bascetin, A. Environmental impact assessment of mining activities. A new approach for mining methods selection. *Gospod. Surowcami Miner.* **2011**, *27*, 113–143.
35. Gwimbi, P.; Nhamo, G. Benchmarking the effectiveness of mitigation measures to the quality of environmental impact statements: Lessons and insights from mines along the Great Dyke of Zimbabwe. *Environ. Dev. Sustain.* **2016**, *18*, 527–546. [CrossRef]
36. Brodny, J.; Tutak, M. The use of artificial neural networks to analyze greenhouse gas and air pollutant emissions from the mining and quarrying sector in the European Union. *Energies* **2020**, *13*, 1925. [CrossRef]
37. Magilligan, F.J.; Goldstein, P.S.; Fisher, G.B.; Bostick, B.C.; Manners, R.B. Late Quaternary hydroclimatology of a hyper-arid Andean watershed: Climate change, floods, and hydrologic responses to the El Niño-Southern Oscillation in the Atacama Desert. *Geomorphology* **2008**, *101*, 14–32. [CrossRef]
38. Santoso, A.; Mcphaden, M.J.; Cai, W. The defining characteristics of ENSO extremes and the strong 2015/2016 El Niño. *Rev. Geophys.* **2017**, *55*, 1079–1129. [CrossRef]
39. Decou, A.; Von Eynatten, H.; Mamani, M.; Sempere, T.; Wörner, G. Cenozoic forearc basin sediments in Southern Peru (15–18 S): Stratigraphic and heavy mineral constraints for Eocene to Miocene evolution of the Central Andes. *Sediment. Geol.* **2011**, *237*, 55–72. [CrossRef]
40. Oerter, E.; Amundson, R.; Heimsath, A.; Jungers, M.; Chong, G.; Renne, P. Early to middle Miocene climate in the Atacama Desert of northern Chile. *Palaeogeogr. Palaeoclimatol. Palaeoecol.* **2016**, *441*, 890–900. [CrossRef]
41. Poveda, G.; Espinoza, J.C.; Zuluaga, M.D.; Solman, S.A.; Garreaud, R.; Van Oevelen, P.J. High impact weather events in the Andes. *Front. Earth Sci.* **2020**, *8*, 162. [CrossRef]
42. Clark, A.H.; Tosdal, R.M.; Farrar, E.; Plazolles V, A. Geomorphologic environment and age of supergene enrichment of the Cuajone, Quellaveco, and Toquepala porphyry copper deposits, southeastern Peru. *Econ. Geol.* **1990**, *85*, 1604–1628. [CrossRef]
43. Masuno, R.K.; Barahona, H.P.A.; Bazán, E.F.; Zenteno, E.G.; Lizárraga, A.M.; Paucara, J.T. *Evaluacion de Peligros de la Ciudad de Moquegua*; INDECI: Arequipa, Peru, 2001.
44. Clesceri, L.S. *Standard Methods for Examination of Water and Wastewater*; American Public Health Association: Washington, DC, USA, 1989; Volume 9.
45. Otto, M. Multivariate methods. In *Analytical Chemistry*; Kellner, R., Mermet, J.M., Otto, M., Widmer, H.M., Eds.; Wiley-VCH: Weinheim, Germany, 1998.
46. Ward, J.H., Jr. Hierarchical grouping to optimize an objective function. *J. Am. Stat. Assoc.* **1963**, *58*, 236–244. [CrossRef]
47. Kazi, T.; Arain, M.; Jamali, M.K.; Jalbani, N.; Afridi, H.; Sarfraz, R.; Baig, J.; Shah, A.Q. Assessment of water quality of polluted lake using multivariate statistical techniques: A case study. *Ecotoxicol. Environ. Saf.* **2009**, *72*, 301–309. [CrossRef]
48. Sokal, R.R.; Rohlf, F.J. The comparison of dendrograms by objective methods. *Taxon* **1962**, *11*, 33–40. [CrossRef]
49. Bhuiyan, M.A.; Islam, M.; Dampare, S.B.; Parvez, L.; Suzuki, S. Evaluation of hazardous metal pollution in irrigation and drinking water systems in the vicinity of a coal mine area of northwestern Bangladesh. *J. Hazard. Mater.* **2010**, *179*, 1065–1077. [CrossRef]
50. Jolliffe, I.; Lovric, M. International encyclopedia of statistical science. In *Principa l Component Analysis*; Springer: Berlin/Heidelberg, Germany, 2011; pp. 1094–1096.
51. Wurts, W.A.; Durborow, R.M. *Interactions of pH, Carbon Dioxide, Alkalinity and Hardness in Fish Ponds*; SRAC Publication: Stoneville, MS, USA, 1992.
52. Talling, J. The depletion of carbon dioxide from lake water by phytoplankton. *J. Ecol.* **1976**, *64*, 79–121. [CrossRef]
53. Möller, D. The Na/Cl ratio in rainwater and the seasalt chloride cycle. *Tellus B* **1990**, *42*, 254–262. [CrossRef]
54. Magaritz, M.; Nadler, A.; Koyumdjisky, H.; Dan, J. The use of Na/Cl ratios to trace solute sources in a semiarid zone. *Water Resour. Res.* **1981**, *17*, 602–608. [CrossRef]
55. Gorchev, H.G.; Ozolins, G. *WHO Guidelines for Drinking-Water Quality*; WHO Chronicle: Washington, DC, USA, 1984; Volume 38.
56. Johnson, D.B.; Hallberg, K.B. Acid mine drainage remediation options: A review. *Sci. Total Environ.* **2005**, *338*, 3–14. [CrossRef]
57. Kalin, M.; Fyson, A.; Wheeler, W.N. The chemistry of conventional and alternative treatment systems for the neutralization of acid mine drainage. *Sci. Total Environ.* **2006**, *366*, 395–408. [CrossRef]

58. Sheoran, A.; Sheoran, V. Heavy metal removal mechanism of acid mine drainage in wetlands: A critical review. *Miner. Eng.* **2006**, *19*, 105–116. [CrossRef]
59. Rösner, U. Effects of historical mining activities on surface water and groundwater-an example from northwest Arizona. *Environ. Geol.* **1998**, *33*, 224–230. [CrossRef]
60. Hudson-Edwards, K.A.; Schell, C.; Macklin, M.G. Mineralogy and geochemistry of alluvium contaminated by metal mining in the Rio Tinto area, southwest Spain. *Appl. Geochem.* **1999**, *14*, 1015–1030. [CrossRef]
61. Miller, J.; Hudson-Edwards, K.; Lechler, P.; Preston, D.; Macklin, M. Heavy metal contamination of water, soil and produce within riverine communities of the Rıo Pilcomayo basin, Bolivia. *Sci. Total Environ.* **2004**, *320*, 189–209. [CrossRef]
62. Pan, Y.; Koopmans, G.F.; Bonten, L.T.; Song, J.; Luo, Y.; Temminghoff, E.J.; Comans, R.N. Influence of pH on the redox chemistry of metal (hydr) oxides and organic matter in paddy soils. *J. Soils Sediments* **2014**, *14*, 1713–1726. [CrossRef]
63. Canfield, D.E. The geochemistry of river particulates from the continental USA: Major elements. *Geochim. Cosmochim. Acta* **1997**, *61*, 3349–3365. [CrossRef]
64. Tipping, E.; Rey-Castro, C.; Bryan, S.E.; Hamilton-Taylor, J. Al (III) and Fe (III) binding by humic substances in freshwaters, and implications for trace metal speciation. *Geochim. Cosmochim. Acta* **2002**, *66*, 3211–3224. [CrossRef]
65. Viers, J.; Dupré, B.; Gaillardet, J. Chemical composition of suspended sediments in World Rivers: New insights from a new database. *Sci. Total Environ.* **2009**, *407*, 853–868. [CrossRef]
66. McPhaden, M.J.; Santoso, A.; Cai, W. Introduction to El Niño Southern Oscillation in a changing climate. In *El Niño Southern Oscillation in a Changing Climate*; American Geophysical Union: Washington, DC, USA, 2020; pp. 1–19.
67. Ayers, R.S.; Westcot, D.W. *Water Quality for Agriculture*; Food and Agriculture Organization of the United Nations Rome: Rome, Italy, 1985; Volume 29.
68. ANA. Modifican los Estándares Nacionales de Calidad Ambiental para Agua y establecen disposiciones complementarias para su aplicación. In *Decreto Supremo, 015–2015–MINAM*; Ministerio del Ambiente: Lima, Peru, 2015.
69. Poshtegal, M.K.; Mirbagheri, S.A. The heavy metals pollution index and water quality monitoring of the Zarrineh river, Iran. *Environ. Eng. Geosci.* **2019**, *25*, 179–188. [CrossRef]
70. Harrison, R.M.; De Mora, S.J. *Introductory Chemistry for the Environmental Sciences*; Cambridge University Press: Cambridge, UK, 1996; Volume 7.
71. Xiao, J.; Jin, Z.D.; Zhang, F.; Wang, J. Solute geochemistry and its sources of the groundwaters in the Qinghai Lake catchment, NW China. *J. Asian Earth Sci.* **2012**, *52*, 21–30. [CrossRef]
72. MINAGRI Peru. Boletín Estadístico de Medios de Producción Agropecuarios. Available online: https://www.midagri.gob.pe/portal/boletin-estadistico-de-medios-de-produccion-agropecuarios (accessed on 20 January 2024).
73. Borrok, D.M.; Engle, M.A. The role of climate in increasing salt loads in dryland rivers. *J. Arid Environ.* **2014**, *111*, 7–13. [CrossRef]
74. LaZerte, B.D.; Burling, K. Manganese speciation in dilute waters of the Precambrian Shield, Canada. *Water Res.* **1990**, *24*, 1097–1101. [CrossRef]
75. Gibbs, R.J. Mechanisms controlling world water chemistry. *Science* **1970**, *170*, 1088–1090. [CrossRef]
76. Kumar, S.K.; Rammohan, V.; Sahayam, J.D.; Jeevanandam, M. Assessment of groundwater quality and hydrogeochemistry of Manimuktha River basin, Tamil Nadu, India. *Environ. Monit. Assess.* **2009**, *159*, 341–351. [CrossRef]
77. US EPA. *National Primary Drinking Water Regulations*; Technical Fact; 2019. Available online: https://www.epa.gov/ground-water-and-drinking-water/national-primary-drinking-water-regulations (accessed on 20 January 2024).
78. Vera Delgado, J. The socio-cultural, institutional and gender aspects of the water transfer-agribusiness model for food and water security. Lessons learned from Peru. *Food Secur.* **2015**, *7*, 1187–1197. [CrossRef]
79. Tessier, A.; Fortin, D.; Belzile, N.; DeVitre, R.; Leppard, G. Metal sorption to diagenetic iron and manganese oxyhydroxides and associated organic matter: Narrowing the gap between field and laboratory measurements. *Geochim. Cosmochim. Acta* **1996**, *60*, 387–404. [CrossRef]
80. Pagenkopf, G.K. Gill surface interaction model for trace-metal toxicity to fishes: role of complexation, pH, and water hardness. *Environ. Sci. Technol.* **1983**, *17*, 342–347. [CrossRef]
81. Pascoe, D.; Evans, S.A.; Woodworth, J. Heavy metal toxicity to fish and the influence of water hardness. *Arch. Environ. Contam. Toxicol.* **1986**, *15*, 481–487. [CrossRef]
82. Zitko, V.; Carson, W. Mechanism of the effects of water hardness on the lethality of heavy metals to fish. *Chemosphere* **1976**, *5*, 299–303. [CrossRef]
83. Bundschuh, J.; Litter, M.I.; Parvez, F.; Román-Ross, G.; Nicolli, H.B.; Jean, J.S.; Liu, C.W.; López, D.; Armienta, M.A.; Guilherme, L.R.; et al. One century of arsenic exposure in Latin America: A review of history and occurrence from 14 countries. *Sci. Total Environ.* **2012**, *429*, 2–35. [CrossRef]
84. Schwartz, M.O. Arsenic in porphyry copper deposits: Economic geology of a polluting element. *Int. Geol. Rev.* **1995**, *37*, 9–25. [CrossRef]
85. Leybourne, M.I.; Cameron, E.M. Source, transport, and fate of rhenium, selenium, molybdenum, arsenic, and copper in groundwater associated with porphyry–Cu deposits, Atacama Desert, Chile. *Chem. Geol.* **2008**, *247*, 208–228. [CrossRef]
86. Byrne, P.; Reid, I.; Wood, P.J. Sediment geochemistry of streams draining abandoned lead/zinc mines in central Wales: The Afon Twymyn. *J. Soils Sediments* **2010**, *10*, 683–697. [CrossRef]

87. Byrne, P.; Taylor, K.G.; Hudson-Edwards, K.A.; Barrett, J.E. Speciation and potential long-term behaviour of chromium in urban sediment particulates. *J. Soils Sediments* **2017**, *17*, 2666–2676. [CrossRef]
88. de agua Moquegua, A.L. Informe del Tercer 1. Monitoreo Participativo de Calidad de Agua Superficial de la Cuenca Moquegua-Ilo. 2013. https://hdl.handle.net/20.500.12543/2771 (accessed on 20 January 2024).
89. de agua Moquegua, A.L. Informe del Cuarto Monitoreo Participativo de Calidad de Agua Superficial de la Cuenca MOQUEGUA-Ilo. 2014. Available online: https://hdl.handle.net/20.500.12543/2830 (accessed on 20 January 2024).
90. de agua Moquegua, A.L. Quinto Monitoreo Participativo de Calidad de Agua Superficial de la Cuenca Moquegua-Ilo. 2014. Available online: https://hdl.handle.net/20.500.12543/2834 (accessed on 20 January 2024).
91. Pincetti-Zúniga, G.; Richards, L.; Daniele, L.; Boyce, A.; Polya, D. Hydrochemical characterization, spatial distribution, and geochemical controls on arsenic and boron in waters from arid Arica and Parinacota, northern Chile. *Sci. Total Environ.* **2022**, *806*, 150206. [CrossRef]
92. Yin, S.; Yang, L.; Wen, Q.; Wei, B. Temporal variation and mechanism of the geogenic arsenic concentrations in global groundwater. *Appl. Geochem.* **2022**, *146*, 105475. [CrossRef]
93. Forbes, W.F.; Hill, G.B. Is exposure to aluminum a risk factor for the development of Alzheimer disease?—Yes. *Arch. Neurol.* **1998**, *55*, 740–741. [CrossRef]
94. Flaten, T.P. Aluminium as a risk factor in Alzheimer's disease, with emphasis on drinking water. *Brain Res. Bull.* **2001**, *55*, 187–196. [CrossRef]
95. Smith, A.H.; Hopenhayn-Rich, C.; Bates, M.N.; Goeden, H.M.; Hertz-Picciotto, I.; Duggan, H.M.; Wood, R.; Kosnett, M.J.; Smith, M.T. Cancer risks from arsenic in drinking water. *Environ. Health Perspect.* **1992**, *97*, 259–267. [CrossRef]
96. Davis, M.A.; Signes-Pastor, A.J.; Argos, M.; Slaughter, F.; Pendergrast, C.; Punshon, T.; Gossai, A.; Ahsan, H.; Karagas, M.R. Assessment of human dietary exposure to arsenic through rice. *Sci. Total Environ.* **2017**, *586*, 1237–1244. [CrossRef]
97. Insel, N.; Poulsen, C.J.; Ehlers, T.A. Influence of the Andes Mountains on South American moisture transport, convection, and precipitation. *Clim. Dyn.* **2010**, *35*, 1477–1492. [CrossRef]
98. Taylor, M.P.; Kesterton, R.G. Heavy metal contamination of an arid river environment: Gruben River, Namibia. *Geomorphology* **2002**, *42*, 311–327. [CrossRef]
99. Arakaki, M.; Cano, A. Floral composition of the Ilo-Moquegua and Lomas de Ilo river basin, Moquegua, Peru. *Peruv. J. Biol.* **2003**, *10*, 5–19. [CrossRef]

Disclaimer/Publisher's Note: The statements, opinions and data contained in all publications are solely those of the individual author(s) and contributor(s) and not of MDPI and/or the editor(s). MDPI and/or the editor(s) disclaim responsibility for any injury to people or property resulting from any ideas, methods, instructions or products referred to in the content.

Article

Runoff Decline Is Dominated by Human Activities

Ping Miao [1], Dagula [1], Xiaojie Li [2], Shahid Naeem [2], Amit Kumar [3,4], Hongli Ma [1], Yenong Ding [1], Ruidong Wang [5] and Jinkai Luan [3,4,*]

1. River and Lake Protection Center, Ordos Water Conservancy Bureau, Ordos 017000, China
2. Key Laboratory of Water Cycle and Related Land Surface Processes, Institute of Geographic Sciences and Natural Resources Research, Chinese Academy of Sciences, Beijing 100101, China
3. Key Laboratory of Hydrometeorological Disaster Mechanism and Warning of Ministry of Water Resources, Nanjing University of Information Science and Technology, Nanjing 210044, China
4. School of Hydrology and Water Resources, Nanjing University of Information Science and Technology, Nanjing 210044, China
5. Ordos Hydrology and Water Resources Subcenter, Ordos 017000, China
* Correspondence: luanjk@nuist.edu.cn

Abstract: Investigations into runoff change and its influencing factors hold immense significance for promoting sustainable development, efficient water resource utilization, and the improvement of the ecological environment. To reduce methodological uncertainties, this study employed six attribution analysis methods, including two statistical approaches, a Budyko equation sensitivity coefficient method, and three hydrology models, to differentiate the contributions of climate change and human activities to the runoff change in the Xiliugou basin. The results indicated an abrupt change point in 2006, and the annual runoff series from 1960 to 2020 demonstrated a significant declining trend. All the six methods revealed that human activities were the major influencing factor. The average contribution rate of climate change was noted to be 24.2%, while that of human activities was 75.8% among the six methods used for this study. The prominent human activities in the Xiliugou basin revolve around soil and water conservation measures. The research findings hold great significance for the comprehensive understanding of runoff formation and its response to the changing environment in the Xiliugou basin. Additionally, these results can provide a foundation for decision-making for water resource management and ecological protection.

Keywords: climate change; human activities; runoff; uncertainty; Xiliugou basin

1. Introduction

The hydrologic cycle system closely links the earth's hydrosphere, atmosphere, biosphere, and lithosphere, forming a mutually coupled subsystem. This system provides an important link between atmospheric, terrestrial, and ecological water [1,2]. Both climatic factors and human activities exert influences on the hydrological cycle within basins [3–5]. Climate change inherently induces alterations in hydrological characteristics across various spatial scales, including in basins and at the regional and global levels. For example, an increase in the global temperature results in an increased evapotranspiration rate and consequently modifies precipitation and hydrological characteristics, and the occurrence and severity of hydrological extreme events, such as floods and droughts, are influenced by their frequency and intensity. These changes affect the total amount of water resources and their redistribution on the spatiotemporal scale [6–8]. Human-induced land use alterations, including the construction of water conservancy projects, deforestation, urbanization, and the restoration of vegetation, have a substantial influence on hydrological processes and the spatiotemporal distribution of water resources within river basins. These modifications make a substantial contribution to the sustainable development of the socio-economy and the ecological environment [9,10].

A river's flow plays a crucial role in maintaining the hydrological balance within a basin, making it highly significant for improving the local ecological environment and fostering economic growth on a larger scale [11]. The accurate evaluation of the individual effects of climate change and human activities on runoff fluctuations provides essential guidance for efficient watershed water resource management, making a substantial contribution to the achievement of sustainable development goals. Hence, it is crucial to regard climate change and human activities as distinct factors that exert influence and to utilize long-term runoff data to quantify and attribute runoff variability for both factors at different time scales [3]. Climate factors, including precipitation, temperature, and potential evapotranspiration, can be easily quantified by simulating a precipitation runoff model with a strong physical mechanism. However, human activity factors indicate an extremely complex mechanism due to their inherent randomness and arbitrariness. Furthermore, aligning the spatiotemporal scale of these activities with the runoff process is challenging, and even the influence of human activities on runoff usually lacks a well-defined physical mechanism.

There are three main approaches to unraveling the impacts of climate change and human activities on the complex processes that regulate runoff. The first one involves a statistical approach based on observed hydrometeorological data [12–18]. This method is straightforward but lacks a physical basis, and it requires highly accurate, long-term statistical data, thus limiting its application. The second one is a sensitivity coefficient method based on the Budyko hydrothermal coupling equilibrium equation [19–23]. This method offers a certain physical mechanism, a simple calculation process, and straightforward parameter calculation. Nevertheless, it is important to note that these methodologies are limited in their applicability solely to multi-year and annual scales, and they thereby can potentially fail to comprehensively quantify the effects of various human activities and the underlying surface modifications on the hydrological dynamics of a given basin. The last approach involves the hydrologic model method [24–27]. This model demonstrates a strong physical mechanism, effectively explaining and attributing runoff change, and it can simulate and restore runoff changes at different time scales from days to years. However, the model structure is relatively complex and may involve a large number of parameters with associated uncertainties. The complex interplay between the underlying surface conditions and climate factors within a basin is intricate and multifaceted, demanding high-quality data, especially for distributed hydrological models. It is important to note that the specific impacts of climate and human activities, as determined by various attribution analysis methods, may exhibit inconsistencies, even within the same basin. These discrepancies highlight the uncertainties associated with the chosen methodology [28].

The Yellow River basin in China stands out as one of the foremost regions grappling with acute water scarcity and significant soil erosion. In recent decades, significant changes have occurred in the hydrological system of this basin, largely due to the combined influence of climate change and human activities [29,30]. To tackle the urgent issues related to soil and water erosion in the Yellow River basin, a wide range of diligent measures for soil and water conservation have been comprehensively implemented. These measures encompass an extensive scope of ecological restoration initiatives alongside the implementation of landscape engineering interventions. For instance, measures like constructing terraces, implementing water diversion projects, and developing reservoirs have been undertaken [29,31,32].

The Xiliugou basin is located within the Ten Kongduis basin, which is situated in the upstream region of the Yellow River basin. It is an ecologically fragile area with an interlacing distribution of sandstone, desert, and floodplain. The region exhibits a significant spatial heterogeneity on a regional scale and is highly sensitive to human disturbance and climate change. The Xiliugou basin experiences a limited availability of water resources. With the development of mineral resources, industrial growth, urban sprawl, and increased groundwater exploitation, the problem of water resource scarcity has intensified. However, limited comprehensive investigations have been conducted on the factors influencing changes in runoff across the Xiliugou basin. Therefore, the main objective of our study is to thoroughly examine the drivers of runoff change in this basin. By employing various methodologies, our aim is to discern the distinct impacts of climate change and human activities on the observed patterns of runoff. This research endeavor is of significant importance, as it can provide valuable insights for informed decision-making in areas such as land use planning, water resource management, and the promotion of sustainable development within the Xiliugou basin.

For the purpose of our investigation, the Xiliugou basin was chosen as the focal area. In our research, we employed the nonparametric Mann–Kendall (MK) test to identify possible abrupt change points in the runoff patterns covering the period from 1960 to 2020. To enhance the investigation, the study period was divided into two separate periods, namely the baseline period and the impact period, considering the identified abrupt change points. Following this division, six separation methods were employed in our research. These approaches encompassed two statistical methods, namely the sensitivity coefficient method utilizing the Budyko equation and the utilization of three hydrology models. The primary aim of employing these methods was to effectively distinguish and quantify the respective contributions of climate variations and human activities to the observed runoff changes. The objectives of this study are as follows: (1) to identify abrupt change points in runoff within the Xiliugou basin; (2) to utilize different methodologies to discern the impacts of climate change and human activities on runoff; and (3) to analyze the discrepancies in the effects of climate change and human activities on runoff changes resulting from the use of different methods and summarize the main factors causing runoff change.

2. Materials and Methods

2.1. Study Region

Situated within the Ten Kongduis basin in Ordos and positioned on the right bank of the Yellow River, the Xiliugou basin occupies a central location, with its specific coordinates being E109°24′~110°00′ and N39°47′~40°30′ (Figure 1). The control area above Longtouguai hydrological station is 1157 km^2, and the altitude varies from 1044 to 1551 m. The Xiliugou basin experiences a semi-arid continental climate, characterized by lengthy cold winters and brief hot summers. The average annual temperature within the basin is approximately 6 °C, the average annual wind speed is 3.7 m·s^{-1}, the annual rainfall is 240–360 mm, and the potential evaporation is 2200 mm. The distribution of rainfall and runoff within this basin exhibits seasonal variations with uneven patterns throughout the year. Notably, during the flood season (June–September), the rainfall accounts for approximately 82% of the total annual precipitation. The rainfall in the flood season tends to be concentrated and takes the form of heavy rain, resulting in sharp changes in the flood level. It serves as a critical source of coarse sand in the Yellow River, China. In recent decades, several strategies have been employed to promote vegetation restoration in the basin, such as the implementation of soil and water conservation techniques and the conversion of agricultural land into forested or grassy areas. These endeavors have led to a remarkable growth in vegetation coverage within the region. As a result, there has been a noticeable alteration in land use and cover, leading to a substantial decrease in both runoff and sediment [33].

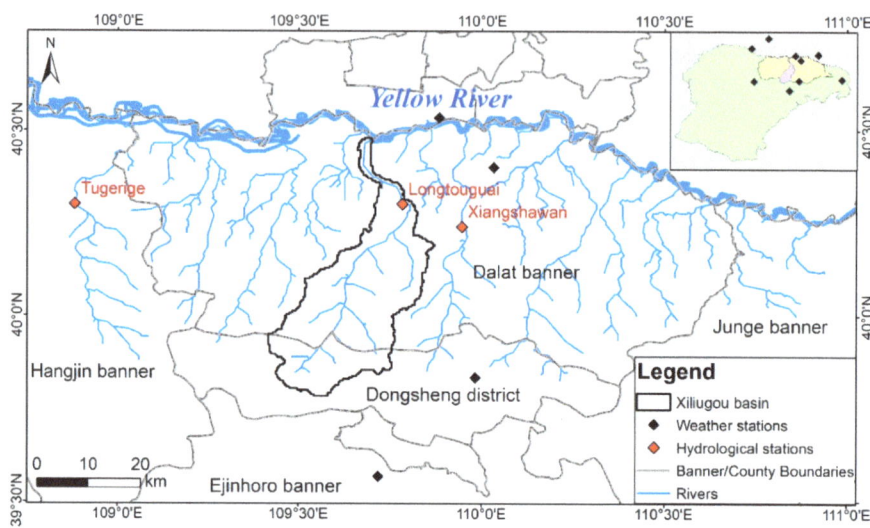

Figure 1. The geographical location of the Xiliugou basin.

2.2. Data

Comprehensive meteorological data, including daily precipitation, temperatures, sunshine duration, specific humidity, and wind speed, were diligently collected from the National Meteorological Information Center of China (https://www.nmic.cn/, 8 August 2022). This comprehensive dataset spans a significant timeframe, ranging from 1960 to 2020. To acquire monthly runoff information specifically for the Longtouguai hydrological station within the aforementioned period, we extensively consulted the esteemed Hydrological Yearbook of the People's Republic of China for accurate references and information. The leaf area index (LAI) data for the period from 1982 to 2020 were acquired from the reputable Global LAnd Surface Satellite (GLASS) and the Advanced Very-High-Resolution Radiometer (AVHRR) LAI product. These datasets were integrated into the proficient Google Earth Engine platform for seamless analysis. These detailed records exhibit an outstanding spatial resolution of $0.05°$ and a temporal resolution of every 8 days. Additionally, we diligently acquired a digital elevation model (DEM) with a resolution of 30 m from the renowned NASA Advanced Spaceborne Thermal Emission and Reflection Radiometer Global Digital Elevation Model (ASTER GDEM) (https://lpdaac.usgs.gov/, 2 August 2022). Subsequently, we precisely extracted the underlying surface data, encompassing crucial elements such as soil properties (including physicochemical aspects), a 1:100,000 soil distribution map, and the crucial land use/land cover (LULC) data, all possessing a remarkable 30 m resolution. These invaluable resources were collected from the renowned Chinese Academy of Sciences' Data Center for Resources and Environmental Sciences (https://www.resdc.cn, 18 August 2022), with a specific focus on the noteworthy time period of the 1980s.

2.3. Trend Analysis and Abrupt Change Detection

The widely employed nonparametric Mann–Kendall (MK) method [34,35] has found extensive application in the analysis of trends and sudden shifts in hydrometeorological time series. In this study, the MK method was applied to estimate annual runoff trends and abrupt points in the Longtouguai hydrological station.

2.4. Six Methods for Attribution Analysis of Runoff Change

2.4.1. Precipitation–Runoff Double Mass Curve (DMC) Method

The double mass curve technique plays a vital role in investigating the coherence and connection between two variables. It involves examining the gradient of cumulatively accumulated values of the two variables over various time periods, which are then plotted on a Cartesian coordinate system [36]. In this study, the variable used as a reference or independent variable, represented by x, signifies precipitation, while the test variable or dependent variable, denoted as y, represents runoff. The notation of x_i and y_i is employed to represent N years of observations, where i ranges from 1 to N.

To commence the analysis, the cumulative yearly values of x and y can be computed using Formulas (1) and (2), respectively. The utilization of these formulas enables the determination of yearly cumulative series, denoted as X_i and Y_i, which is achieved through the following calculations:

$$X_i = \sum_{i=1}^{N} x_i \tag{1}$$

$$Y_i = \sum_{i=1}^{N} y_i \tag{2}$$

Subsequently, a relationship curve is constructed within a rectangular coordinate system to graphically depict the cumulative values of the two variables. This curve serves as a visual representation of the association between the accumulated values of the respective variables. In the absence of any systematic bias in the dependent variable or test variable, the cumulative curve exhibits a linear pattern. An upward bias indicates an increase, while a downward bias suggests a decrease.

2.4.2. Slope Change Ratio of Cumulative Quantity (SCRCQ) Method

The SCRCQ method [17] serves as a valuable tool for quantifying and distinguishing the relative influences exerted by these two factors on the observed changes in runoff. This approach can be used by two ways: (1) reflecting climate change only by precipitation to estimate its contribution rate, and (2) reflecting climate change comprehensively by precipitation and temperature. Within the scope of this study, the climate change factors under investigation encompassed precipitation and temperature. These variables were carefully chosen as key indicators to better comprehend the dynamics of climate change within the analyzed context.

To commence the analysis, the calculation of the slope for cumulative runoff, cumulative precipitation, and cumulative temperature was conducted individually for both the baseline period and the period of change. These slopes are denoted as S_{Ra}, S_{Pa}, and S_{Ta} for the baseline period and S_{Rb}, S_{Pb}, and S_{Tb} for the change period. By calculating the difference between the slopes of the baseline period and the change period, we can determine the rates of change for cumulative runoff (R_R), cumulative precipitation (R_P), and cumulative temperature (R_T). It is important to note that in this method, precipitation and runoff were assumed to exhibit a positive correlation, while temperature and runoff considered to exhibit a negative correlation. Furthermore, we defined the attribution of precipitation (η_P) and temperature (η_T) to the variability in runoff as follows:

$$\eta_P = (R_P/R_R) \times 100\% \tag{3}$$

$$\eta_T = -(R_T/R_R) \times 100\% \tag{4}$$

We calculated and quantified the respective impacts of climate change (η_C) and human activities (η_H) on the observed alterations in runoff through the following calculations:

$$\eta_C = \eta_P + \eta_T \tag{5}$$

$$\eta_H = 1 - \eta_C \tag{6}$$

2.4.3. Sensitivity Coefficient Methods by the Budyko Equation

The Budyko theory provides a comprehensive framework for understanding the distribution mechanism of precipitation between runoff and evapotranspiration, offering insights into the intricate coupling between water and energy within a given basin. This theory delves into the fundamental principles that govern the interplay of water and energy dynamics, shedding light on the intricate relationship between these two crucial components. This study used the Budyko equation for Fu's formula [37] to study the attribution analysis of runoff change:

$$\frac{E}{P} = 1 - \frac{Q}{P} = 1 + \frac{E_P}{P} - \left[1 + \left(\frac{E_P}{P}\right)^\omega\right]^{1/\omega} \quad (7)$$

where Q represents the average annual runoff depth (mm), P signifies the average annual precipitation (mm), E denotes the average annual evapotranspiration (mm), and E_P corresponds to the average annual potential evapotranspiration (mm). Furthermore, the basin-specific parameters are represented by the symbol ω.

To accurately assess the influence of climate elements, specifically precipitation (P) and potential evapotranspiration (E_P) on runoff, we utilized the sensitivity coefficient technique. This approach facilitates the computation of runoff sensitivity to changes in P and E_P, enabling a rigorous and quantitative assessment of their respective influences:

$$\Delta Q_C = \frac{\partial Q}{\partial P}\Delta P + \frac{\partial Q}{\partial E_P}\Delta E_P \quad (8)$$

where the runoff change caused by this phenomenon is represented by ΔQ_C. The sensitivity coefficients of runoff with respect to P and E_P are denoted by $\frac{\partial Q}{\partial P}$ and $\frac{\partial Q}{\partial E_P}$, respectively, and ΔP and ΔE_P represent the changes in precipitation and potential evapotranspiration, respectively.

The calculation of sensitivity coefficients was conducted as follows:

$$\frac{\partial Q}{\partial P} = P^{\omega-1}(E_P^\omega + P^\omega)^{\frac{1}{\omega}-1} \quad (9)$$

$$\frac{\partial Q}{\partial E_P} = E_P^{\omega-1}(E_P^\omega + P^\omega)^{\frac{1}{\omega}-1} - 1 \quad (10)$$

When evaluating the influence of human activities on runoff, it is possible to quantify the alterations in runoff caused by these activities. This calculation involves determining the alteration in runoff attributable to human interventions:

$$\Delta Q_H = \Delta Q - \Delta Q_C \quad (11)$$

where the alteration in runoff attributed specifically to human activities is represented as ΔQ_H. It is crucial to distinguish ΔQ_H from the overall actual change in runoff, represented as ΔQ, in order to separate and quantify the distinct impact of human interventions on changes in runoff.

2.4.4. HBV and SIMHYD Hydrological Model

The Hydrologiska Byrans Vattenbalansavdelning (HBV) hydrological model [38] and SIMulation of HYDrology (SIMHYD) [39–43] are two conceptual precipitation runoff models. These models exhibit their proficiency in establishing the connections between precipitation, evaporation, and runoff. Moreover, they integrate modifications in soil, water and groundwater processes using a set of precise mathematical equations. These equations form the basis for accurately simulating and analyzing the intricate dynamics of a hydrological system.

The HBV model structure is mainly composed of a snow melting module, a soil moisture module, and a production confluence module. The specific model structure is shown in Figure 2.

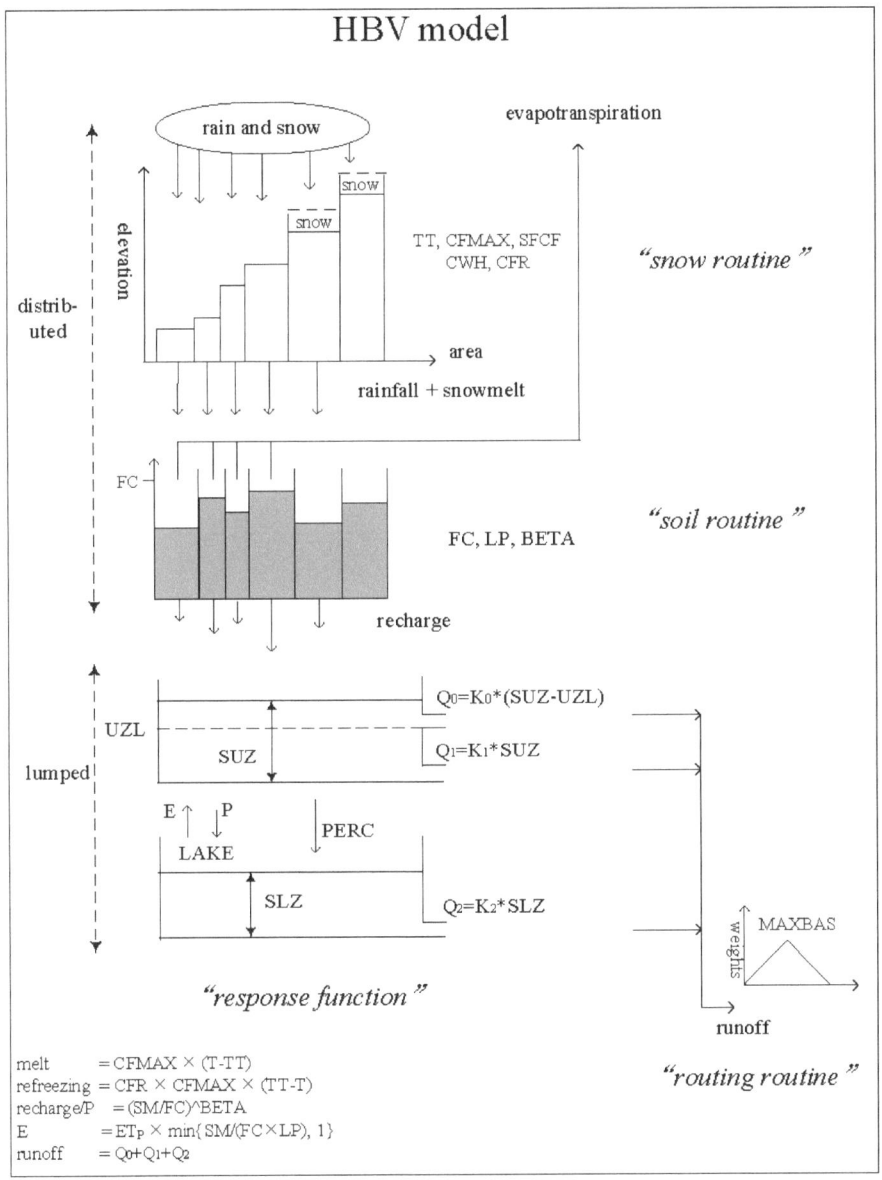

Figure 2. HBV hydrological model structure.

The SIMHYD model comprises three main elements: surface runoff, soil water movement, and subsurface flow. The fundamental structure of the model is depicted in Figure 3.

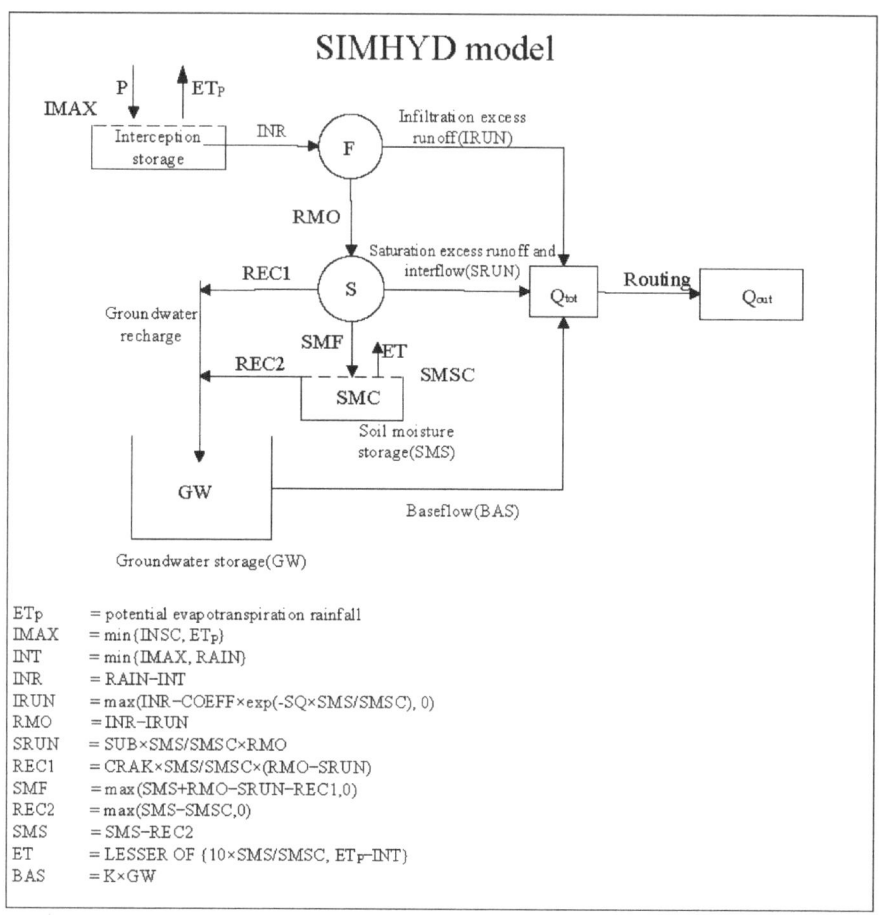

Figure 3. SIMHYD hydrological model structure.

2.4.5. SWAT Hydrological Model

The Soil and Water Assessment Tool (SWAT) model, established by the United States Department of Agriculture, is a widely acknowledged and extensively employed physically based model at a large scale [44,45]. SWAT incorporates a comprehensive set of physical principles and mathematical algorithms to accurately represent the complex dynamics of a soil–water system. It includes several modules, such as a hydrology module, a sediment transport module, and a nutrient transport module, etc. This model has a wide range of applicability, including basin water balance accounting. Additionally, it facilitates explorations into the impact of management strategies, as well as variations in climate and land use, on both the quantity and quality of water within basins. The versatility and robustness of this model make it an invaluable tool for the comprehensive and accurate analysis of hydrological processes and their associated environmental impacts. The model corresponds each process of the hydrological cycle to different sub-modules through modular modeling, and each module can be run independently or combined, which is conducive to expansion and application. The hydrological module includes numerous variables, such as evapotranspiration, river discharge, surface runoff, soil water transport, groundwater recharge, etc., realizing the calculation and simulation of the hydrological cycle. The SWAT model is divided into a sub-basin calculation module and a channel production and confluence module influenced by the underlying surface properties and rainfall processes.

The basis of this model is built upon the water balance equation, which can be represented as follows:

$$SW_t = SW_0 + \sum_{i=1}^{t} \left(P_{day,i} - Q_{surf,i} - E_{a,i} - W_{seep,i} - Q_{gw,i} \right) \tag{12}$$

where SW_t (mm) represents the soil water storage at a specific time, SW_0 (mm) indicates the initial soil water storage, t denotes the time, $P_{day,i}$ (mm) signifies precipitation on the ith day, $Q_{surf,i}$ (mm) denotes the surface runoff on the ith day, $E_{a,i}$ (mm) represents the evapotranspiration on the ith day, $W_{seep,i}$ (mm) indicates the amount of water infiltration on the ith day, and $Q_{gw,i}$ (mm) represents the return flow from the groundwater on the ith day.

2.5. Calibration and Verification of HBV and SIMHYD Models

To optimize the parameters of the HBV and SIMHYD hydrological models, a widely recognized global optimizer called the particle swarm optimization (PSO) toolbox [46] was employed. These specific values were selected to ensure an effective exploration of the solution space, leading to the identification of the most optimal solution.

The two hydrological models underwent a rigorous calibration and validation process, utilizing the monthly runoff data from the baseline period. For calibration purposes, 70% of the available data were employed, while the remaining 30% were reserved for validation. For the purpose of model calibration in this study, the Nash–Sutcliffe efficiency (NSE) [47] was chosen as the objective function. This widely recognized metric serves as a reliable indicator of model performance and enabled a robust assessment of the model's ability to accurately replicate the observed runoff patterns:

$$F = (1 - NSE) + 5|LN(1 + Bias)|^{2.5} \tag{13}$$

$$NSE = 1 - \frac{\sum_{i=1}^{N}(Q_{obs,i} - Q_{sim,i})^2}{\sum_{i=1}^{N}(Q_{obs,i} - \overline{Q_{obs}})^2} \tag{14}$$

$$Bias = \frac{\sum_{i=1}^{N} Q_{obs,i} - \sum_{i=1}^{N} Q_{sim,i}}{\sum_{i=1}^{N} Q_{obs,i}} \tag{15}$$

where the NSE (Nash–Sutcliffe efficiency) was calculated by comparing the simulated monthly runoff (Q_{sim}) with the observed monthly runoff (Q_{obs}). The NSE was determined using the average value of the observed monthly runoff, with "i" representing the ith month and "N" denoting the total number of months in the calibration period. The F value was minimized during the calibration to enhance the model's performance and its ability to accurately replicate the observed runoff patterns. The NSE value serves as a robust metric to evaluate and compare the results of the calibration and verification processes, providing valuable insights into the accuracy and reliability of the hydrological models.

2.6. Calibration and Verification of SWAT Model

In this study, the calibration and verification of the model parameters were carried out utilizing the SWAT-CUP Premium tool (https://www.2w2e.com/, 10 October 2022). This tool offers advanced features for rigorous model evaluation. The calibration and validation of the SWAT model were proficiently performed using the monthly runoff data from the baseline period. To ensure an accurate representation of the hydrological processes, 70% of the available data were allocated for calibration, while the remaining 30% were dedicated to validation. This systematic approach enabled the identification of optimal parameter values and ensured the model's ability to effectively reproduce the observed runoff patterns.

During the calibration and validation processes, a comprehensive objective function, denoted as F, was implemented in SWAT-CUP Premium. This function incorporated multiple evaluation metrics, including R^2, the Nash–Sutcliffe efficiency (NSE) [47], and Bias. By integrating these diverse criteria, the objective function facilitated a thorough assessment of the model's performance and enhanced the reliability of the results. Simultaneously considering multiple aspects of model accuracy, precision, and bias, this approach provided a robust framework for both the calibration and verification stages:

$$F = w_3 R^2 + w_5 NSE - |w_8 Bias| \tag{16}$$

$$NSE = 1 - \frac{\sum_{i=1}^{N}(Q_{obs,i} - Q_{sim,i})^2}{\sum_{i=1}^{N}(Q_{obs,i} - \overline{Q_{obs}})^2} \tag{17}$$

$$Bias = \frac{\sum_{i=1}^{N} Q_{obs,i} - \sum_{i=1}^{N} Q_{sim,i}}{\sum_{i=1}^{N} Q_{obs,i}} \tag{18}$$

$$R^2 = \frac{\left[\sum_{i=1}^{N}(Q_{obs,i} - \overline{Q_{obs}})(Q_{sim,i} - \overline{Q_{sim}})\right]^2}{\sum_{i=1}^{N}(Q_{obs,i} - \overline{Q_{obs}})^2 \sum_{i=1}^{N}(Q_{sim,i} - \overline{Q_{sim}})^2} \tag{19}$$

where the weights (w_3, w_5, w_8) used in the objective function were calculated following the prescribed equations outlined in the SWAT-CUP Premium handbook:

$$w_3 = 1, w_5 = |avg\_goal\_R^2/avg\_goal\_NSE|, w_8 = avg\_goal\_R^2/avg\_goal\_Bias \tag{20}$$

The findings validated the suitability of these three hydrological models for assessing the attribution on runoff variations in the Xiliugou basin.

2.7. Calculation of Contribution of Climate Factors and Human Activities to Runoff Variations

To quantify the variations in runoff resulting from climate change and human activities, the following equations were utilized for the computation:

$$\Delta Q = Q_{past} - Q_{pre} = \Delta Q_C + \Delta Q_H \tag{21}$$

$$\Delta Q_H = Q_{past} - Q_{sim} \tag{22}$$

$$\Delta Q_C = \Delta Q - \Delta Q_H = Q_{sim} - Q_{pre} \tag{23}$$

where the change in runoff (ΔQ) can be attributed to two primary factors: climate variability (ΔQ_C) and human activities (ΔQ_H). To determine their relative contributions, the observed annual average runoff during the baseline period (Q_{past}) and the change period (Q_{pre}) were utilized. Additionally, the multiyear average simulated runoff (Q_{sim}), which only accounts for the impact of climate factors, was obtained. The relative contributions of climate factors (η_C) and human disturbance (η_H) towards the overall runoff variations were calculated as follows:

$$\eta_C = \frac{|\Delta Q_C|}{|\Delta Q_C| + |\Delta Q_H|} \times 100\% \tag{24}$$

$$\eta_H = \frac{|\Delta Q_H|}{|\Delta Q_C| + |\Delta Q_H|} \times 100\% \tag{25}$$

3. Results

3.1. Abrupt Change analysee of Annual Runoff

The outcomes of the Mann–Kendall (MK) test indicated a statistically significant decrease in the annual runoff ($Z = -4.25$, $p < 0.01$). Additionally, a sudden shift was identified in 2006, as depicted in Figure 4.

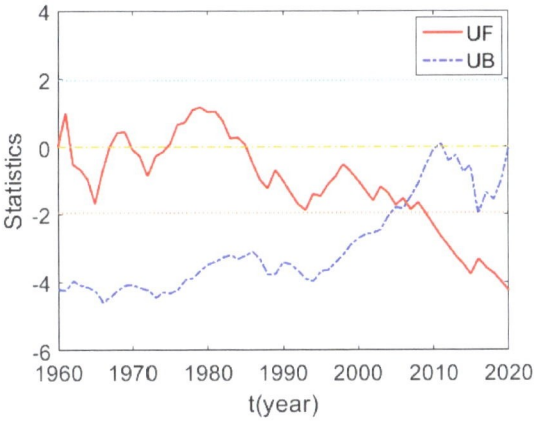

Figure 4. Analysis of the long-term trend and identification of abrupt changes in the annual runoff sequence within the Xiliugou basin spanning from 1960 to 2020.

The precipitation–runoff DMC in the Xiliugou basin was generated based on the identified abrupt change points (refer to Figure 5). The analysis confirmed the validity of the detected abrupt change point in 2006.

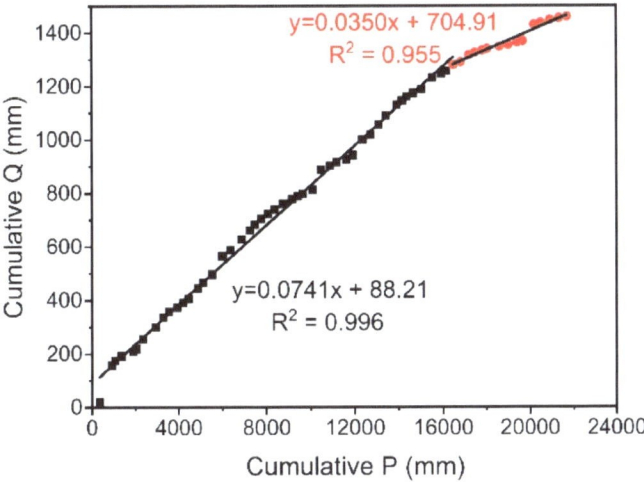

Figure 5. Precipitation–runoff double mass curve relationship in Xiliugou basin from 1960 to 2020.

3.2. Calibration and Validation Results of Three Hydrological Models

In this study, the calibration and validation of the model were performed using the baseline period. The calibration period ranged from 1960 to 1992, while the validation period spanned from 1993 to 2006. The hydrology models were evaluated based on multiple criteria, and their performance is illustrated in Figure 6. The assessment indicated satisfactory results, with NSE and R^2 values exceeding 0.5 and bias values falling within the acceptable range of ±0.15 during both the calibration and validation periods [43,48]. Consequently, these three hydrology models were considered appropriate for analyzing the effects of climate change and human activities on runoff changes in the Xiliugou basin.

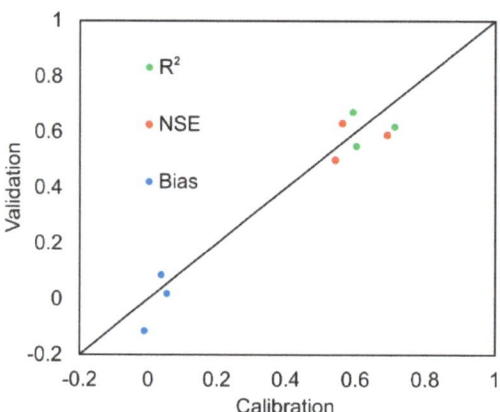

Figure 6. The calibration and validation periods of the three hydrology models were evaluated using NSE, R^2, and Bias values.

3.3. The Impact of Climate Change and Human Activities on Changes in Runoff

In this study (refer to Figure 7), significant variations were observed in the results obtained from the six different methodologies, indicating the presence of methodological uncertainties. The DMC method calculated that climate change contributed to 40.4% of the runoff change, while human activities accounted for the remaining 59.6%. Similarly, the SCRCQ method indicated a contribution rate of 39.9% for climate change and 60.1% for human activities. By employing the Budyko equation, the runoff change was attributed to climate change at a rate of 12.9%, whereas human activities contributed to 87.1%. When considering the three hydrological models, their respective contributions to climate change were 6.9%, 16%, and 28.9%, while the contributions of human activities were 93.1%, 84%, and 71.1%, respectively. According to the six methods used, the average contribution of climate change was determined to be 24.2%, while human activities accounted for an average contribution of 75.8%. Consistently, all the methods indicated that human activities significantly influenced the modification of runoff patterns in the Xiliugou basin. Figure 8 presents the simulated monthly runoff process lines by the three hydrological models and the measured monthly runoff process line during the impact/change period (2007–2020). It is evident that the simulated monthly runoff processes by all three hydrological models exhibited consistency during the impact period. Notably, the SWAT model produced the highest simulation results. However, the three hydrological models demonstrated limited capability in capturing extreme runoff events, with all simulations underestimating the magnitude of extreme runoff occurrences in 2016.

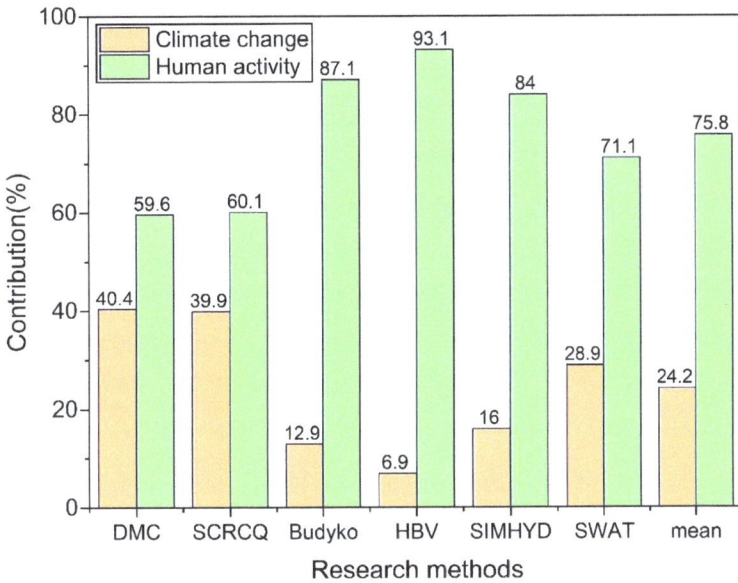

Figure 7. The assessment of runoff change was conducted using six research methods to determine the respective contributions of climate change and human activities.

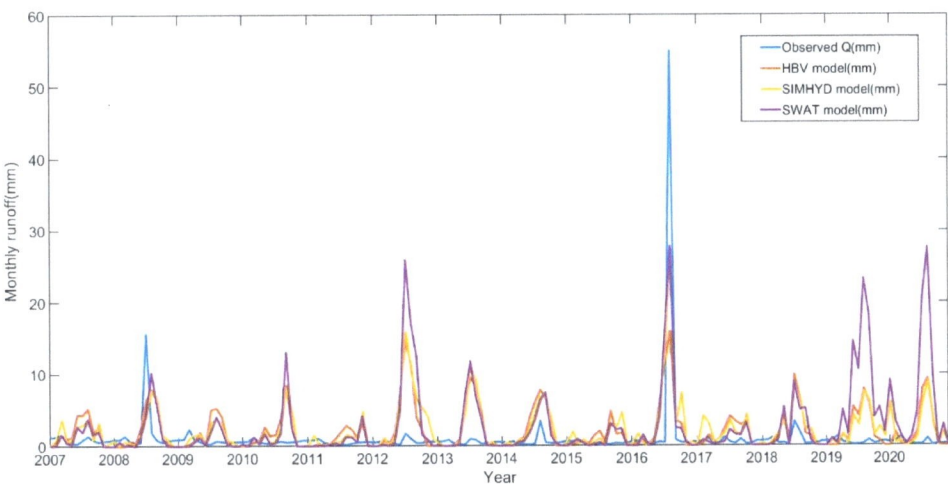

Figure 8. The simulated monthly runoff process lines by the three hydrological models and measured monthly runoff process line during 2007–2020.

4. Discussion

In this study, the MK test was utilized as a statistical tool to identify the point of abrupt change in the runoff series. The analysis indicated that the year 2006 was identified as the observed abrupt change point. This finding was further substantiated by examining the precipitation–runoff double mass curve, which displayed a gradual increase in cumulative runoff around the aforementioned year, thereby corroborating the results obtained from the MK test. The results showed that the contribution rates calculated by the two statistical analysis methods were extremely close, which could be enforced by the similar principle of mechanism of both methods. The results calculated by the three hydrological models were slightly different, which may have been caused by the different model structures and physical mechanisms of the three hydrological models.

The collective findings from all the utilized methodologies consistently demonstrated that the variations in runoff within the Xiliugou basin were predominantly influenced by human activities. This observation is consistent with prior studies conducted in the Yellow River basin, which further strengthens the understanding that human activities have a notable impact on the formation of runoff patterns in these areas [7,19,24,28,49]. The typical human activities entail the implementation of comprehensive soil and water conservation management practices in the Xiliugou basin. These activities include engineering and vegetation ecological measures. The implementation of comprehensive soil and water conservation practices in the Xiliugou basin commenced in the 1960s. However, due to the low investment, small management scale, low conservation rate, and fewer gully dam projects for flood and sand control, the problem has not been taken care thoroughly. During the 1990s, the Ministry of Water Resources, together with the Inner Mongolia Autonomous Region and relevant departments in Ordos, demonstrated a strong dedication to executing the initial phase of the soil and water conservation project in the Dalate banner, which received financial support from the World Bank. The basic aims of that project were afforestation, grass plantation, soil and water conservation, and cancelling all sloping farmland above 25°. To bolster vegetation coverage and combat soil erosion in the Loess Plateau, a collaborative endeavor spanning over two decades has been undertaken. These sustained endeavors have led to significant improvements in the overall vegetation condition within the region. Several vegetation restoration programs were introduced in the late 1990s. The "Natural Forest Conservation Program" and the "Grain for Green Program" [5,11,31] are the most prominent projects introduced for vegetation restoration in the region. Over a span of more than two decades, the implementation of these projects has resulted in substantial improvements in the vegetation coverage of the Loess Plateau [50,51]. Figure 9 shows the annual changes in the leaf area index (LAI) in the Xiliugou basin from 1982 to 2020. It can be observed that the implementation of large-scale vegetation restoration measures after 1997 significantly increase the LAI trends from 0.12 to 0.36 between 1997 and 2020.

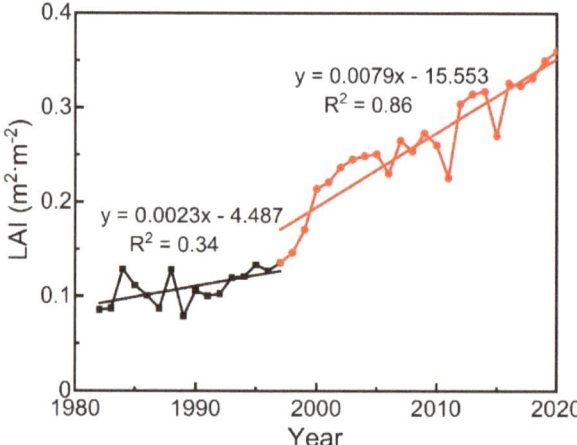

Figure 9. Annual LAI changes in Xiliugou basin from 1982 to 2020.

In addition to the extensive vegetation restoration initiatives, the construction of check dams also plays a crucial role in conserving soil and water resources. In recent years, a substantial number of check dams have been erected in the Xiliugou basin to mitigate soil erosion. Figure 10 illustrates the annual variations in the count and storage of check dams in the Xiliugou basin. The construction of these dams primarily commenced after

2000, with a particular focus between 2005 and 2010. By 2018, the Xiliugou basin hosted a total of 105 check dams, including 39 large/backbone dams, 31 medium-sized dams, and 35 small dams. The total storage capacity corresponding to check dams is 0.493×10^8 m^3. These intense human activities have substantially affected on the runoff distributions and patterns in the studied basin.

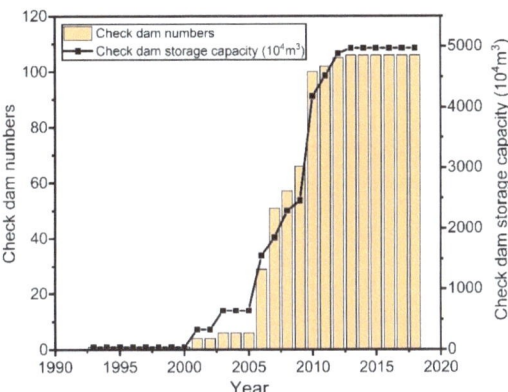

Figure 10. The temporal variation in the annual count and storage of check dams in the Xiliugou basin.

Because there is an absence of meteorological stations within the investigated basin (Figure 1), the meteorological data (such as precipitation, temperature, etc.) used in this study could be a little different from the actual meteorological data in the basin. Therefore, there could be some uncertainties in the interannual variation in the hydrometeorological series and the role of climate change and human activities in the runoff change. To enhance the scientific rigor and accuracy of the research findings, it is imperative to intensify the collection and assessment of meteorological data within the basin during subsequent research endeavors.

The current research approaches for distinguishing the impacts of climate change and human activities on basin runoff changes rely on the assumption that climate change and human activities are relatively independent factors. The contribution of climate variations to runoff change is evaluated using various methodologies, while the remaining portion is attributed to human activities. However, the precise quantitative assessment of the interplay between climate change and human activities in runoff changes is still not fully understood due to the complex multi-factorial influence mechanisms and interactions involved. Therefore, isolating the impacts of climate factors and human activities cannot be achieved simplistically. For instance, climate change can influence human activities, and vice versa. Given the uncertainties associated with the methodology and limitations of hydrological data, future studies should adopt a comprehensive approach that considers various factors to accurately attribute the response of runoff to climate change and human activities. This holistic approach will contribute to ensuring water resource security and facilitate a more informed regulation and control of water resources.

5. Conclusions

Six distinct methods were employed to discern the respective impacts of climate change and human activities on runoff changes within the Xiliugou basin in this study. By conducting the Mann–Kendall test, an abrupt change point in the runoff series was identified in 2006. The findings indicated a significant decrease in annual runoff from 1960 to 2020. To account for this abrupt change point, the research period was divided into two distinct periods: 1960–2006 as the baseline period and 2007–2020 as the change period/impact period. The results obtained from all the methodologies consistently demonstrated that human activities predominantly drove the observed runoff changes. The average contribution rates attributed to climate change and human activities across all six methods were determined to be 24.2% and 75.8%, respectively. Notably, the Xiliugou basin exhibits typical human activities characterized by soil and water conservation measures. In recent years, the implementation of vegetation restoration programs and check dam construction, aimed at enhancing vegetation coverage and promoting soil and water conservation, have significantly impacted the runoff processes in the region. This study provides valuable insights into the medium-to-long-term consequences of soil and water conservation measures on water availability, spanning from the catchment to regional scales. The gained insights hold immense significance in fostering sustainable water resource management within river basins.

Author Contributions: P.M.: investigation, formal analysis, data curation, writing—original draft, visualization. D.: data curation and supervision. X.L.: data curation, methodology, revising manuscript. S.N.: supervision and revising manuscript. A.K.: supervision and revising manuscript. H.M., Y.D. and R.W.: supervision. J.L.: conceptualization, methodology, supervision, writing—revising manuscript. All authors have read and agreed to the published version of the manuscript.

Funding: This work was supported by the "Science for a Better Development of Inner Mongolia" Program (KJXM-EEDS-2020005) of the Bureau of Science and Technology of the Inner Mongolia Autonomous Region. The "Quantifying Groundwater Changes and Development of Better Agricultural Water Saving Techniques in the Western Ordos" Program was funded by the Bureau of Science and Technology of the Erdos. The Natural Science Foundation of Inner Mongolia (2021BS04009).

Data Availability Statement: Data are contained within the article.

Conflicts of Interest: The authors declare no conflict of interest.

References

1. Kleidon, A.; Renner, M. Thermodynamic Limits of Hydrologic Cycling within the Earth System: Concepts, Estimates and Implications. *Hydrol. Earth Syst. Sci.* **2013**, *17*, 2873–2892. [CrossRef]
2. Yang, D.; Yang, Y.; Xia, J. Hydrological Cycle and Water Resources in a Changing World: A Review. *Geogr. Sustain.* **2021**, *2*, 115–122. [CrossRef]
3. Dey, P.; Mishra, A. Separating the Impacts of Climate Change and Human Activities on Streamflow: A Review of Methodologies and Critical Assumptions. *J. Hydrol.* **2017**, *548*, 278–290. [CrossRef]
4. Li, C.; Wang, L.; Wanrui, W.; Qi, J.; Linshan, Y.; Zhang, Y.; Lei, W.; Cui, X.; Wang, P. An Analytical Approach to Separate Climate and Human Contributions to Basin Streamflow Variability. *J. Hydrol.* **2018**, *559*, 30–42. [CrossRef]
5. Zhang, L.; Nan, Z.; Yu, W.; Zhao, Y.; Xu, Y. Comparison of Baseline Period Choices for Separating Climate and Land Use/Land Cover Change Impacts on Watershed Hydrology Using Distributed Hydrological Models. *Sci. Total Environ.* **2018**, *622–623*, 1016–1028. [CrossRef]
6. Schewe, J.; Heinke, J.; Gerten, D.; Haddeland, I.; Arnell, N.W.; Clark, D.B.; Dankers, R.; Eisner, S.; Fekete, B.M.; Colón-González, F.J.; et al. Multimodel Assessment of Water Scarcity under Climate Change. *Proc. Natl. Acad. Sci. USA* **2014**, *111*, 3245–3250. [CrossRef]
7. Yin, J.; Gentine, P.; Zhou, S.; Sullivan, S.C.; Wang, R.; Zhang, Y.; Guo, S. Large Increase in Global Storm Runoff Extremes Driven by Climate and Anthropogenic Changes. *Nat. Commun.* **2018**, *9*, 4389. [CrossRef]
8. Zhang, W.; Villarini, G.; Vecchi, G.A.; Smith, J.A. Urbanization Exacerbated the Rainfall and Flooding Caused by Hurricane Harvey in Houston. *Nature* **2018**, *563*, 384–388. [CrossRef] [PubMed]
9. Jaramillo, F.; Destouni, G. Local Flow Regulation and Irrigation Raise Global Human Water Consumption and Footprint. *Science* **2015**, *350*, 1248–1251. [CrossRef]
10. Wang, G.; Zhang, J.; Yang, Q. Attribution of Runoff Change for the Xinshui River Catchment on the Loess Plateau of China in a Changing Environment. *Water* **2016**, *8*, 267. [CrossRef]
11. Li, L.; Ni, J.; Chang, F.; Yue, Y.; Frolova, N.; Magritsky, D.; Borthwick, A.G.L.; Ciais, P.; Wang, Y.; Zheng, C.; et al. Global Trends in Water and Sediment Fluxes of the World's Large Rivers. *Sci. Bull.* **2020**, *65*, 62–69. [CrossRef]
12. Shi, H.; Hu, C.; Wang, Y.; Liu, C.; Li, H. Analyses of Trends and Causes for Variations in Runoff and Sediment Load of the Yellow River. *Int. J. Sediment Res.* **2017**, *32*, 171–179. [CrossRef]
13. Du, J.; Shi, C. Effects of Climatic Factors and Human Activities on Runoff of the Weihe River in Recent Decades. *Quat. Int.* **2012**, *282*, 58–65. [CrossRef]
14. Miao, C.; Ni, J.; Borthwick, A.G.L.; Yang, L. A Preliminary Estimate of Human and Natural Contributions to the Changes in Water Discharge and Sediment Load in the Yellow River. *Glob. Planet. Change* **2011**, *76*, 196–205. [CrossRef]

15. Gao, P.; Li, P.; Zhao, B.; Xu, R.; Zhao, G.; Sun, W.; Mu, X. Use of Double Mass Curves in Hydrologic Benefit Evaluations. *Hydrol. Process.* **2017**, *31*, 4639–4646. [CrossRef]
16. Xue, L.; Yang, F.; Yang, C.; Chen, X.; Zhang, L.; Chi, Y.; Yang, G. Identification of Potential Impacts of Climate Change and Anthropogenic Activities on Streamflow Alterations in the Tarim River Basin, China. *Sci Rep* **2017**, *7*, 8254. [CrossRef] [PubMed]
17. Wang, S.; Yan, M.; Yan, Y.; Shi, C.; He, L. Contributions of Climate Change and Human Activities to the Changes in Runoff Increment in Different Sections of the Yellow River. *Quat. Int.* **2012**, *282*, 66–77. [CrossRef]
18. Sun, W.; Song, X.; Zhang, Y.; Chiew, F.; Post, D.; Zheng, H.; Song, S. Coal Mining Impacts on Baseflow Detected Using Paired Catchments. *Water Resour. Res.* **2020**, *56*, e2019WR025770. [CrossRef]
19. Liang, K. Comparative Investigation on the Decreased Runoff between the Water Source and Destination Regions in the Middle Route of China's South-to-North Water Diversion Project. *Stoch. Environ. Res. Risk Assess.* **2018**, *32*, 369–384. [CrossRef]
20. Xu, X.; Yang, D.; Yang, H.; Lei, H. Attribution Analysis Based on the Budyko Hypothesis for Detecting the Dominant Cause of Runoff Decline in Haihe Basin. *J. Hydrol.* **2014**, *510*, 530–540. [CrossRef]
21. Xu, X.; Liu, W.; Scanlon, B.R.; Zhang, L.; Pan, M. Local and Global Factors Controlling Water-Energy Balances within the Budyko Framework. *Geophys. Res. Lett.* **2013**, *40*, 6123–6129. [CrossRef]
22. Yang, Y.; Zhang, S.; McVicar, T.R.; Beck, H.E.; Zhang, Y.; Liu, B. Disconnection between Trends of Atmospheric Drying and Continental Runoff. *Water Resour. Res.* **2018**, *54*, 4700–4713. [CrossRef]
23. Zhou, X.; Yang, Y.; Sheng, Z.; Zhang, Y. Reconstructed Natural Runoff Helps to Quantify the Relationship between Upstream Water Use and Downstream Water Scarcity in China's River Basins. *Hydrol. Earth Syst. Sci.* **2019**, *23*, 2491–2505. [CrossRef]
24. Feng, D.; Zheng, Y.; Mao, Y.; Zhang, A.; Wu, B.; Li, J.; Tian, Y.; Wu, X. An Integrated Hydrological Modeling Approach for Detection and Attribution of Climatic and Human Impacts on Coastal Water Resources. *J. Hydrol.* **2018**, *557*, 305–320. [CrossRef]
25. Jehanzaib, M.; Shah, S.A.; Yoo, J.; Kim, T.-W. Investigating the Impacts of Climate Change and Human Activities on Hydrological Drought Using Non-Stationary Approaches. *J. Hydrol.* **2020**, *588*, 125052. [CrossRef]
26. Li, X.; Zhang, Y.; Ma, N.; Li, C.; Luan, J. Contrasting Effects of Climate and LULC Change on Blue Water Resources at Varying Temporal and Spatial Scales. *Sci. Total Environ.* **2021**, *786*, 147488. [CrossRef]
27. Yan, R.; Zhang, X.; Yan, S.; Zhang, J.; Chen, H. Spatial Patterns of Hydrological Responses to Land Use/Cover Change in a Catchment on the Loess Plateau, China. *Ecol. Indic.* **2018**, *92*, 151–160. [CrossRef]
28. Luan, J.; Zhang, Y.; Ma, N.; Tian, J.; Li, X.; Liu, D. Evaluating the Uncertainty of Eight Approaches for Separating the Impacts of Climate Change and Human Activities on Streamflow. *J. Hydrol.* **2021**, *601*, 126605. [CrossRef]
29. Fu, B.; Wang, S.; Liu, Y.; Liu, J.; Liang, W.; Miao, C. Hydrogeomorphic Ecosystem Responses to Natural and Anthropogenic Changes in the Loess Plateau of China. *Annu. Rev. Earth Planet. Sci.* **2017**, *45*, 223–243. [CrossRef]
30. Li, P.; Mu, X.; Holden, J.; Wu, Y.; Irvine, B.; Wang, F.; Gao, P.; Zhao, G.; Sun, W. Comparison of Soil Erosion Models Used to Study the Chinese Loess Plateau. *Earth-Sci. Rev.* **2017**, *170*, 17–30. [CrossRef]
31. Cao, S.; Chen, L.; Shankman, D.; Wang, C.; Wang, X.; Zhang, H. Excessive Reliance on Afforestation in China's Arid and Semi-Arid Regions: Lessons in Ecological Restoration. *Earth-Sci. Rev.* **2011**, *104*, 240–245. [CrossRef]
32. Gao, P.; Deng, J.; Chai, X.; Mu, X.; Zhao, G.; Shao, H.; Sun, W. Dynamic Sediment Discharge in the Hekou–Longmen Region of Yellow River and Soil and Water Conservation Implications. *Sci. Total Environ.* **2017**, *578*, 56–66. [CrossRef]
33. Qian, L.; Dang, S.; Bai, C.; Wang, H. Variation in the Dependence Structure between Runoff and Sediment Discharge Using an Improved Copula. *Theor. Appl. Climatol.* **2021**, *145*, 285–293. [CrossRef]
34. Mann, H.B. Nonparametric Tests Against Trend. *Econometrica* **1945**, *13*, 245. [CrossRef]
35. Mavromatis, T.; Stathis, D. Response of the Water Balance in Greece to Temperature and Precipitation Trends. *Theor. Appl. Climatol.* **2011**, *104*, 13–24. [CrossRef]
36. Kohler, M.A. On the Use of Double-Mass Analysis for Testing the Consistency of Meteorological Records and for Making Required Adjustments. *Bull. Am. Meteorol. Soc.* **1949**, *30*, 188–195. [CrossRef]
37. Fu, B. The calculation of the evaporation from land surface. *Chin. J. Atmos. Sci.* (In Chinese). **1981**, *5*, 23–31.
38. Krysanova, V.; Bronstert, A.; Müller-Wohlfeil, D.-I. Modelling River Discharge for Large Drainage Basins: From Lumped to Distributed Approach. *Hydrol. Sci. J.* **1999**, *44*, 313–331. [CrossRef]
39. Chiew, F.H.S.; Kirono, D.G.C.; Kent, D.M.; Frost, A.J.; Charles, S.P.; Timbal, B.; Nguyen, K.C.; Fu, G. Comparison of Runoff Modelled Using Rainfall from Different Downscaling Methods for Historical and Future Climates. *J. Hydrol.* **2010**, *387*, 10–23. [CrossRef]
40. Li, F.; Zhang, Y.; Xu, Z.; Teng, J.; Liu, C.; Liu, W.; Mpelasoka, F. The Impact of Climate Change on Runoff in the Southeastern Tibetan Plateau. *J. Hydrol.* **2013**, *505*, 188–201. [CrossRef]
41. Li, F.; Zhang, Y.; Xu, Z.; Liu, C.; Zhou, Y.; Liu, W. Runoff Predictions in Ungauged Catchments in Southeast Tibetan Plateau. *J. Hydrol.* **2014**, *511*, 28–38. [CrossRef]
42. Zhang, Y.; Vaze, J.; Chiew, F.H.S.; Teng, J.; Li, M. Predicting Hydrological Signatures in Ungauged Catchments Using Spatial Interpolation, Index Model, and Rainfall–Runoff Modelling. *J. Hydrol.* **2014**, *517*, 936–948. [CrossRef]
43. Zhang, Y.; Chiew, F.H.S. Relative Merits of Different Methods for Runoff Predictions in Ungauged Catchments: RUNOFF PREDICTIONS IN UNGAUGED CATCHMENT. *Water Resour. Res.* **2009**, *45*, 1–13. [CrossRef]
44. Arnold, J.G.; Srinivasan, R.; Muttiah, R.S.; Williams, J.R. Large Area Hydrologic Modeling and Assessment PART I: Model Development. *J. Am. Water Resour. Assoc.* **1998**, *34*, 73–89. [CrossRef]

45. Arnold, J.G.; Allen, P.M. Estimating Hydrologic Budgets forstand Three Illinois Watersheds. *J. Hydrol.* **1996**, *176*, 57–77. [CrossRef]
46. Eberhart, R.; Kennedy, J. A New Optimizer Using Particle Swarm Theory. In Proceedings of the MHS'95. Proceedings of the Sixth International Symposium on Micro Machine and Human Science, Nagoya, Japan, 4–6 October 1995; pp. 39–43.
47. Nash, J.E.; Sutcliffe, J.V. River forcasting using conceptual models. Part I: A discussion of principles. *J. Hydrol.* **1970**, *10*, 280–290. [CrossRef]
48. Moriasi, D.N.; Arnold, J.G.; Van Liew, M.W.; Bingner, R.L.; Harmel, R.D.; Veith, T.L. Model Evaluation Guidelines for Systematic Quantification of Accuracy in Watershed Simulations. *Trans. ASABE* **2007**, *50*, 885–900. [CrossRef]
49. Luan, J.; Miao, P.; Tian, X.; Li, X.; Ma, N.; Abrar Faiz, M.; Xu, Z.; Zhang, Y. Estimating Hydrological Consequences of Vegetation Greening. *J. Hydrol.* **2022**, *611*, 128018. [CrossRef]
50. Feng, X.; Fu, B.; Piao, S.; Wang, S.; Ciais, P.; Zeng, Z.; Lü, Y.; Zeng, Y.; Li, Y.; Jiang, X.; et al. Revegetation in China's Loess Plateau Is Approaching Sustainable Water Resource Limits. *Nat. Clim. Change* **2016**, *6*, 1019–1022. [CrossRef]
51. Piao, S.; Yin, G.; Tan, J.; Cheng, L.; Huang, M.; Li, Y.; Liu, R.; Mao, J.; Myneni, R.B.; Peng, S.; et al. Detection and Attribution of Vegetation Greening Trend in China over the Last 30 Years. *Glob. Change Biol.* **2015**, *21*, 1601–1609. [CrossRef]

Disclaimer/Publisher's Note: The statements, opinions and data contained in all publications are solely those of the individual author(s) and contributor(s) and not of MDPI and/or the editor(s). MDPI and/or the editor(s) disclaim responsibility for any injury to people or property resulting from any ideas, methods, instructions or products referred to in the content.

Article

The Response of NDVI to Drought at Different Temporal Scales in the Yellow River Basin from 2003 to 2020

Wen Liu [1,2]

1 College of Urban and Environmental Sciences, Northwest University, Xi'an 710127, China; liuwen@nwu.edu.cn
2 Shaanxi Key Laboratory of Earth Surface System and Environmental Carrying Capacity, College of Urban and Environmental Sciences, Northwest University, Xi'an 710127, China

Abstract: Ecological protection in the Yellow River Basin (YRB) is a major strategy for China's sustainable development. Amid global warming, droughts have occurred more frequently, severely affecting vegetation growth. Based on the Standardized Precipitation Evapotranspiration Index (SPEI) and Normalized Difference Vegetation Index (NDVI) at different time scales from 2003 to 2020, this study employed the linear trend method and the Spearman correlation coefficient method to calculate the trends and correlation coefficients of NDVI and SPEI at different scales at the pixel scale and explored the spatial distribution pattern of the sensitivity of vegetation growth in the YRB to drought. The results show that: (1) NDVI and SPEI are positively correlated in 77% of the area, negatively correlated in 9%, and are positively correlated in the arid and semi-arid areas, while negatively correlated in the humid and subhumid areas. The significant negative correlation between NDVI and drought at high altitudes may be due to the fact that Gramineae vegetation is more sensitive to drought, with heat being more affected than water. (2) Urbanization has a relatively obvious impact on the distribution of drought. Extreme drought mainly occurs in the middle and upper reaches of the Wei River; severe drought mainly occurs in the central area of the Guanzhong Plain centered on Xi'an; the central area of the Loess Plateau; and the surrounding areas of the Zhengzhou-centered Central Plains City Group. (3) The NDVI showed an upward trend from 2003 to 2020, indicating an increase in vegetation density or an expansion of vegetation coverage. From the temporal trend, SPEI decreased at a rate of -0.17/decade, indicating that the entire watershed has a drought trend on an annual scale. (4) Spring NDVI is more sensitive to the water supply provided by SPEI-1, while the positive correlation between SPEI and NDVI begins to rise in June and reaches its peak in July, then starts to decline in August. In autumn and winter, NDVI is more sensitive to 3–6-month accumulated drought. (5) From the dynamic transmission laws of different levels of positive correlation, the positive impact of the 3-month accumulated drought on NDVI is most significant, and the influence of SPEI-1 on the negative correlation between SPEI and NDVI is most significant. This paper aims to clarify the sensitivity of vegetation to different time-scale droughts, provide a basis for alleviating drought in the YRB, and promote sustainable development of ecological environmental protection. The research findings enable us to gain a profound insight into the responsiveness of vegetation growth to drought in the context of global warming and offer a valuable theoretical foundation for devising pertinent measures to alleviate stress on vegetation growth in regions prone to frequent droughts.

Keywords: Standardized Precipitation Evapotranspiration Index (SPEI); Normalized Difference Vegetation Index (NDVI); different time scales; spatial difference; Yellow River Basin (YRB)

Citation: Liu, W. The Response of NDVI to Drought at Different Temporal Scales in the Yellow River Basin from 2003 to 2020. *Water* **2024**, *16*, 2416. https://doi.org/10.3390/w16172416

Academic Editor: Steven G. Pueppke

Received: 3 July 2024
Revised: 22 August 2024
Accepted: 24 August 2024
Published: 27 August 2024

Copyright: © 2024 by the author. Licensee MDPI, Basel, Switzerland. This article is an open access article distributed under the terms and conditions of the Creative Commons Attribution (CC BY) license (https://creativecommons.org/licenses/by/4.0/).

1. Introduction

The dependence of vegetation growth on water fully reflects the water, carbon, and energy exchange between land and atmosphere [1–4], and drought has a significant negative impact on the function of terrestrial ecosystems [5]. Extreme drought reduces the strength of terrestrial carbon sinks [6] and leads to imbalance in terrestrial ecosystems [7]. The accelerated water cycle brought by global warming is limited by soil moisture, resulting in a weak water supply in the atmosphere [8–11], and drought is intensifying in frequency, intensity, and duration, triggering drought self-propagation [12], with more vegetation growth affected by water scarcity [13]. However, whether vegetation has the ability to adapt to changes in water availability under the background of climate warming is unknown, which hinders our deep understanding of the physiological mechanisms by which drought affects vegetation [6].

The rise in global vegetation coverage and enhanced net primary productivity has led to an increased demand for water by plants [14,15], creating a positive feedback loop in ecosystem water scarcity [16,17]. This results in a mutual constraint between plant growth and water availability [18,19]. Factors such as vegetation types, sudden droughts, urban heat waves, and prolonged low water supply all play significant roles in influencing regional vegetation cover and plant growth [20,21], with drought exerting a particularly detrimental impact on ecosystems' vegetation [22]. However, non-water-limited high-latitude regions may have beneficial effects on vegetation function within the context of climate change [23]. Therefore, studying the relationship between vegetation dynamics and drought can enhance our understanding of plant sensitivity and resilience to drought under changing conditions.

The development of drought in the Yellow River basin (YRB) is often accompanied by warm periods. Even if precipitation occurs at the same time, it is difficult to prevent the drought from causing destructive impacts on ecosystems during the plant-growing period [21]. The YRB encompasses a large area of arid and semi-arid regions as well as semi-humid areas with diverse vegetation cover, including grasslands, sandy areas, original forests, artificial forests, and construction land. Compared with grasslands and forests, arid and semi-arid areas with higher agricultural or construction coverage experience shorter drought propagation times, while drought propagation speed in forested areas is slower [24]. Research indicates that agricultural land exhibits the greatest response to drought, followed by urban and rural land; meanwhile, forests have the weakest response to the drought index due to their strong resistance and resilience [25]. Therefore, vegetation sensitivity to drought is not only related to water scarcity but also associated with vegetation type and water-heat combination in different seasons.

The seasonal dynamics and extent of dry spells play a crucial role in shaping how these events affect plant life over time [26], with winter dry spells generally lasting longer than those occurring in summer [27]. As such, investigating these phenomena across various time scales becomes essential for accurately gauging their impact on plant communities [28]. Drought-induced water scarcity places considerable stress on plants; initially, they exhibit some resistance but eventually display resilience following relief from or cessation of prolonged dry conditions [29,30], resulting in delayed responses with cumulative effects [28]. Consequently, both spatial distribution patterns as well as temporal responses vary among plant communities affected by drought, underscoring the importance of conducting detailed assessments at finer scales. The standardized precipitation evapotranspiration index (SPEI), which accounts for both precipitation levels and evaporation rates, proves highly sensitive to natural environmental fluctuations [24]; it enables assessment of varying degrees of drought severity across different time frames while highlighting accumulated impacts and delayed vegetative responses [28]. Nevertheless, there remains limited research focusing on finely-scaled evaluations assessing plant sensitivities to drought, particularly regarding dynamic responses within individual seasons.

The main objective of this study is as follows: (1) to analyze the temporal and spatial trends of SPEI and NDVI to evaluate the drought-vegetation evolution pattern of YRB from 2003 to 2020; (2) to conduct a spatial analysis of the correlation between annual scale NDVI and SPEI from 2003 to 2020 at the pixel level to elucidate the response of NDVI to SPEI at different temporal scales; and (3) to investigate the regularity of the correlation between SPEI and NDVI in space and time. Consequently, this study conducts a temporal and spatial dynamic analysis of vegetation responses across different seasons at various temporal scales. It identifies the season and month with the highest vegetation sensitivity, investigates the cumulative effect of drought and the delayed response of vegetation, and presents a research case for addressing the contradiction between regional drought and vegetation. These findings have significant implications for advancing ecological protection and promoting high-quality development in the YRB.

2. Materials and Methods

2.1. Study Area

The YRB, characterized by its west-high-east-low topography (Figure 1), is esteemed as the cradle of Chinese civilization. Originating from the lofty Qinghai-Tibet Plateau at an average elevation of 4000 m, it intricately meanders through the Inner Mongolia Plateau and the Loess Plateau at an average elevation of 1000–2000 m before gracefully descending onto the North China Plain with an average elevation of 50 m or less [31]. Finally, it merges into the Bohai Sea. Encompassing Shaanxi, Gansu, Ningxia, Qinghai, Henan, and Shanxi provinces within its embrace [32], this region boasts a diverse climate ranging from plateau to temperate zones and encompasses arid deserts as well as semi-humid areas. Notably, China's highest annual evaporation rate exceeds 2500 mm while receiving annual precipitation between 200 and 600 mm [33]. Covering a staggering 38% of China's land area and supporting one-third of its population, which contributes to one-quarter of China's GDP. This basin is home to seven city clusters, including mature ones such as Guanzhong Plain and developing ones like Hohhot-Baotou-Yinchuan [34]. However, the water resources in the YRB are characterized by significant contradictions and frequent droughts, which have severely constrained local social and economic development. The proposal and implementation of a strategy for ecological protection and high-quality development in the YRB hold far-reaching significance for China's ecological and environmental security as well as sustained high-speed economic growth.

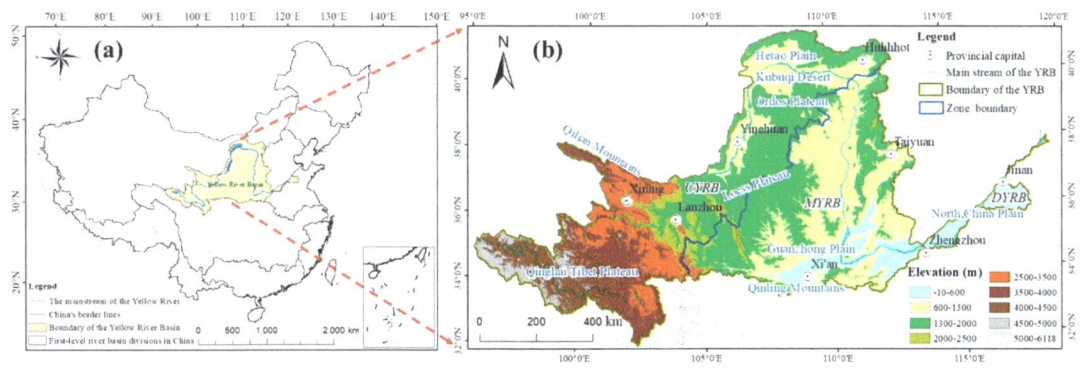

Figure 1. The geographical location (**a**) and digital elevation model (**b**) of the YRB.

2.2. Method and Materials

2.2.1. Data Source

The Normalized Difference Vegetation Index (NDVI) data come from the MOD13A2 product provided by NASA (https://ladsweb.modaps.eosdis.nasa.gov/, accessed on 1 December 2023), which is collected by the MODIS satellite and then stitched and projected to convert from sinusoidal projection to WGS84 projection coordinates, with a spatial resolution of 1 km, to obtain the monthly NDVI data for the YRB from 2003 to 2020.

The SPEI data are calculated based on meteorological data, which are sourced from the China Meteorological Data Sharing Network (https://www.cma.gov.cn/en/, accessed on 30 June 2021). Interpolate meteorological data to obtain gridded meteorological data, calculate the monthly scale SPEI data, and align the projection, resolution, and temporal scale of the SPEI data with the NDVI data. The SPEI classification is based on Table 1 [35,36], with the calculation method described in the references [37–39].

Table 1. Classification of SPEI.

Drought Grade	SPEI Value
Extreme drought	SPEI ≤ -2.00
Severe drought	$-2 < \text{SPEI} \leq -1.5$
Moderate drought	$-1.5 < \text{SPEI} \leq -1$
Mild drought	$-1 < \text{SPEI} \leq -0.5$
Normal	SPEI ≥ -0.5

2.2.2. Linear Trend Method

The trend analysis in this paper is carried out using linear least squares regression [31]. Calculate the trends of SPEI and NDVI at a resolution of 1000 m per pixel for the period 2003–2020 to fully reflect the temporal and spatial evolution characteristics of NDVI and SPEI [40]. The calculation formula is as follows:

$$Slope = \frac{n \times \sum_{i=1}^{n} (i \times y_i) - \sum_{i=1}^{n} i \sum_{i=1}^{n} y_i}{n \times \sum_{i=1}^{n} i^2 - (\sum_{i=1}^{n} i)^2} \qquad (1)$$

In the formula, n represents the time series for the period 2003–2020 under study, and y_i denotes the value of y at time i. $Slope > 0$ indicates an upward trend in y, while $Slope < 0$ indicates a downward trend. Perform a t-test on the trend values, and if the p-value is less than 0, the result is significant.

2.2.3. Correlation Analysis

Calculate the Pearson Correlation Coefficient (CC) between the 2003–2020 NDVI sequence and different temporal scales of SPEI to measure the correlation between vegetation and drought. Using Pearson's CC, the relationship between SPEI and NDVI at different temporal scales can be evaluated. The higher the CC, the better the correlation between SPEI and NDVI, and it can also clarify the lag time of vegetation's response to drought [41]. The CC classification is shown in Table 2 [42,43]. The calculation formula is as follows:

$$R_{xy} = \frac{\sum_{i=1}^{n} [(x_i - \bar{x})(y_i - \bar{y})]}{\sqrt{\sum_{i=1}^{n} (x_i - \bar{x})^2} \times \sqrt{\sum_{i=1}^{n} (y_i - \bar{y})^2}} \qquad (2)$$

Table 2. Correlation coefficient grading.

Level of CC	CC
High positive correlation	$0.8 \leq \text{CC} < 1.0$
Strong positive correlation	$0.6 \leq \text{CC} < 0.8$
Moderate positive correlation	$0.4 \leq \text{CC} < 0.6$
Weak positive correlation	$0.2 \leq \text{CC} < 0.4$
No correlation	$-0.2 < \text{CC} < 0.2$
Weak negative correlation	$-0.4 < \text{CC} \leq -0.2$
Moderate negative correlation	$-0.6 < \text{CC} \leq -0.4$
Strong negative correlation	$-0.8 < \text{CC} \leq -0.6$
High negative correlation	$-1 < \text{CC} \leq -0.8$

In this equation, n represents the number of years in the time series; R_{xy} is the CC between two influencing factors x and y, while x_i and y_i represent the values of x and y in the i-th year. Additionally, $\bar{x}$ and $\bar{y}$ denote the average values of these two influencing factors over n years.

The significance test for linear trend and correlation analysis is completed by a *t*-test. In this article, *P* = 0.05 is used as the dividing line. *P* > 0.05 is considered non-significant, while *P* ≤ 0.05 is considered significant. The formula for the *t*-test is as follows:

$$P = (x - \mu)/(s/\sqrt{n}) \tag{3}$$

In this equation, *P* represents the significance value of the *t*-test, *x* is the sample mean, *μ* is the population mean, *s* is the sample standard deviation, and *n* is the sample size.

3. Results

3.1. Spatio-Temporal Distribution Characteristics of SPEI

From 2003 to 2020, the spatial pattern of SPEI in the YRB displayed remarkable heterogeneity (Figure 2a). The SPEI exhibited lower values in the middle reaches of the Yellow River Basin (MYRB) and downstream of the Yellow River Basin (DYRB), while higher values were observed in the upper reaches of the Yellow River Basin (UYRB). Extreme drought affected only 0.11% of the total area (Figure 2b), primarily concentrated in the upper reaches of the Weihe River. Severe drought impacted 13.34% of the region, mainly encompassing the central plain area centered on Xi'an as well as areas within the loess plateau and southern part of the North China Plain, including cities such as Zhengzhou, western cities in Shanxi, and the Central Plain City Group. Moderate drought covered 39.1% of the area, predominantly found in the MYRB and DYRB and surrounding regions near Xining and Lanzhou in its upper reaches. Mild drought affected approximately 45.29% of landmasses, mainly distributed across its upper reaches, whereas areas free from drought constituted about 2.16%, primarily located at the source of YRB and Hetao Plain.

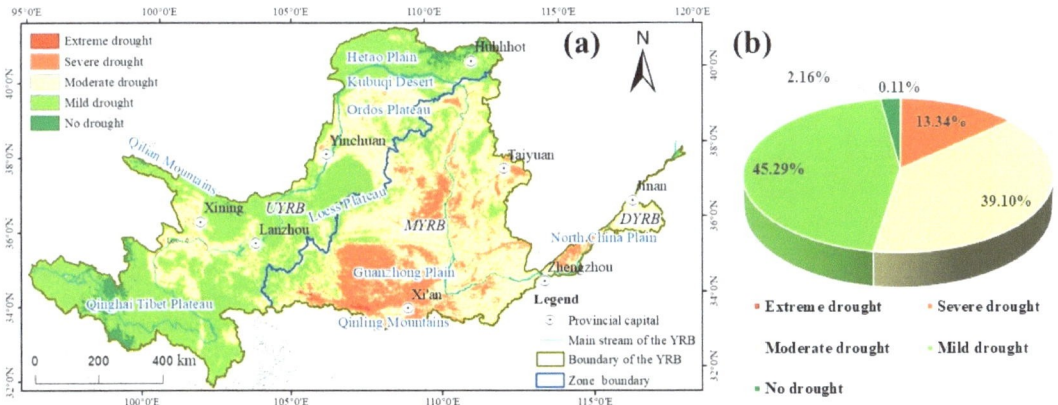

Figure 2. Spatial distribution of SPEI (**a**) and percentage of different SPEI grades (**b**).

An analysis of the spatial trends of SPEI in the YRB from 2003 to 2020 (Figure 3a) reveals that the rate of change decreases from the middle to the periphery of the basin, with five levels of change rate. The area with the highest rate of decrease is concentrated in the Guanzhong Plain and the Loess Plateau of North Shaanxi, with a rate of change of −0.7 to −1.2/decade, accounting for 8% of the total basin area. The rate of change in the surrounding areas gradually decreases until it reaches zero. In the source areas of YRB, the Hetao Plain, and DYRB, the trend of SPEI reverses to an increase, with an increase rate of up to 0.7/decade. The trend of drought relief and moisture increase is observed, accounting for 12% of the total basin area.

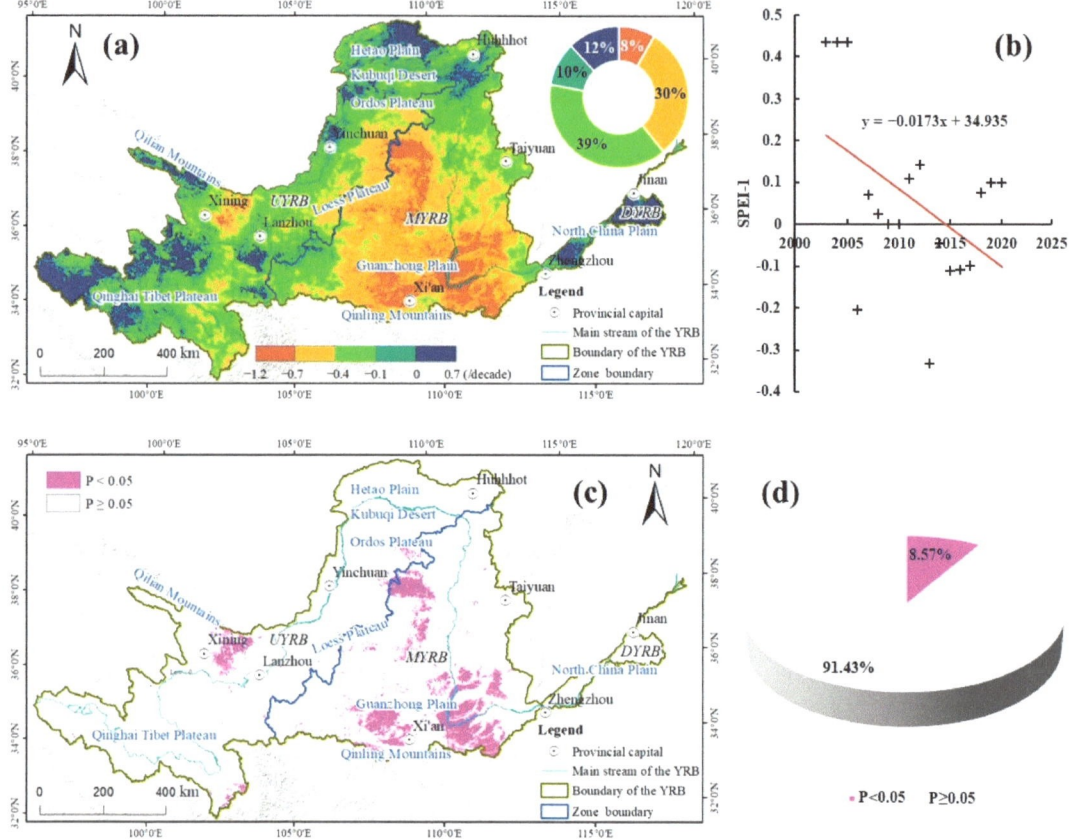

Figure 3. Spatial distribution of SPEI trend (**a**), trend of SPEI (**b**), significance test of SPEI trend (**c**), and percentage of significance test (**d**).

Observing the temporal evolution depicted in Figure 3b, it is evident that SPEI exhibits a declining trend at a rate of −0.17 per decade, signifying a drying tendency on an annual scale throughout the entire watershed. The range of SPEI variations spans from −0.33 to 0.43. Among them, 2006 and 2013 were the most severe drought years, with SPEI values of −0.21 and −0.33, respectively, indicating extreme drought and severe drought. By conducting a significance test on the spatial trend of SPEI change (Figure 3c), it was found that the trend of SPEI change passed the significance test ($P < 0.05$) in the Guanzhong Plain, the mountain and valley basins around Xining, and the loess plateau of north Shaanxi, accounting for 8.57% of the watershed area (Figure 3d).

3.2. Spatial and Temporal Distribution Characteristics of NDVI

The spatial arrangement of yearly changes in NDVI within the YRB exhibits striking diversity with a discernible correlation to elevation (Figure 4a). Approximately 40.77% of these regions demonstrate statistically significant alterations (Figure 4b), among which approximately 34.04% display an increasing tendency primarily observed at altitudes below 2000 m with a minimum significance level set at or above 0.05; conversely, around 6.73% exhibit a declining pattern predominantly situated at higher elevations within the YRB's origin area as well as sporadically across Guanzhong Plain, Hetao Plain, and Kubuqi Desert—all passing statistical tests with *p*-values less than 0.05.

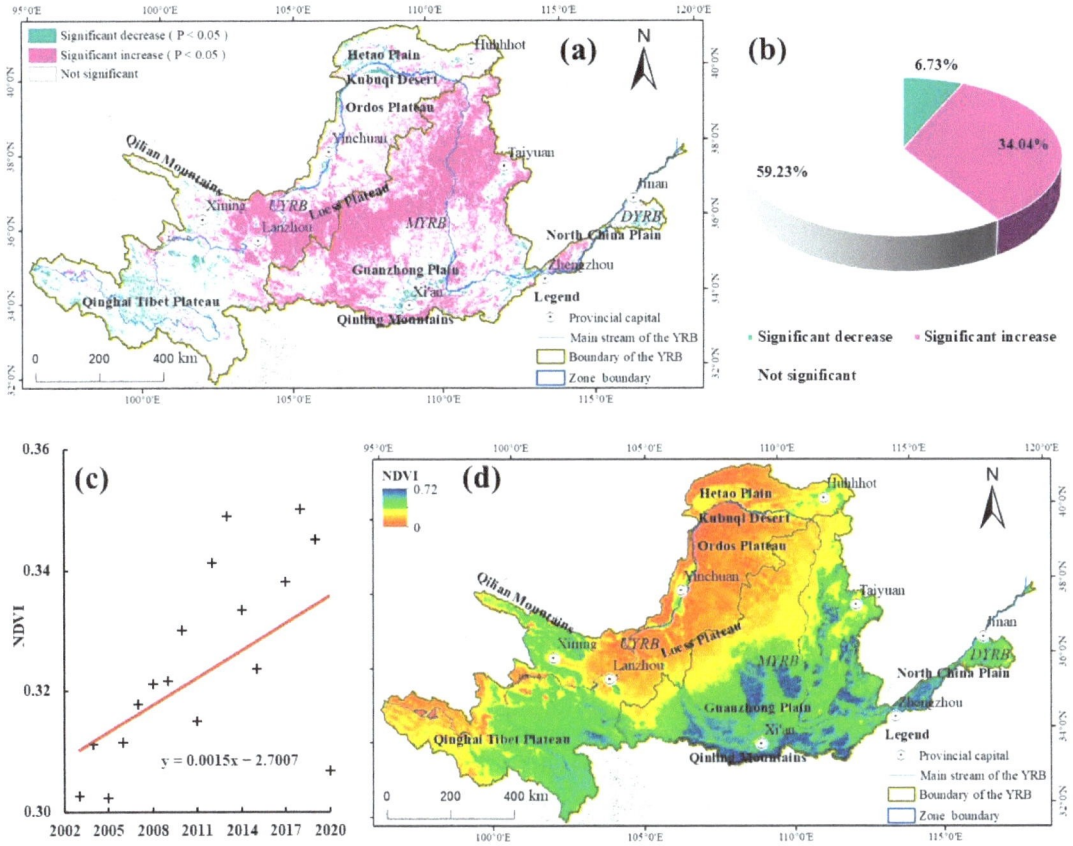

Figure 4. The spatial distribution of significant changes in NDVI trend from 2003 to 2020 (**a**), the percentage of significant changes (**b**), the trend of NDVI (**c**), and the spatial distribution of NDVI (**d**).

From 2003 to 2020, the NDVI exhibited a continuous upward trend at a rate of 0.015/decade, indicating an increase in vegetation density or expansion of vegetation coverage. Based on the current status of vegetation coverage, NDVI ranges from 0 to 0.72 within the basin. The southern part of the basin shows relatively good vegetation coverage, including the source areas of YRB, Guanzhong Plain, western Shanxi, and southern North China Plain. In contrast, the northwestern part of the basin exhibits lower vegetation coverage, encompassing the highest altitude areas of the source region of YRB and the northwestern Loess Plateau.

The spatial distribution of monthly NDVI trends in the YRB is not the same. During the entire winter season, from December to February of the following year, vegetation is in a dormant state and grows slowly, so the areas with significant NDVI decline in December and January are larger than those with significant NDVI increase, with the areas of significant decline accounting for 15%, 7%, and 17%, respectively. In February, the temperature rises, causing the area of NDVI increase to recover to 16%. However, the overall area of non-significant change is at least 77% or more (Figure 5k,l,a). During the spring period, from March to May, the areas with significant NDVI increase gradually expand from 22% to 26%, eventually covering 44% of the entire basin, mainly distributed in the middle and lower reaches of the basin (Figure 5b–d). During the summer, the areas with significant NDVI increase decreased from 56% to 36% and finally decreased to 15% in August. The areas with significant NDVI increase in the autumn are relatively stable, accounting for 4–5%, while the areas with significant NDVI decreases account for 5–14%.

It can be seen that the areas with significant NDVI increase in May, June, and July of 2003–2020 were the largest, and vegetation growth and vegetation coverage improved.

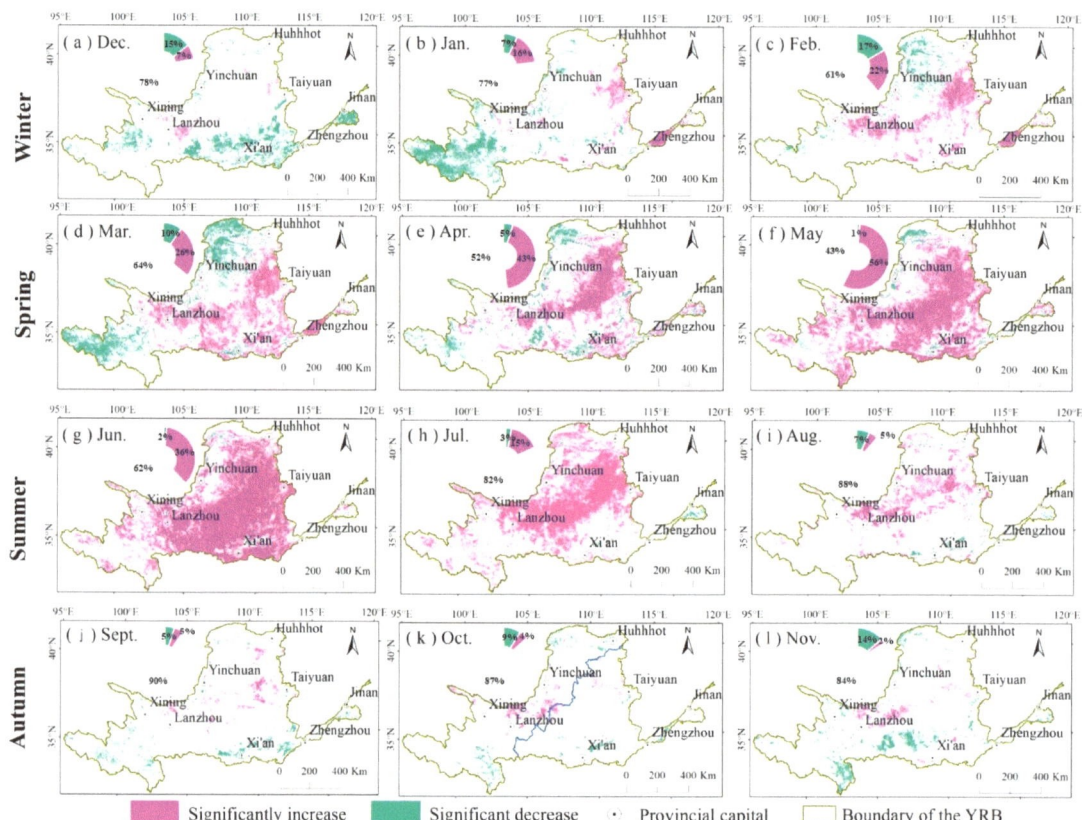

Figure 5. Significance test of monthly scale NDVI trend from 2003 to 2020.

3.3. The Response of NDVI to SPEI

An intricate spatial examination of NDVI-SPEI correlation on an annual scale spanning from 2003 to 2020 (Figure 6a) unveils that approximately three-quarters of this region exhibits positive correlations (Figure 6b). This phenomenon is predominantly concentrated within elevations below 2000 m along the lower stretches of YRB, encompassing vast portions of its MYRB and DYRB. Notably, over one-tenth of this domain showcases a CC surpassing 0.6, primarily clustered around Lanzhou while sporadically dispersed across the Hexi Corridor and vicinity to Hohhot; meanwhile, an additional 16% manifests correlations falling between 0.4 and 0.6 across widely scattered locations situated on peripheries adjoining strongly correlated zones. The region exhibiting a weak positive correlation between NDVI and SPEI is the most extensive, encompassing 50.74% of the area and spanning across the middle reaches of the Loess Plateau and the Guanzhong Plain in the YRB. Conversely, areas where NDVI and SPEI display a negative correlation are relatively limited, accounting for only 9% of the total area and primarily located in the headwaters of the YRB as well as its lower reaches. These findings indicate that in arid and semi-arid regions, NDVI and SPEI tend to exhibit a positive correlation, while in humid and subhumid areas such as those found in both the headwaters of YRB and DYRB, this relationship is reversed. Furthermore, approximately 12.19% of these correlations are statistically significant (Figure 6c,d), predominantly distributed from the upper reaches to

Lanzhou, Yinchuan, and Hetao Plain within YRB as well as the southern North China Plain. This suggests that NDVI in these specific areas is significantly influenced by SPEI.

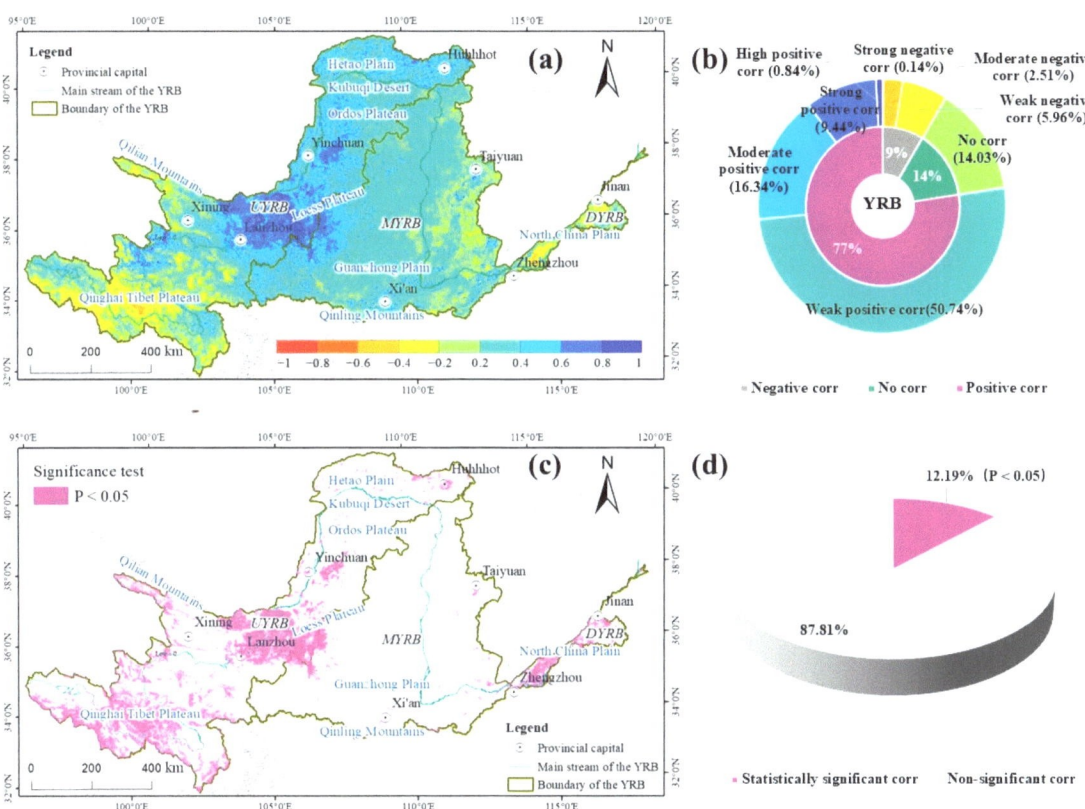

Figure 6. The spatial distribution of correlation between NDVI and SPEI from 2003 to 2020 (**a**), percentage of correlation grade (**b**), the spatial distribution of CC significance of NDVI and SPEI (**c**), and percentage of CC significance of NDVI and SPEI (**d**).

To further understand the impact of SPEI on NDVI, we analyzed the spatial correlation between monthly SPEI-1, SPEI-3, and SPEI-6 and NDVI. During the three-month winter period (Figure 7a–c), the spatial distribution of the CC between SPEI-1 and NDVI showed significant differences, with fewer regions having a CC greater than 0.6 (0.15%, 0.08%, and 0.45%) and fewer regions having a CC less than −0.6 (0.02%, 0.03%, and 0.20%). The area of moderate negative correlation was 4.57%, 5.19%, and 13.19%, respectively. The area of negative correlation gradually increased in winter and reached its highest value in February, mainly distributed in the source area of the YRB, which may be related to the vegetation characteristics and semi-humid climate environment of the source area of the YRB. The vegetation growth is less controlled by water and more affected by heat. At the same time, we can see that the negative correlation area between SPEI-3 and NDVI expanded further in winter, and the area of moderate negative correlation increased most significantly, reaching 6.78%, 7.94%, and 18.19% (Figure 8a–c). However, the correlation between SPEI-6 and NDVI in winter has shown more positive correlation, with the area of moderate positive correlation accounting for 15.51%, 4.73%, and 8.81% (Figure 9a–c). The study found that SPEI-3 has the most significant impact on NDVI in the winter.

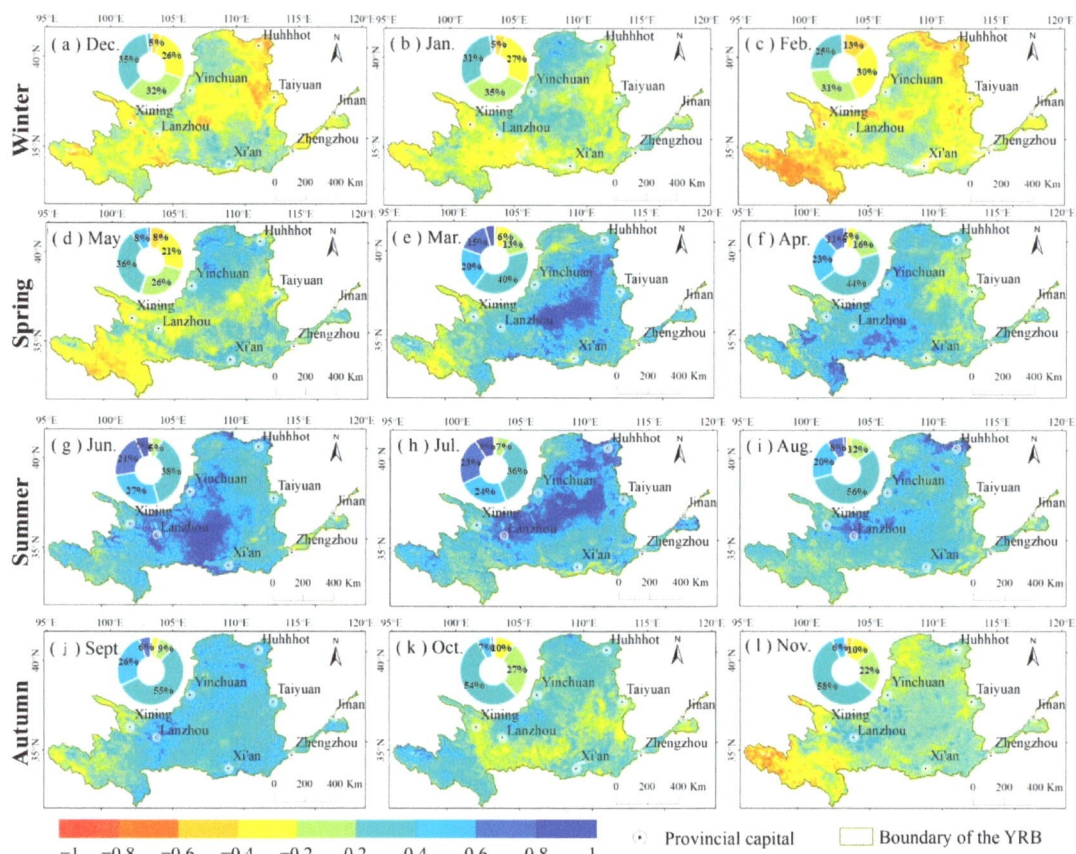

Figure 7. The spatial distribution of the CC between SPEI-1 and NDVI.

The positive correlation strength between spring SPEI and NDVI further intensifies and expands in the area. Throughout the spring, from March to May, the dominant relationship between SPEI-1 and NDVI is a positive correlation, with the areas of CC greater than 0.4 being 9.33%, 39.45%, and 35.35% (Figure 7d–f). The dominant relationship between SPEI-3 and NDVI, as well as the relationship between SPEI-6 and NDVI, is also a positive correlation, but the areas of CC greater than 0.4 decrease, being 2.04%, 3.97%, and 13.91% (Figure 8d–f), and 14.75%, 23.59%, and 38.49% (Figure 9d–f), possibly because spring temperatures have a greater impact on vegetation growth and SPEI-1 scale and SPEI-6 water supply are more sensitive to vegetation growth.

After the maximum positive correlation region between SPEI and NDVI in the summer, it began to decline in August, and the change in the region where the CC between SPEI-1 and NDVI was greater than 0.4 in each month of the summer was 54.57%, 55.93%, and 29.29%. The change in the region where the CC between SPEI-3 and NDVI was greater than 0.4 in each month of the summer was 39.61%, 40.49%, and 27.35%. The region where the CC between SPEI-6 and NDVI was greater than 0.4 rose to its maximum, at 62.31%, 66.85%, and 51.59% in each month of the summer. The results show that although there are significant differences in the correlation between different scales of SPEI and NDVI, the internal fluctuation trend in the summer is consistent, i.e., it began to rise in June, reached its peak in July, and then began to decline in August. The significant positive correlation regions are concentrated in the Loess Plateau region after the significance test.

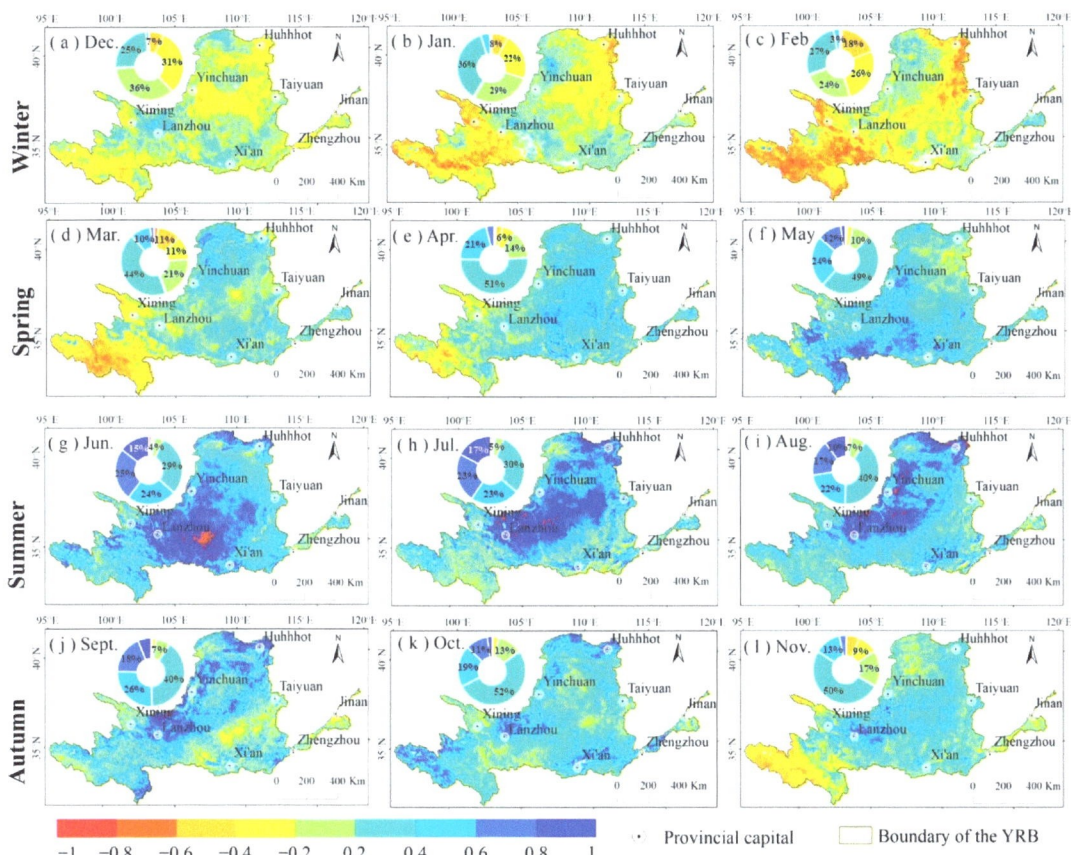

Figure 8. The spatial distribution of the CC between SPEI-3 and NDVI.

The correlation between SPEI and NDVI in autumn continued to decline, with more than 40% of the area falling into a weak correlation. Upon observation, the decline in the correlation between SPEI-1 and NDVI was the most obvious, with weak correlation areas accounting for about 60% (Figure 9g–l). The weak correlation areas of SPEI-3 and NDVI, as well as SPEI-6 and NDVI, decreased to about 50%, while the corresponding strong correlation areas increased. The results show that SPEI-3 and SPEI-6 have a more significant impact on NDVI in the autumn.

The above results indicate that there are certain regular patterns in the correlation between different temporal scales of SPEI and NDVI, both spatially and temporally. Figure 10 illustrates the trends in the number of positive correlations at various levels. Strong and moderate positive correlations exhibit clear seasonal patterns, with peak values occurring in July. SPEI-3 has the most significant impact on NDVI, suggesting that a three-month accumulated drought has the greatest influence (Figure 10a,b). When expanding the area of negative correlation between different temporal scales of SPEI and NDVI, SPEI-1 has the most notable effect, indicating that a one-month accumulated drought has the strongest impact (Figure 10c,d). SPEI and NDVI show a negative correlation in certain regions (Figure 11), and from the negative correlation propagation process, the negative correlation between SPEI-3 and NDVI is the strongest, indicating that the impact of an accumulated 3-month drought on vegetation is the highest.

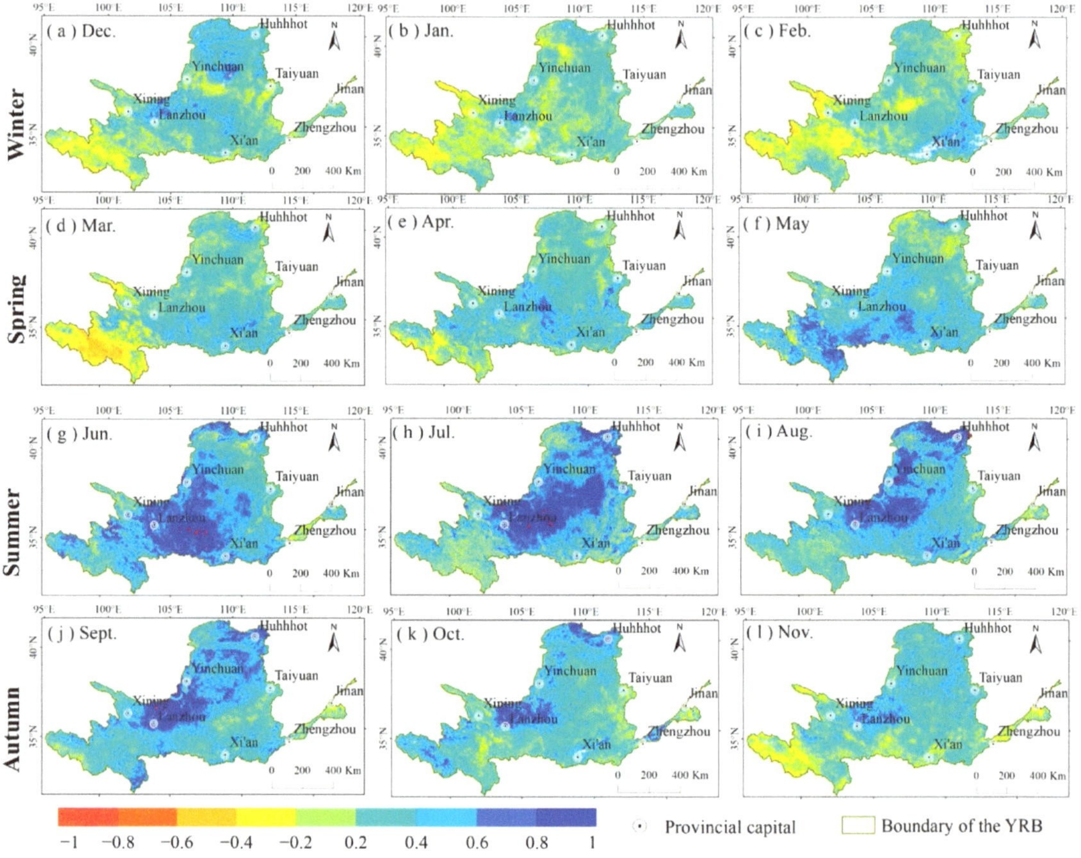

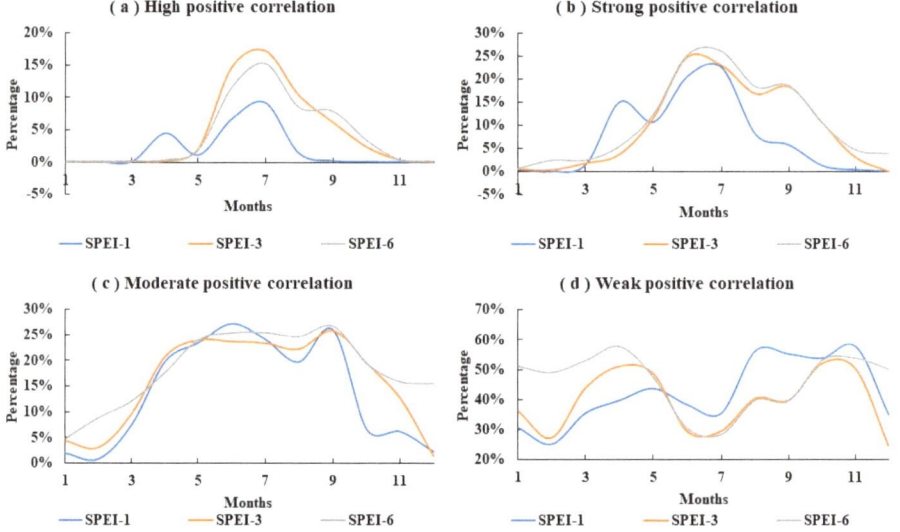

Figure 9. The spatial distribution of the CC between SPEI-6 and NDVI.

Figure 10. The process of spreading the area of positive correlation between SPEI and NDVI at different temporal scales.

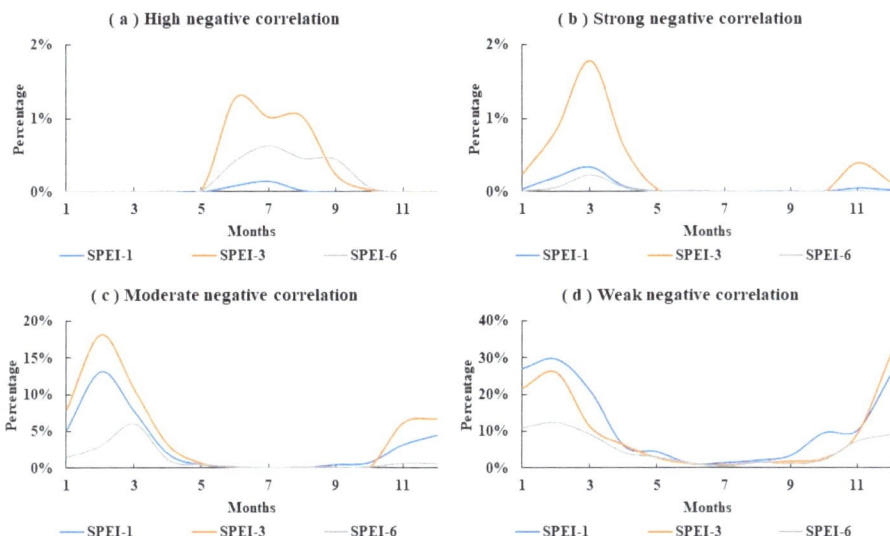

Figure 11. The process of spreading the area of negative correlation between SPEI and NDVI at different temporal scales.

4. Discussion

4.1. The Impact Mechanism of SPEI on NDVI

The impact of drought on vegetation is both severe and far-reaching [21,44–47]. As a result of water scarcity, the process of plant photosynthesis decelerates [48–51], leading to desiccation and shedding of foliage [52,53], as well as stunted or halted growth [54]. Prolonged drought significantly hampers vegetation productivity [55–57] and serves as a pivotal factor in precipitating plant mortality [58–61]. Research indicates that various ecosystems (such as grasslands and forests) exhibit distinct sensitivities to water deficits during different seasons. The intricate response of vegetation to drought is shaped by the combined influences of water availability and thermal energy [62], while also being intricately linked with diverse regional climates and types of flora [63,64].

At the same time, drought causes soil to become dry and infertile, impacting plant growth and increasing the likelihood of vegetation diseases and pest infestations, thereby affecting the survival and reproduction of plants. In ecosystems, drought can also lead to a reduction in vegetation [65], soil structure damage, and ultimately land desertification. Studies have indicated that the sustained increase in transpiration demand over the past decade has shortened the growing season of ecosystems. Insufficient soil moisture levels cannot support increased atmospheric water demand during the summer, resulting in transpiration suppression that prevents vegetation from meeting higher transpiration demands; water scarcity inhibits vegetation growth [66]. Therefore, drought has a severe impact on vegetation, with significant consequences for both ecosystems and human society [67].

The relationship between SPEI and NDVI mainly reflects the impact of water conditions on plant growth and health. In general, when SPEI is high, it indicates that there is sufficient water supply, which is beneficial to plant growth and leads to an increase in NDVI. On the other hand, when SPEI is low, it indicates drought or water scarcity, which limits plant growth, resulting in a decrease in NDVI [68]. At the same time, NDVI reflects the health status of plant growth. Water scarcity can lead to plant stress [69], which in turn affects its photosynthesis and growth, resulting in a decrease in NDVI. The availability of water in different growing seasons can have a significant impact on the plant's photosynthesis and growth rate. For example, in the dry season, NDVI may decrease due to insufficient water; in the wet season, when SPEI is high, NDVI will increase accordingly. The water conditions reflected by SPEI are also closely related to soil moisture [70]. When soil moisture is sufficient, plant roots can obtain more water and nutrients, which can promote a higher NDVI. Different ecosystems respond differently to water changes [71], such as forests, grasslands, and farmlands, and changes in SPEI may also affect the availability of water in these ecosystems and subsequently affect NDVI. At the same time, NDVI is also affected by human activities, such as climate change having a greater impact on vegetation turning green than human activities and human activities having a greater impact on vegetation degradation than climate change [72].

4.2. Response of NDVI to SPEI

Vegetation's response to drought is reflected in its resilience and recovery capacity [30]. Short-term responses (e.g., 3–6 months) of vegetation to drought are more sensitive, while long-term responses (e.g., 12–24 months) of vegetation show resilience and recovery capacity [73]. High altitude and steep slopes enhance vegetation's drought resistance, while vegetation in plain areas shows stronger recovery capacity after drought [74]. The order of vegetation death after drought events is roughly: temperate, subtropical, tropical, temperate, alpine climate zone, i.e., the alpine climate zone is less affected by drought events [74]. The YRB has an alpine climate zone, where most of the highland areas have an altitude of over 4600 m, with an extremely fragile ecological environment and decreasing evapotranspiration [75]. Drought causes the growth rate of the dominant and low-abundance species on high altitudes and steep slopes to be lower [76], which is consistent with the conclusions of this study. In the high-altitude areas of the YRB, vegetation has stronger resistance to drought, and its growth is less restricted by water, with the average CC in the region being approximately −0.3 to 0.4 (Figure 12). Some studies have indicated that the increase in atmospheric water demand will accelerate evaporation in the region, especially in temperate areas, causing vegetation growth to be impeded [77]. While vegetation also plays a role in mitigating drought, the changes in drought are still primarily driven by climate change [78], especially under the control of precipitation variability [79]. On the other hand, climate warming leads to an extension of the growing season in early spring and late autumn, which reduces the benefits of carbon sequestration [56,80]. Meanwhile, temperature rise and afforestation will further exacerbate the risk of drought spreading [81].

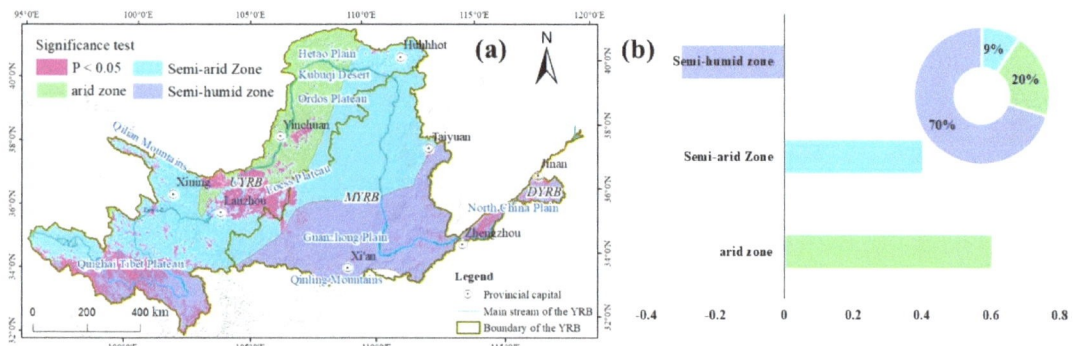

Figure 12. The zonal division of aridity in the YRB (**a**), the percentage area of different arid and humid zones, as well as the CC between SPEI and NDVI (**b**).

4.3. Uncertainty and Vegetation Management Recommendations

NDVI is a widely used vegetation index in remote sensing and ecological research, which mainly evaluates ecological environment changes by reflecting vegetation coverage and health status. SPEI is an index that evaluates climate drought and moisture, combining the effects of precipitation and temperature. The NDVI data are composed of 16-day-resolution remote sensing data, and the remote sensing data will be affected by factors such as cloud cover, atmosphere, solar altitude, surface characteristics of different objects, and sensor characteristics [28]. We use the maximum synthesis method to the greatest extent possible to eliminate these errors, but these errors cannot be avoided, so in the calculation of SPEI, we used measured data to calculate it as much as possible to ensure the accuracy of drought monitoring data [72].

Urbanization has a noticeable impact on drought distribution [82], therefore cities and surrounding areas in the Huanghe River Basin should supplement artificial irrigation measures to alleviate the inhibition of extreme drought on vegetation growth and development. The negative correlation between SPEI and NDVI in high-altitude arid and semi-arid areas is because these regions are adapted to drought, and temperature may have a more significant impact on vegetation growth. Global warming may have a certain degree of promotion in these areas [46]. For sensitive high-altitude areas, appropriate measures can be taken to promote the migration of species that are adapted to climate change, and the implementation of measures to increase species diversity is needed. For the humid and subhumid regions, SPEI and NDVI show a significant positive correlation. It is beneficial to strengthen drought monitoring in these areas for the protection of the ecological environment in the YRB.

5. Conclusions

This paper analyzes the impact of drought at different temporal scales on vegetation growth in the YRB, and the main conclusions are as follows:

(1) The process of urbanization exerts a discernible influence on the distribution of drought. Extreme drought predominantly occurs in the middle and upper reaches of the Weihe River, while severe drought is primarily concentrated in areas surrounding Xi'an, the Guanzhong Plain, the central part of the Loess Plateau, and the southern region of the North China Plain. These regions are characterized by city clusters such as the Chengdu-Chongqing metropolitan area, the Xi'an-Wuhan-Chengdu city cluster, and the Chengdu-Chongqing metropolitan area.

(2) The spatial trend of SPEI exhibits a gradual decline from the central and southern regions towards the boundaries of the river basin, with the most pronounced decrease concentrated in the Guanzhong Plain and the Loess Plateau, where the rate of decline reaches -0.7 to -1.2/decade. The drought trend is notably significant. Conversely, there is an upward trajectory in SPEI observed in the source regions of YRB, Hetao Plain, and DYRB, with the highest rate of increase reaching 0.7/decade. This indicates a mitigation of drought trends and a shift towards increased moisture.

(3) From 2003 to 2020, the NDVI showed an upward trend, indicating an increase in vegetation density or an expansion of vegetation coverage. From the current vegetation coverage, the southern part of the watershed has better vegetation coverage, including the source areas of YBB, Guanzhong Plain, western Shanxi, and the southern North China Plain. The vegetation coverage in the northwestern part of the watershed is low, including the highest altitude area of the source area in YRB and the northwestern part of the Loess Plateau.

(4) The impact of spring temperature on vegetation growth is profound, while the water supply from both the SPEI-1 scale and SPEI-6 exhibits heightened sensitivity towards vegetation growth. Furthermore, the initial expansion of a positive correlation between drought and NDVI occurs during June before reaching its zenith in July, followed by a gradual decline through August. Notably, areas demonstrating significant positive correlations successfully pass rigorous significance tests, with a concentration observed within the Loess Plateau region. As autumn progresses, there is a continued decrease in correlation between SPEI and NDVI, with over 40% of regions regressing into weaker correlations. Additionally, during this season, both SPEI-3 and SPEI-6 exert a more pronounced influence on NDVI compared to other periods, and it is worth noting that during the winter months, it is specifically SPEI-3 that exerts maximal influence on NDVI.

(5) According to the dynamic transmission laws of positive CC at various levels, it is evident that strong and moderate positive correlations exhibit distinct transmission patterns, reaching their peak values in July. The most significant positive impact on NDVI

comes from SPEI-3. Among the areas showing a negative correlation between SPEI and NDVI, the influence of SPEI-1 is the most prominent.

Funding: This research was funded by the National Natural Science Foundation of China (Grant No. 41901110).

Data Availability Statement: The data acquisition method has been explained in the article.

Conflicts of Interest: The author declares no conflicts of interest.

References

1. Novick, K.A.; Ficklin, D.L.; Stoy, P.C.; Williams, C.A.; Bohrer, G.; Oishi, A.C.; Papuga, S.A.; Blanken, P.D.; Noormets, A.; Sulman, B.N.; et al. The increasing importance of atmospheric demand for ecosystem water and carbon fluxes. *Nat. Clim. Chang.* **2016**, *6*, 1023–1027. [CrossRef]
2. Huang, K.; Xia, J.; Wang, Y.; Ahlström, A.; Chen, J.; Cook, R.B.; Cui, E.; Fang, Y.; Fisher, J.B.; Huntzinger, D.N.; et al. Enhanced peak growth of global vegetation and its key mechanisms. *Nat. Ecol. Evol.* **2018**, *2*, 1897–1905. [CrossRef]
3. Ciais, P.; Reichstein, M.; Viovy, N.; Granier, A.; Ogée, J.; Allard, V.; Aubinet, M.; Buchmann, N.; Bernhofer, C.; Carrara, A.; et al. Europe-wide reduction in primary productivity caused by the heat and drought in 2003. *Nature* **2005**, *437*, 529–533. [CrossRef] [PubMed]
4. Wang, K.; Dickinson, R.E. A review of global terrestrial evapotranspiration: Observation, modeling, climatology, and climatic variability. *Rev. Geophys.* **2012**, *50*. [CrossRef]
5. Piao, S.; Zhang, X.; Chen, A.; Liu, Q.; Lian, X.; Wang, X.; Peng, S.; Wu, X. The impacts of climate extremes on the terrestrial carbon cycle: A review. *Sci. China Earth Sci.* **2019**, *62*, 1551–1563. [CrossRef]
6. Jiao, W.; Wang, L.; Smith, W.K.; Chang, Q.; Wang, H.; D'Odorico, P. Observed increasing water constraint on vegetation growth over the last three decades. *Nat. Commun.* **2021**, *12*, 3777. [CrossRef]
7. Peterson, T.J.; Saft, M.; Peel, M.C.; John, A. Watersheds may not recover from drought. *Science* **2021**, *372*, 745–749. [CrossRef]
8. Shu, Z.; Jin, J.; Zhang, J.; Wang, G.; Lian, Y.; Liu, Y.; Bao, Z.; Guan, T.; He, R.; Liu, C.; et al. 1.5 °C and 2.0 °C of global warming intensifies the hydrological extremes in China. *J. Hydrol.* **2024**, *635*, 131229. [CrossRef]
9. Furtak, K.; Wolińska, A. The impact of extreme weather events as a consequence of climate change on the soil moisture and on the quality of the soil environment and agriculture—A review. *Catena* **2023**, *231*, 107378. [CrossRef]
10. Cao, X.; Zheng, Y.; Lei, Q.; Li, W.; Song, S.; Wang, C.; Liu, Y.; Khan, K. Increasing actual evapotranspiration on the Loess Plateau of China: An insight from anthropologic activities and climate change. *Ecol. Indic.* **2023**, *157*, 111235. [CrossRef]
11. Liu, Y.; Lin, Z.; Wang, Z.; Chen, X.; Han, P.; Wang, B.; Wang, Z.; Wen, Z.; Shi, H.; Zhang, Z.; et al. Discriminating the impacts of vegetation greening and climate change on the changes in evapotranspiration and transpiration fraction over the Yellow River Basin. *Sci. Total Environ.* **2023**, *904*, 166926. [CrossRef]
12. Schumacher, D.L.; Keune, J.; Dirmeyer, P.; Miralles, D.G. Drought self-propagation in drylands due to land–atmosphere feedbacks. *Nat. Geosci.* **2022**, *15*, 262–268. [CrossRef]
13. Yuan, W.; Zheng, Y.; Piao, S.; Ciais, P.; Lombardozzi, D.; Wang, Y.; Ryu, Y.; Chen, G.; Dong, W.; Hu, Z.; et al. Increased atmospheric vapor pressure deficit reduces global vegetation growth. *Sci. Adv.* **2019**, *5*, eaax1396. [CrossRef]
14. Lucht, W.; Prentice, I.C.; Myneni, R.B.; Sitch, S.; Friedlingstein, P.; Cramer, W.; Bousquet, P.; Buermann, W.; Smith, B. Climatic Control of the High-Latitude Vegetation Greening Trend and Pinatubo Effect. *Science* **2002**, *296*, 1687–1689. [CrossRef]
15. Zhu, Z.; Piao, S.; Myneni, R.B.; Huang, M.; Zeng, Z.; Canadell, J.G.; Ciais, P.; Sitch, S.; Friedlingstein, P.; Arneth, A.; et al. Greening of the Earth and its drivers. *Nat. Clim. Chang.* **2016**, *6*, 791–795. [CrossRef]
16. Wang, Y.; Li, J.; Hou, J.; Zhao, K.; Wu, R.; Sun, B.; Lu, J.; Liu, Y.; Cui, C.; Liu, J. Enhanced evapotranspiration induced by vegetation restoration may pose water resource risks under climate change in the Yellow River Basin. *Ecol. Indic.* **2024**, *162*, 112060. [CrossRef]
17. Li, X.; Xu, X.; Tian, W.; Tian, J.; He, C. Contribution of climate change and vegetation restoration to interannual variability of evapotranspiration in the agro–pastoral ecotone in northern China. *Ecol. Indic.* **2023**, *154*, 110485. [CrossRef]
18. Kreuzwieser, J.; Rennenberg, H. Molecular and physiological responses of trees to waterlogging stress. *Plant Cell Environ.* **2014**, *37*, 2245–2259. [CrossRef]
19. Lan, X.; Xie, Y.; Liu, Z.; Yang, T.; Huang, L.; Chen, X.; Chen, X.; Lin, K.; Cheng, L. Vegetation greening accelerated hydrological drought in two-thirds of river basins over China. *J. Hydrol.* **2024**, *637*, 131436. [CrossRef]
20. Zhao, M.; Running, S.W. Drought-Induced Reduction in Global Terrestrial Net Primary Production from 2000 Through 2009. *Science* **2010**, *329*, 940–943. [CrossRef]
21. Xu, W.; Yuan, W.; Wu, D.; Zhang, Y.; Shen, R.; Xia, X.; Ciais, P.; Liu, J. Impacts of record-breaking compound heatwave and drought events in 2022 China on vegetation growth. *Agric. For. Meteorol.* **2024**, *344*, 109799. [CrossRef]
22. Xu, H.-j.; Wang, X.-p.; Zhao, C.-y.; Yang, X.-m. Diverse responses of vegetation growth to meteorological drought across climate zones and land biomes in northern China from 1981 to 2014. *Agric. For. Meteorol.* **2018**, *262*, 1–13. [CrossRef]
23. Ye, J.; Gao, Z.; Wu, X.; Lu, Z.; Li, C.; Wang, X.; Chen, L.; Cui, G.; Yu, M.; Yan, G.; et al. Impact of increased temperature on spring wheat yield in northern China. *Food Energy Secur.* **2021**, *10*, 368–378. [CrossRef]

24. Li, Y.; Huang, Y.; Li, Y.; Zhang, H.; Fan, J.; Deng, Q.; Wang, X. Spatiotemporal heterogeneity in meteorological and hydrological drought patterns and propagations influenced by climatic variability, LULC change, and human regulations. *Sci. Rep.* **2024**, *14*, 5965. [CrossRef]
25. Wu, J.; Miao, C.; Zheng, H.; Duan, Q.; Lei, X.; Li, H. Meteorological and Hydrological Drought on the Loess Plateau, China: Evolutionary Characteristics, Impact, and Propagation. *J. Geophys. Res. Atmos.* **2018**, *123*, 11–569. [CrossRef]
26. Apurv, T.; Sivapalan, M.; Cai, X. Understanding the Role of Climate Characteristics in Drought Propagation. *Water Resour. Res.* **2017**, *53*, 9304–9329. [CrossRef]
27. Van Loon, A.F.; Tijdeman, E.; Wanders, N.; Van Lanen, H.A.J.; Teuling, A.J.; Uijlenhoet, R. How climate seasonality modifies drought duration and deficit. *J. Geophys. Res. Atmos.* **2014**, *119*, 4640–4656. [CrossRef]
28. Zhan, C.; Liang, C.; Zhao, L.; Jiang, S.; Niu, K.; Zhang, Y. Drought-related cumulative and time-lag effects on vegetation dynamics across the Yellow River Basin, China. *Ecol. Indic.* **2022**, *143*, 109409. [CrossRef]
29. Vicente-Serrano, S.M. Response of vegetation to drought time-scales across global land biomes. *Proc. Natl. Acad. Sci. USA* **2013**, *110*, 52–57. [CrossRef]
30. Wu, D.; Vargas, G.G.; Powers, J.S.; McDowell, N.G.; Becknell, J.M.; Pérez-Aviles, D.; Medvigy, D.; Liu, Y.; Katul, G.G.; Calvo-Alvarado, J.C.; et al. Reduced ecosystem resilience quantifies fine-scale heterogeneity in tropical forest mortality responses to drought. *Glob. Chang. Biol.* **2021**, *28*, 2081–2094. [CrossRef]
31. Jiang, X.; Fang, X.; Zhu, Q.; Jin, J.; Ren, L.; Jiang, S.; Yan, Y.; Yuan, S.; Liao, M. Time-series satellite images reveal abrupt changes in vegetation dynamics and possible determinants in the Yellow River Basin. *Agric. For. Meteorol.* **2024**, *355*, 110124. [CrossRef]
32. Niu, H.; Xiu, Z.; Xiao, D. Impact of land-use change on ecological vulnerability in the Yellow River Basin based on a complex network model. *Ecol. Indic.* **2024**, *166*, 112212. [CrossRef]
33. Shen, J.; Zhao, M.; Tan, Z.; Zhu, L.; Guo, Y.; Li, Y.; Wu, C. Ecosystem service trade-offs and synergies relationships and their driving factor analysis based on the Bayesian belief Network: A case study of the Yellow River Basin. *Ecol. Indic.* **2024**, *163*, 112070. [CrossRef]
34. Ci, F.; Wang, Z.; Hu, Q. Spatial pattern characteristics and optimization policies of low-carbon innovation levels in the urban agglomerations in the Yellow River Basin. *J. Clean. Prod.* **2024**, *439*, 140856. [CrossRef]
35. Wu, J.; Chen, X. Spatiotemporal trends of dryness/wetness duration and severity: The respective contribution of precipitation and temperature. *Atmos. Res.* **2019**, *216*, 176–185. [CrossRef]
36. Vicente-Serrano, S.M.; Beguería, S.; López-Moreno, J.I. A Multiscalar Drought Index Sensitive to Global Warming: The Standardized Precipitation Evapotranspiration Index. *J. Clim.* **2010**, *23*, 1696–1718. [CrossRef]
37. Liang, R.G. Spatial-Temporal Variation Characteristics and Influencing Factors of Drought in Karst Region of Southwest China Based on Remote Sensing. Master's Thesis, Guizhou Normal University, Guiyang, China, 2022.
38. Xue, H.; Li, Y.; Dong, G. Analysis of Spatial-temporal Variation Characteristics of Meteorological Drought in the Hexi Corridor Based on SPEI Index. *Chin. J. Agrometeorol.* **2022**, *43*, 932–934. [CrossRef]
39. Zhang, X.; Duan, Y.; Duan, J.; Jian, D.; Ma, Z. A daily drought index based on evapotranspiration and its application in regional drought analyses. *Sci. China Earth Sci.* **2021**, *65*, 317–336. [CrossRef]
40. Li, Y.; Qin, Y. The Response of Net Primary Production to Climate Change: A Case Study in the 400 mm Annual Precipitation Fluctuation Zone in China. *Int. J. Environ. Res. Public Health* **2019**, *16*, 1497. [CrossRef]
41. Ding, Y.; Xu, J.; Wang, X.; Cai, H.; Zhou, Z.; Sun, Y.; Shi, H. Propagation of meteorological to hydrological drought for different climate regions in China. *J. Environ. Manag.* **2021**, *283*, 111980. [CrossRef]
42. Ding, Y.; Xu, J.; Wang, X.; Peng, X.; Cai, H. Spatial and temporal effects of drought on Chinese vegetation under different coverage levels. *Sci. Total Environ.* **2020**, *716*, 137166. [CrossRef] [PubMed]
43. Xu, Y.; Zhang, X.; Hao, Z.; Singh, V.P.; Hao, F. Characterization of agricultural drought propagation over China based on bivariate probabilistic quantification. *J. Hydrol.* **2021**, *598*, 126194. [CrossRef]
44. Jiang, T.; Su, X.; Singh, V.P.; Zhang, G. Spatio-temporal pattern of ecological droughts and their impacts on health of vegetation in Northwestern China. *J. Environ. Manag.* **2022**, *305*, 114356. [CrossRef]
45. Ge, W.; Han, J.; Zhang, D.; Wang, F. Divergent impacts of droughts on vegetation phenology and productivity in the Yungui Plateau, southwest China. *Ecol. Indic.* **2021**, *127*, 107743. [CrossRef]
46. Wang, C.-P.; Huang, M.-T.; Zhai, P.-M. Change in drought conditions and its impacts on vegetation growth over the Tibetan Plateau. *Adv. Clim. Chang. Res.* **2021**, *12*, 333–341. [CrossRef]
47. Chen, Z.; Shao, Z.; Huang, X.; Zhuang, Q.; Dang, C.; Cai, B.; Zheng, X.; Ding, Q. Assessing the impact of drought-land cover change on global vegetation greenness and productivity. *Sci. Total Environ.* **2022**, *852*, 158499. [CrossRef]
48. Wang, S.; Zhang, Y.; Ju, W.; Porcar-Castell, A.; Ye, S.; Zhang, Z.; Brümmer, C.; Urbaniak, M.; Mammarella, I.; Juszczak, R.; et al. Warmer spring alleviated the impacts of 2018 European summer heatwave and drought on vegetation photosynthesis. *Agric. For. Meteorol.* **2020**, *295*, 108195. [CrossRef]
49. Frank, D.; Reichstein, M.; Bahn, M.; Thonicke, K.; Frank, D.; Mahecha, M.D.; Smith, P.; van der Velde, M.; Vicca, S.; Babst, F.; et al. Effects of climate extremes on the terrestrial carbon cycle: Concepts, processes and potential future impacts. *Glob. Chang. Biol.* **2015**, *21*, 2861–2880. [CrossRef]
50. Sippel, S.; Reichstein, M.; Ma, X.; Mahecha, M.D.; Lange, H.; Flach, M.; Frank, D. Drought, Heat, and the Carbon Cycle: A Review. *Curr. Clim. Chang. Rep.* **2018**, *4*, 266–286. [CrossRef]

51. Li, D.; Li, X.; Li, Z.; Fu, Y.; Zhang, J.; Zhao, Y.; Wang, Y.; Liang, E.; Rossi, S. Drought limits vegetation carbon sequestration by affecting photosynthetic capacity of semi-arid ecosystems on the Loess Plateau. *Sci. Total Environ.* **2024**, *912*, 168778. [CrossRef]
52. Cao, S.; He, Y.; Zhang, L.; Chen, Y.; Yang, W.; Yao, S.; Sun, Q. Spatiotemporal characteristics of drought and its impact on vegetation in the vegetation region of Northwest China. *Ecol. Indic.* **2021**, *133*, 108420. [CrossRef]
53. Song, W.; Song, R.; Zhao, Y.; Zhao, Y. Research on the characteristics of drought stress state based on plant stem water content. *Sustain. Energy Technol. Assess.* **2023**, *56*, 103080. [CrossRef]
54. Li, Q.; Xue, Y. Simulated impacts of land cover change on summer climate in the Tibetan Plateau. *Environ. Res. Lett.* **2010**, *5*, 015102. [CrossRef]
55. Jiang, W.; Wang, L.; Zhang, M.; Yao, R.; Chen, X.; Gui, X.; Sun, J.; Cao, Q. Analysis of drought events and their impacts on vegetation productivity based on the integrated surface drought index in the Hanjiang River Basin, China. *Atmos. Res.* **2021**, *254*, 105536. [CrossRef]
56. Jha, S.; Srivastava, R. Impact of drought on vegetation carbon storage in arid and semi-arid regions. *Remote Sens. Appl. Soc. Environ.* **2018**, *11*, 22–29. [CrossRef]
57. Zheng, L.; Lu, J.; Chen, X. Drought offsets the vegetation greenness-induced gross primary productivity from 1982 to 2018 in China. *J. Hydrol.* **2024**, *632*, 130881. [CrossRef]
58. Gouveia, C.M.; Trigo, R.M.; Beguería, S.; Vicente-Serrano, S.M. Drought impacts on vegetation activity in the Mediterranean region: An assessment using remote sensing data and multi-scale drought indicators. *Glob. Planet. Chang.* **2017**, *151*, 15–27. [CrossRef]
59. Scheiter, S.; Kumar, D.; Pfeiffer, M.; Langan, L. Modeling drought mortality and resilience of savannas and forests in tropical Asia. *Ecol. Model.* **2024**, *494*, 110783. [CrossRef]
60. Holtmann, A.; Huth, A.; Bohn, F.; Fischer, R. Assessing the impact of multi-year droughts on German forests in the context of increased tree mortality. *Ecol. Model.* **2024**, *492*, 110696. [CrossRef]
61. Socha, J.; Hawryło, P.; Tymińska-Czabańska, L.; Reineking, B.; Lindner, M.; Netzel, P.; Grabska-Szwagrzyk, E.; Vallejos, R.; Reyer, C.P.O. Higher site productivity and stand age enhance forest susceptibility to drought-induced mortality. *Agric. For. Meteorol.* **2023**, *341*, 109680. [CrossRef]
62. Shen, C.; Ma, R. Estimating suitable hydrothermal conditions for vegetation growth for land use cover across China based on maximum-probability-density monthly NDVI. *Remote Sens. Appl. Soc. Environ.* **2023**, *30*, 100958. [CrossRef]
63. Weng, Z.; Niu, J.; Guan, H.; Kang, S. Three-dimensional linkage between meteorological drought and vegetation drought across China. *Sci. Total Environ.* **2023**, *859*, 160300. [CrossRef]
64. Jiang, L.; Guli·Jiapaer; Bao, A.; Guo, H.; Ndayisaba, F. Vegetation dynamics and responses to climate change and human activities in Central Asia. *Sci. Total Environ.* **2017**, *599–600*, 967–980. [CrossRef]
65. Yuan, M.; Zhao, L.; Lin, A.; Wang, L.; Li, Q.; She, D.; Qu, S. Impacts of preseason drought on vegetation spring phenology across the Northeast China Transect. *Sci. Total Environ.* **2020**, *738*, 140297. [CrossRef]
66. Rahmati, M.; Graf, A.; Poppe Terán, C.; Amelung, W.; Dorigo, W.; Franssen, H.-J.H.; Montzka, C.; Or, D.; Sprenger, M.; Vanderborght, J.; et al. Continuous increase in evaporative demand shortened the growing season of European ecosystems in the last decade. *Commun. Earth Environ.* **2023**, *4*, 236. [CrossRef]
67. Li, Z.; Bai, X.; Tan, Q.; Zhao, C.; Li, Y.; Luo, G.; Chen, F.; Li, C.; Ran, C.; Zhang, S.; et al. Dryness stress weakens the sustainability of global vegetation cooling. *Sci. Total Environ.* **2024**, *909*, 168474. [CrossRef]
68. Wu, C.; Zhong, L.; Yeh, P.J.F.; Gong, Z.; Lv, W.; Chen, B.; Zhou, J.; Li, J.; Wang, S. An evaluation framework for quantifying vegetation loss and recovery in response to meteorological drought based on SPEI and NDVI. *Sci. Total Environ.* **2024**, *906*, 167632. [CrossRef]
69. Akula, N.N.; Abdelhakim, L.; Knazovický, M.; Ottosen, C.-O.; Rosenqvist, E. Plant responses to co-occurring heat and water deficit stress: A comparative study of tolerance mechanisms in old and modern wheat genotypes. *Plant Physiol. Biochem.* **2024**, *210*, 108595. [CrossRef]
70. Zhao, H.; Huang, Y.; Wang, X.; Li, X.; Lei, T. The performance of SPEI integrated remote sensing data for monitoring agricultural drought in the North China Plain. *Field Crop. Res.* **2023**, *302*, 109041. [CrossRef]
71. Liu, S.; Xue, L.; Xiao, Y.; Yang, M.; Liu, Y.; Han, Q.; Ma, J. Dynamic process of ecosystem water use efficiency and response to drought in the Yellow River Basin, China. *Sci. Total Environ.* **2024**, *934*, 173339. [CrossRef] [PubMed]
72. Xu, Y.; Dai, Q.-Y.; Lu, Y.-G.; Zhao, C.; Huang, W.-T.; Xu, M.; Feng, Y.-X. Identification of ecologically sensitive zones affected by climate change and anthropogenic activities in Southwest China through a NDVI-based spatial-temporal model. *Ecol. Indic.* **2024**, *158*, 111482. [CrossRef]
73. Zhang, Q.; Kong, D.; Singh, V.P.; Shi, P. Response of vegetation to different time-scales drought across China: Spatiotemporal patterns, causes and implications. *Glob. Planet. Chang.* **2017**, *152*, 1–11. [CrossRef]
74. Xiao, L.; Zhou, J.; Wu, X.; Anas Khan, M.; Zhao, S.; Wu, X. The dominant influence of terrain and geology on vegetation mortality in response to drought: Exploring resilience and resistance. *Catena* **2024**, *243*, 108156. [CrossRef]
75. Zhang, F.; Geng, M.; Wu, Q.; Liang, Y. Study on the spatial-temporal variation in evapotranspiration in China from 1948 to 2018. *Sci. Rep.* **2020**, *10*, 17139. [CrossRef]

76. Hollunder, R.K.; Mariotte, P.; Carrijo, T.T.; Holmgren, M.; Luber, J.; Stein-Soares, B.; Guidoni-Martins, K.G.; Ferreira-Santos, K.; Scarano, F.R.; Garbin, M.L. Topography and vegetation structure mediate drought impacts on the understory of the South American Atlantic Forest. *Sci. Total Environ.* **2021**, *766*, 144234. [CrossRef]
77. Konapala, G.; Mishra, A.K.; Wada, Y.; Mann, M.E. Climate change will affect global water availability through compounding changes in seasonal precipitation and evaporation. *Nat. Commun.* **2020**, *11*, 3044. [CrossRef]
78. Padrón, R.S.; Gudmundsson, L.; Decharme, B.; Ducharne, A.; Lawrence, D.M.; Mao, J.; Peano, D.; Krinner, G.; Kim, H.; Seneviratne, S.I. Observed changes in dry-season water availability attributed to human-induced climate change. *Nat. Geosci.* **2020**, *13*, 477–481. [CrossRef]
79. Liu, M.; Huang, J.; Sun, A.Y.; Wang, K.; Chen, H. What roles can water-stressed vegetation play in agricultural droughts? *Sci. Total Environ.* **2022**, *803*, 149810. [CrossRef]
80. Buermann, W.; Forkel, M.; O'Sullivan, M.; Sitch, S.; Friedlingstein, P.; Haverd, V.; Jain, A.K.; Kato, E.; Kautz, M.; Lienert, S.; et al. Widespread seasonal compensation effects of spring warming on northern plant productivity. *Nature* **2018**, *562*, 110–114. [CrossRef] [PubMed]
81. Li, Y.; Huang, S.; Wang, H.; Huang, Q.; Li, P.; Zheng, X.; Wang, Z.; Jiang, S.; Leng, G.; Li, J.; et al. Warming and greening exacerbate the propagation risk from meteorological to soil moisture drought. *J. Hydrol.* **2023**, *622*, 129716. [CrossRef]
82. Huang, S.; Zhang, X.; Yang, L.; Chen, N.; Nam, W.-H.; Niyogi, D. Urbanization-induced drought modification: Example over the Yangtze River Basin, China. *Urban Clim.* **2022**, *44*, 101231. [CrossRef]

Disclaimer/Publisher's Note: The statements, opinions and data contained in all publications are solely those of the individual author(s) and contributor(s) and not of MDPI and/or the editor(s). MDPI and/or the editor(s) disclaim responsibility for any injury to people or property resulting from any ideas, methods, instructions or products referred to in the content.

Review

Research Progress and Application Analysis of the Returning Straw Decomposition Process Based on CiteSpace

Yitong Wang [1,*], Qiujie Shan [1], Chuan Wang [1,2], Shaoyuan Feng [1] and Yan Li [1]

[1] College of Hydraulic Science and Engineering, Yangzhou University, Yangzhou 225009, China; shanqj0128@126.com (Q.S.); wangchuan198710@126.com (C.W.); syfeng@yzu.edu.cn (S.F.); liyan7986@126.com (Y.L.)

[2] International Shipping Research Institute, Gongqing Institute of Science and Technology, Jiujiang 332020, China

* Correspondence: wangyitong@yzu.edu.cn; Tel.: +86-17761970122

Abstract: Straw returning is an important measurement to determine the utilization of straw resources. Understanding the decomposition process and nutrient release process of straw is of great significance to the efficient utilization of straw resources and the sustainable development of the agricultural economy. In this study, the literature published in the CNKI and WOS from 2002 to 2022 was used as the data pool, and a keyword co-occurrence network map was drawn with the CiteSpace (6.2.R4) software. Visual analyses were based on the straw returning literature (1998 articles) and straw decomposition agent literature (125 articles), and the decomposition and nutrient release of straw under the action of a decomposition agent were analyzed using a straw decomposition characterization experiment. In general, returning straw can effectively improve soil fertility conditions and provide nutrients for crop growth, and the use of a straw decomposition agent can further improve soil conditions and increase crop yield. The straw decomposition characterization experiment further showed that *Pseudomonas* could effectively increase the decomposition rate and increase the nutrient release rate of straw. According to the above results, determining how to improve the utilization efficiency of straw resources via decomposable bacteriological agents according to local conditions will become a research hotspot in the future.

Keywords: utilization of straw resources; straw decomposition; straw decomposition agent; CiteSpace analysis

Citation: Wang, Y.; Shan, Q.; Wang, C.; Feng, S.; Li, Y. Research Progress and Application Analysis of the Returning Straw Decomposition Process Based on CiteSpace. *Water* **2023**, *15*, 3426. https://doi.org/10.3390/w15193426

Academic Editor: Arturo Alvino

Received: 30 June 2023
Revised: 19 September 2023
Accepted: 27 September 2023
Published: 29 September 2023

Copyright: © 2023 by the authors. Licensee MDPI, Basel, Switzerland. This article is an open access article distributed under the terms and conditions of the Creative Commons Attribution (CC BY) license (https://creativecommons.org/licenses/by/4.0/).

1. Introduction

With the continuous improvement of agricultural technology, agricultural production is growing. The amount of crop straw has soared, and the annual output of crop straw has reached 965 million tons [1]. Crop straw is a type of easily dispersed and low-density agricultural waste that is prone to regional, seasonal, and structural surplus problems [2]. The decomposition rate of crop straw is slow, and large amounts of crop straw accumulation will create a soil C/N imbalance [3]. These problems have limited the enthusiasm of farmers in the use of in situ straw returning. At present, the utilization of excess crop straw is in a critical stage of transitioning from "quantitative change" to "qualitative change". Crop straw used as fertilizer, raw material energy, agricultural raw materials, and industrial raw materials is of great significance to the sustainable development of the agricultural economy [4,5]. Thus, how to use crop straw more effectively and avoid farmland ecological nonpoint source pollution during crop straw utilization are becoming research hotspots.

Crop straw, as an important soil fertilizer, can return some of the nutrients absorbed during the crop period to the soil, improve soil fertility conditions, and promote the growth of crops in later periods. Firstly, straw returning can change the physical and chemical properties of the soil. Straw can help improve soil water retention performance, reduce soil bulk density, increase soil total porosity, improve soil aeration conditions, and promote soil deep root penetration [6,7]. Straw returning can improve the nutrient absorption of crops by promoting the soil permeability of deep roots and increasing crop yield [8]. Secondly, straw returning provides better breeding conditions for microorganisms and improves soil biodiversity by increasing the number of microorganisms in the soil. For example, Zhang et al. [9] effectively increased the diversity and abundance of fungi in the 0–10 cm soil layer by returning sugarcane straw to the field in 14 months. Thirdly, straw returning also has a positive effect on soil pollution by heavy metals. Greenhouse pot experiments showed that rice straw could reduce nickel bioavailability by 68% with a 2% return application [10]. However, the slow decomposition rate of returning straw will affect the germination and rooting of the next crop, and long-term straw mulching may increase the risk of blight and lead to crop death [11]. Thus, using a decomposition agent to accelerate straw decomposition is an effective way to solve these problems.

Straw decomposition agents are composed of microorganisms that can degrade straw rapidly, including fungi, bacteria, and actinomyces. The principle of these agents is to convert the cellulose, hemicelluloses, and lignin in straw into glucose, amino acids, fatty acids, and other small molecular organic compounds and minerals via the microbial metabolism [12]. The results show that adding a decomposition agent can significantly improve the decomposition rate of straw and prevent or reduce the adverse effects of excessive straw in the field on crop growth [13]. In addition, adding a decomposition agent can improve the soil microbial activity, kill pathogenic microorganisms in straw, provide more nutrients for crop growth, increase yield, and improve quality [14]. At present, most studies on the application of straw decomposition agents focus on the influence of decomposition agents on soil properties. However, nutrient release due to changes in the straw decomposition rate remains an understudied area.

Bibliometric analysis is a popular technical tool for quantitatively analyzing published literature on a specific topic in scientific research. The CiteSpace (6.2.R4) software is often used in academic research. This software was developed by the Chinese scholar Dr. Chen Chaomei using mathematical and statistical methods [15] and was introduced in China in early 2007. This software has been widely applied by scholars from various fields [16]. CiteSpace can draw a visual map through literature data retrieval; concisely and intuitively display the publication time, author institutions, research hotspots, and frontiers of literature in related fields [17]; and help further interpret and analyze the research status and development direction in this field [18,19].

With the continuous improvement of straw resource utilization and the gradual maturity of decomposition technology, the research results in the field of straw returning are becoming increasingly comprehensive, which provides strong guidance for the practical application of straw returning. In view of the above research background, the CiteSpace software was used to summarize the relevant research progress and research trends in straw returning and the application of straw decomposition agents in this paper, as well as to clarify the research context and frontiers to timely and comprehensively understand the technical level and application direction of straw returning. In addition, the improvement effects of straw returning were also analyzed in this paper, and the application effects of straw decomposition agents on straw returning are discussed. A simulation test was used to explore changes in the decomposition rate and nutrient release of straw returning under different decomposition agents. The purpose of this paper is to help researchers quickly understand the research development direction and existing achievements in the field of straw returning, providing reference for further related research.

2. Research Vein and Trend Analysis of Straw Returning

2.1. Statistical Analysis of Straw Returning Studies

In total, 1999 studies published from 2002 to 2022 were retrieved from the China National Knowledge Infrastructure (CNKI) with the keyword "straw returning". In total, 995 studies were retrieved from the international database Web of Science (WOS) with the keyword "straw mulching".

2.1.1. Number of Published Papers

Figure 1 presents the number of papers related to "straw returning" published from 2002 to 2022, showing a general trend of fluctuating growth. From 2002 to 2013, a total of 514 papers were published, with an annual average of 43. This rate indicates a slow growth trend with a small fluctuation. From 2014 to 2022, the total number of published papers was 1484, with an annual average of 165. The annual growth rate reached 188.72%, much higher than that of the first stage (2002–2013). Notably, the number of papers published in 2014 exceeded 100. Starting in that year, the related studies on straw returning experienced a small upsurge. It can be predicted that straw returning will remain a research hotspot in the future.

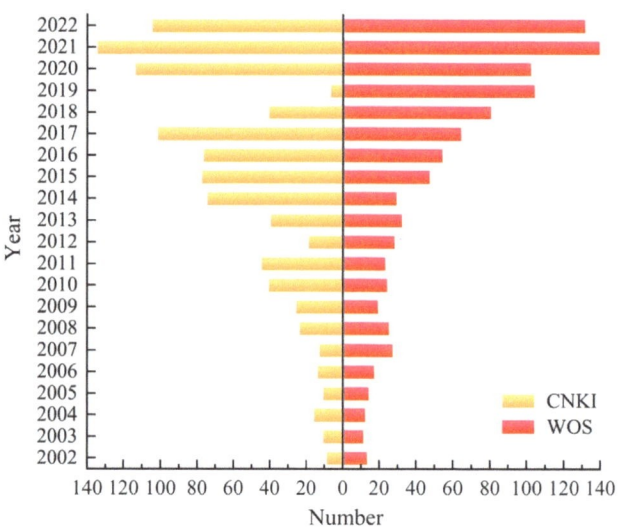

Figure 1. Number of papers in the field of straw returning from 2002 to 2022.

Figure 2 shows a co-occurrence network map of the keywords. In general, soil fertility, soil type, crop nutrition, and tillage methods were the main foci of current research on straw returning. The papers in the CNKI encompass 537 nodes and 713 lines, while the papers in the WOS encompass 545 nodes and 4127 lines. The size of each graph node represents the frequency of the keywords, and the betweenness represents the importance of the node in the network. Here, a node with a betweenness exceeding 0.1 is considered a key node. As shown in Figure 2a, in the CNKI, the most frequent keywords were "straw return" and "straw returning"; the frequency was 444 occurrences, and the betweenness was 1.04. The keyword "soil nutrient" ranked second, with a frequency of 55 and a betweenness of 0.11. Ranked next were the keywords related to planted crops, soil indexes, soil types, and tillage, such as "winter wheat", "grain yield", "crop yield", "soil enzyme activity", "soil organic carbon", "soil respiration", "paddy soil", "black soil", and "conservation tillage". As shown in Figure 2b, the most frequent keyword was "yield"; the frequency was 220 occurrences, and the betweenness was 0.07. The keyword "water use efficiency" ranked next, with a frequency of 186 occurrences and a betweenness of 0.04. Ranked next were keywords related to straw, soil indicators, crops, and tillage, including "straw mulch", "straw mulching", "soil temperature", "organic carbon", "growth", "winter wheat", "grain yield", "crop yield", "management", and "tillage". The keywords of CNKI and WOS were clustered on this basis. The modularity value of CNKI was $Q_{CNKI} = 0.5739$ ($Q > 0.3$), the mean silhouette of CNKI was $S_{CNKI} = 0.8708$ ($S > 0.5$), the modularity value of WOS was $Q_{WOS} = 0.3588$ ($Q > 0.3$), and the mean silhouette of CNKI was $S_{CNKI} = 0.6906$ ($S > 0.5$). The results show that the clustering structure was reasonably divided.

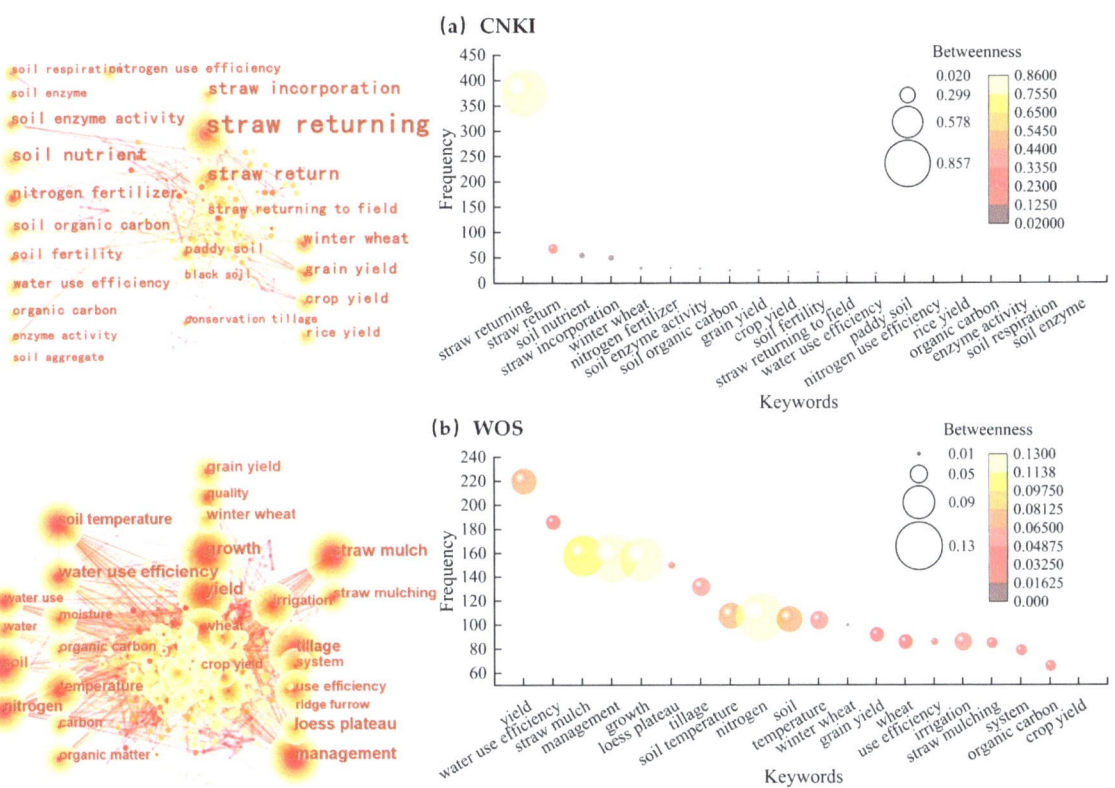

Figure 2. Keywords co-occurrence network map in the field of straw returning.

2.1.2. Research Trends of Straw Returning

Figure 3 shows the time and intensity of the keywords' emergence. The CiteSpace software was used to create the burst detection of the papers. The results show that the use of conservation tillage technology to improve soil fertility conditions will be a future development trend in the straw returning field. The keyword "summer maize" first appeared in 2004. The initial research object in the field of straw returning was mainly maize, which corresponds to a large volume of straw. With the advancement of research, crop straw resources, such as rice (2007) and wheat (2010), have gradually been fully introduced to the field. From 2007 to 2009, the keywords "soil K" and "soil fertility" suddenly appeared, indicating that straw returning research had become an effective way to measure soil improvements and that the importance of soil fertility in agricultural production was gradually becoming recognized. From 2009 to 2011, "CH4 flux" and "greenhouse gas" appeared suddenly, indicating that straw returning research was further extended to explore the air environment. From 2013 to 2014, the emergence of "grain yield", "crop yield", and "lime concretion black soil" suggests that the research scope of straw returning extended to soil improvement and crop yield. Notably, the keyword "soil aggregate" first appeared in 2018, ranking first in strength (strength = 4.17). Soil aggregates are the basic units of soil's structure, composition, and stability, which play important roles in influencing soil pore structure and coordinating water, fertilizer, gas, and heat in soil. These factors determine the sustainability of soil utilization [20]. In addition, the emergence of "microbial community" in 2019 indicates that the research direction of straw returning was shifting from improving soil toward restoring the topsoil's ecological structure. This shift indicates that research and testing in the field of straw returning is becoming increasingly complex.

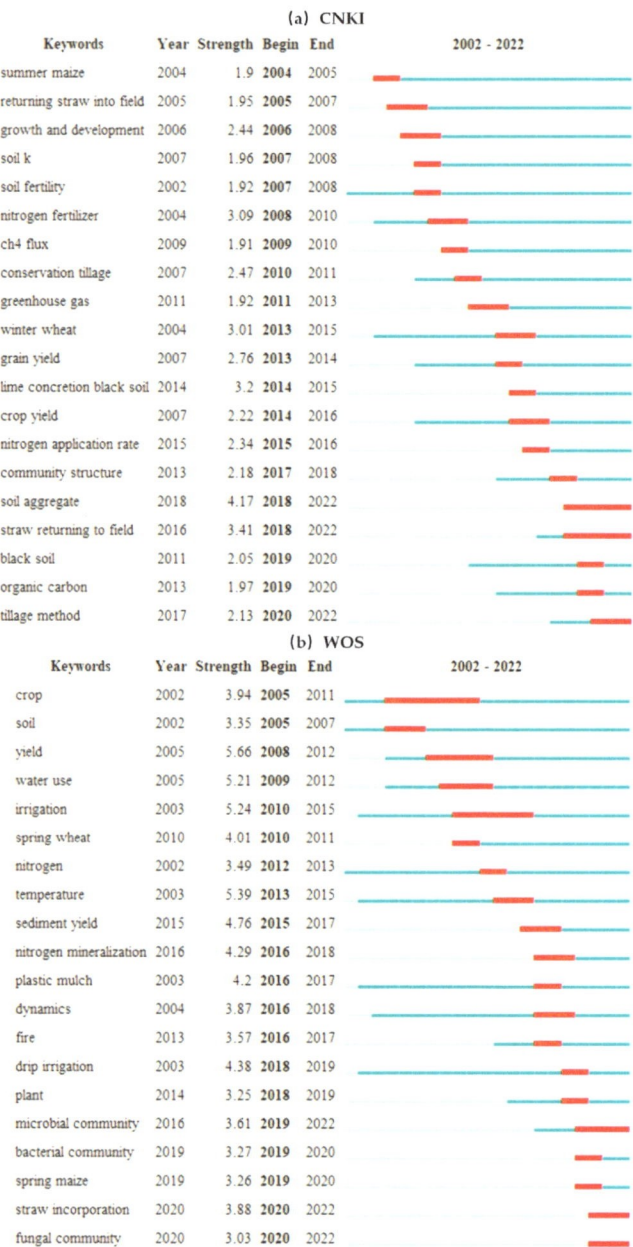

Figure 3. The first 20 burst terms in the field of straw returning. The blue lines are the periods in which the keywords appear, and the red lines are the sudden periods of the keywords.

2.2. Statistical Analysis of Straw Decomposition Agent Studies

In the CNKI, a total of 58 studies were retrieved with the keywords "maturation agent" and "promoting agent" from 2002 to 2022. In the WOS, a total of 72 studies were retrieved with the keywords "straw decomposing agent" and "straw decomposition agent" from 2002 to 2022.

2.2.1. Statistical Analysis of Straw Decomposition Agent Studies

Figure 4 shows the number of papers related to straw decomposition agents from 2002 to 2022. In general, this chart indicates a fluctuating growth trend year by year, which was divided into two stages. The first stage spanned 2004–2013: the total number of published papers was 48, and the average annual number of published papers was 5. The second phase covered 2014–2022: the total number of publications was 77, with an average annual number of 6, of which only 5 were published in 2017. As can be seen from the visible change trend of the papers in Figure 4, studies on applying decomposition agents to straw returning are mainly concentrated after 2019, with the peak years being 2019 and 2021 (13 articles). Therefore, it can be predicted that research on the application of decomposition agents to straw returning will remain an academic hotspot in the future.

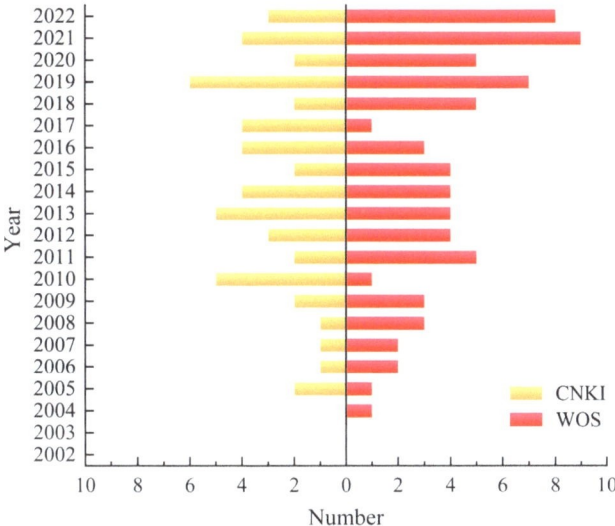

Figure 4. Number of papers in the field of straw decomposition agents from 2002 to 2022.

Figure 5 presents a co-occurrence network map of the major keywords. In general, soil fertility and straw type were the main concerns of current research on straw decomposition agents. The documents in the CNKI encompassed 102 nodes and 197 lines, and the documents in the WOS encompassed 344 nodes and 1318 lines. As shown in Figure 5a, "straw returning" appeared eight times, with the highest frequency in the CNKI in the field of straw decomposition agents, and the betweenness was 0.63. The keywords "decomposing agent", "decomposition agent", and "decomposing inoculant" ranked second, with a frequency of 17 and a betweenness of 0.65, followed by keywords related to soil indicators and straw types such as "soil nutrient", "soil enzyme activity", "nutrient release", "rice straw", and "wheat straw". As shown in Figure 5b, the highest occurrence frequency among the keywords was "decomposition" in the WOS, with 220 occurrences and a betweenness of 0.07. The next most common keywords were related to soil properties and straw properties, such as "carbon", "organic matter", "nitrogen", "microbial community", "bacterial", "CH4 emission", "bioma", "lignin", "lignocellulosic bioma", and "mineralization". The keywords were clustered on this basis. The modularity value of CNKI was $Q_{CNKI} = 0.7351$ ($Q > 0.3$), the mean silhouette of CNKI was $S_{CNKI} = 0.9496$ ($S > 0.5$), the modularity value of WOS was $Q_{WOS} = 0.726$ ($Q > 0.3$), and the mean silhouette of CNKI was $S_{WOS} = 0.9267$ ($S > 0.5$). The results show that the clustering structure was reasonably divided.

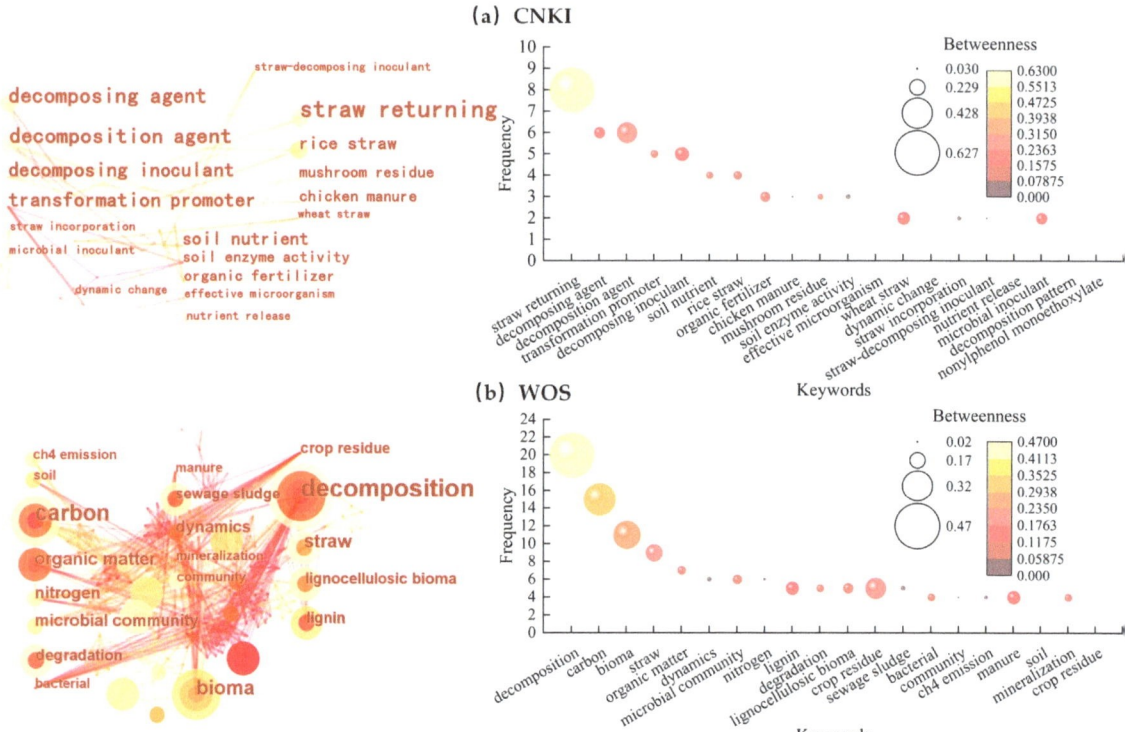

Figure 5. Keywords co-occurrence network map in the field of straw decomposition agents.

2.2.2. Research Trend of Straw Decomposition Agent

Figure 6 shows the time and intensity of keywords' emergence. The results indicate that the development trends of straw decomposition agents focused on how to improve soil fertility and promote crop growth by accelerating straw decomposition. The keyword "transformation promoter" first appeared in 2005 and remained commonplace for seven years, with strength = 2.42. From 2011 to 2016, keywords for soil properties and straw properties emerged, such as "organic matter", "soil enzyme activity", and "lignin". This result indicates that earlier research on straw maturation agents focused on soil fertility and straw decomposition effects. From 2016 to 2019, the keywords "bioma" and "crop yield" appeared suddenly, and related research was further extended toward crop physiological and biochemical indexes and yield. In 2019, the emergent keywords "CH4 emissions" and "biochar" indicate that the research was extended to soil carbon and air-quality effects. The keywords of "decomposing agent", "decomposition agent", and "straw returning" emerged in 2020, indicating the great significance of straw decomposing agents in straw returning.

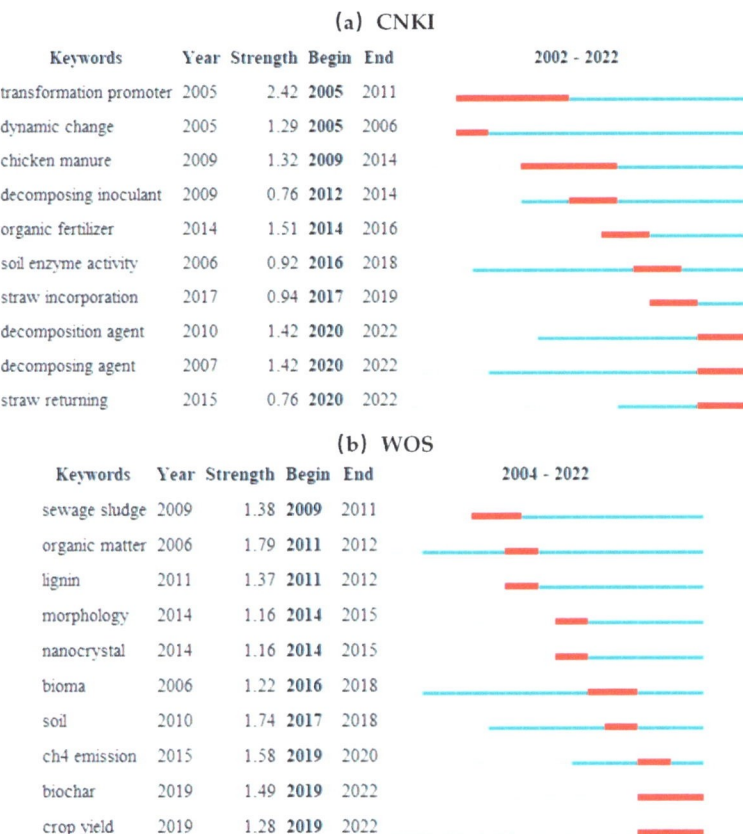

Figure 6. The first 10 burst terms in the field of straw decomposition agents. The blue lines are the periods in which the keywords appeared, and the red lines are the sudden periods of the keywords.

3. Study on the Ecological Effects of Straw Returning to Farmland

3.1. Effects of Straw Returning on Soil Physical and Chemical Properties

Straw returning is one of the most important agronomic measures to improve soil's physical and chemical properties and can help reduce soil bulk density, increase soil porosity, and prevent soil compaction. Lu et al. [21] found that straw returning significantly reduced the soil bulk of surface soil (5–25 cm) by 9.15–23.99%, increased the total porosity by 4.6–15.4%, and increased the capillary porosity by 8.78–58.8%. Straw returning prevents soil water from escaping and evaporating by changing the contact between the soil surface and the external environment. At the same time, straw returning on the soil surface can reduce surface runoff and increase rainwater seepage into the soil layer, impacting evaporation inhibition, water collection, and water retention [22,23]. Straw returning can significantly increase the soil moisture of crops (such as wheat, corn, and potato) by 6.3–61.0% during the most critical period of water demand [24–27].

Soil organic matter is the foundation of farmland fertility, and organic carbon is one of the main elements of organic matter, which is of great significance to crop yield and agricultural environmental sustainability. Research showed that the soil organic carbon content of monoculture with dry crop residues incorporation (i.e., conventional monoculture) (7.59 t/ha) was significantly higher than that of monoculture with residues removal (i.e., alternative monoculture) (1.02 t/ha) over the 2006–2016 period [28]. An increase in soil organic matter content can provide a better environment for the growth and propagation of microorganisms. Hu et al. [29] demonstrated that 3-year straw returning could increase soil organic matter content by 0.07%. The combined application of returned straw and inorganic fertilizer was found to have positive effects on the diversity and population of soil fungi in paddy fields and significantly changed the function of fungi [30]. The combined application of returned straw and fertilizer, the bacterial richness (Shannon index), and the AMF bacterial richness (Shannon index) of maize root increased by 63.6% (27.9%) and 40.1% (35.7%), respectively [31]. Soil enzyme activity, which is a potential index to maintain soil fertility, shows positive correlations with soil TC and TN content and also plays an important role in the ecological cycle [32,33]. Straw returning combined with lower fertilizer use of a reduction of 20% compared to conventional fertilization significantly increased urease activity in the rice season by 20.31% and in the rape season by 24.33% [34].

With appropriate environmental conditions and soil microorganisms, N, P, K, and other mineral nutrients released by straw decomposition can be reabsorbed by crops. Returning straw has already become an important nutrition source in place of chemical fertilizer [35]. Many studies have shown that straw returning can effectively reduce not only fertilizer use but also soil nutrient loss [36–38]. The potential for crop straw to replace K_2O, N, and P_2O_5 fertilizers was 33.08–285.95 kg/hm^2, 9.52–82.32 kg/hm^2, and 4.91–28.71 kg/hm^2, respectively [39]. In addition, the nutrient release rate of straw returning was significantly different during decomposition, when the release rate of K was the highest, followed by P. The release rate of N was the lowest [40].

3.2. Effects of Straw Returning on Saline Soil Properties

Salinized soil is widely distributed globally, covering 424 million ha of topsoil (0–30 cm) and 833 million ha of subsoil (30–100 cm) [41], with an annual growth rate of 1.0×10^6–1.5×10^6 hm^2 [42,43]. The character of saline soil includes a heavy texture, poor permeability, high salt content, lack of nutrients, and low microbial activity, which seriously affects crop growth and ecosystem functions [44,45]. However, the saline soil area is flat, which is suitable for mechanical cultivation and has great fertility potential. Therefore, this soil plays a key role in alleviating the shortages of cultivated land resources and promoting the sustainable development of agriculture through scientific improvements and the exploitation of salinized soil.

As an effective form of conservation tillage, straw returning is considered to be an effective measure for improving saline soil. As a barrier for water and salt migration in the soil layer, returning straw can promote soil salt leaching during precipitation and irrigation, inhibit soil salt return, and reduce soil salt accumulation in the topsoil [46,47]. Zhang et al. [48] found that the soil salt content in the 0–40 cm soil layer decreased by 3.07–36.82% after a three-year straw returning. A soil column experiment conducted by Liao et al. [49] showed that treatment with desulfurized gypsum and decomposed straw could significantly reduce the soil salinity of sodic saline soil after leaching. The simultaneous application of straw returning and nitrogen fertilizer improved the properties of the Shandong coastal saline soil and reduced the soil salinity by 27.08% [50]. Straw returning can also optimize the root growth environment by changing water and salt migration, soil porosity, and aggregate size distribution in the soil layer [51,52]. Straw returning can provide sufficient organic matter for microbial proliferation [53,54]. Additionally, organic matter mineralization releases nutrients for crop growth and increases soil nitrogen, phosphorus, and potassium contents [55,56]. A 4-year straw-return cotton experiment showed that soil salt decreased by 6.8–11.9% in the 0–20 cm and 20–40 cm soil layers [57].

3.3. Effects of Straw Returning on Greenhouse Gases

Straw returning has an important effect on controlling greenhouse gas emissions. CO_2 is the main source of soil carbon pool input. Straw returning has a good carbon sequestration effect, which is of great significance for soil organic carbon (SOC) fixation and CO_2 emission reductions. After returning to the field, 8% to 35.7% of the organic carbon in straw is converted into soil organic carbon and stored in the soil carbon bank [58], which enables the conversion of farmland from a carbon source into a carbon sink and indirectly reduces CO_2 emissions. Therefore, how to quantitatively estimate the carbon emission and carbon sink effects of straw with different disposal methods, how to optimize the efficient utilization of straw resources, and how to maximize the potential of carbon emission reductions are current research hotspots in agricultural sustainable development. Using the emission factor method, Ma et al. [12] calculated that the average annual carbon sink from straw returning in China is 271 tons of CO_2, contributing 10–31% of the global terrestrial carbon sink [59]. A meta-analysis found that straw returning significantly increased CO_2, CH_4, and N_2O emissions by 31.7%, 130.9%, and 12.2%, respectively [60], and the annual carbon sequestration could reach 0.597 tons/ha [61]. Taking Jiangsu Province in China as an example, the carbon emission reduction potential of straw returning was evaluated based on the amount of recoverable straw resources. It was found that the carbon emission reduction potential of straw returning totaled 362,000 tons of CO_2e, equivalent to 0.18% of the greenhouse gas emissions in Jiangsu Province [62]. In addition, straw returning can significantly increase CH_4 emissions (21.1–39.6%) and enhance the soil's ability to absorb CH_4 [63,64]. The increase is 21.1–39.6% but has no effect on the emissions of N_2O [65]. Recent studies have shown that the utilization of biogas produced from agricultural wastes has reduced greenhouse gas (GHG) emissions [66]. Biogas is produced through anaerobic digestion of various biomass energy sources. Anaerobic digestion (AD) technology allows for the bioconversion of organic matter from agricultural wastes, as well as the recovery of biogas for the generation of electricity or the production of biomethane [67,68] and renewable fertilizers biogas residue [69]. *Cynara cardunculus* L. residue has been widely used as a source of biomass for biogas production process because of its high biomass content and renewable characteristics [70]. De Menna et al. [71] investigated the biogas production potential of five different varieties of artichoke and found methane production of 292 LCH4/kg VS.

4. Study on Accelerating Decomposition of Retuning Straw

4.1. Mechanism of Straw Decomposition Agent

Cellulose, hemicellulose, lignin, protein, and soluble sugar are the main components of straw, and cellulose, hemicellulose, and lignin account for approximately 80% of straw's total dry matter mass [72]. Straw is also rich in C, N, P, K, Ca, Mg, and other elements. However, because of the slow decomposition of straw under natural conditions [73], only 25% of the mass of straw is effectively returned to the field [74]. Therefore, the rational application of a straw decomposition agent is one of the most important measures to promote the use of straw returning technology.

A straw decomposition agent is mainly composed of bacteria, fungi, actinomyces, and bioenzymes. These microorganisms can effectively degrade straw components in the process of growth and reproduction and convert straw components into mineral elements such as N, P, K, Ca, and Mg, which are required for crop growth [12]. The application of a decomposition agent can effectively improve the decomposition rate and degree of straw [75].

However, the effects of straw decomposition agents on straw decomposition were found to be different under different field temperatures. When the temperature was below 3 °C, the decomposition rate of rice straw after 25 days was basically 0. When the temperature was higher than 3 °C, the decomposition rate of rice straw increased rapidly with an increase in temperature [76]. During the wheat season, the field temperature was low for 30 to 60 days of rice straw decomposition, so the effect of the decomposition agent was not obvious [77]. Therefore, it is necessary to analyze the climatic characteristics of the cultivation area and select a suitable decomposition agent to provide ideal environmental conditions for a straw decomposition agent and, thus, ensure effective straw decomposition. Adding a straw decomposition agent also increased the diversity of the microbial community in the soil [78] and significantly enhanced the activities of hydrolase and other soil enzymes [79], thus effectively improving soil texture, alleviating soil nutrient loss, improving soil fertility, and improving crop quality and yield [80,81].

4.2. Effects of Accelerating Straw Decomposition on Soil Properties

Adding a straw decomposition agent to straw returning can further improve the soil structure and increase soil fertility. Compared with the control soil, the soil water content of the soil with a straw decomposition agent increased by 1.88–10.80% [82], and soil N, P, and K concentrations also increased significantly [83]. In rape cultivation, soil total nitrogen and available nitrogen levels increased by 3% and 4%, respectively, under straw returning treatment with a microbial agent [84]. Microorganisms play a vital role in the straw decomposition and soil nutrient cycle, which can accelerate straw decomposition [85,86]. The straw returning treatment, which involved rotary tillage with microbial agents, yielded a higher decomposition rate of cellulose, hemicellulose, and lignin at 35.49%, 84.23%, and 85.50%, respectively, and the soil microbial biomass carbon and soluble carbon under these treatments increased by 14.22 mg/kg and 25.10 mg/kg compared with the levels under other treatments [87]. Under the condition of no-tillage straw returning, the straw was exposed to the surface. Because of insufficient contact between the straw and soil, the decomposition agent was affected by sunlight and water shortage. For this reason, microorganisms could not fully play their respective roles, thereby affecting the decomposition and nutrient release of the straw [88,89]. Microbial agents can also enhance the functional diversity of soil microorganisms and increase the content of enzymes in the microbial community, thus accumulating more enzymes in the soil. Soil enzyme activity is involved in nutrient cycling and the decomposition of organic matter and has a synergistic correlation with soil nutrients [90–92].

4.3. Effects of Accelerating Straw Decomposition on Crop Growth

Yield and quality are important indicators related to the economic and social benefits of agricultural products. Straw returning combined with a decomposition agent can promote the growth of crops and increase yield. In this study, straw returning combined with a microbial agent significantly increased crop water utilization rates by 7.9–8.4% and rice yield by 7.3–7.7% [93], possibly because microbial agents can rapidly release nutrient elements in crop straws and improve soil properties to promote the growth of rice [94,95]. Zhang et al. [96] found that when rape straw was decomposed and returned to the field, the plant height, ear length, and other agronomic traits of rice from the subsequent crop were improved compared with the traits of rice from crops not exposed to a decomposition agent, and the yield further increased by 14.73%. A meta-analysis showed that the effects of straw-decomposing microbial inoculant (SDMI) on crop yield were significantly different under different soil conditions [97]. Compared with neutral soil, SDMI significantly increased the straw decomposition rate and crop yield in acidic soil and alkaline soil. This result may reflect the competition between SDMI and the microbial community in neutral soil, which limited the efficacy of SDMI [98].

5. Quantitative Characterization Experiment of Straw Decomposition

5.1. Materials and Methods

5.1.1. Experimental Sample

(1) Wheat straw: The straw was taken from a wheat field in Lianyungang, with a length of 3–5 cm. The nutrient content of the straw included total N of 2.82 mg/kg, total P of 2.23 mg/kg and total K of 18.94 mg/kg.

(2) Straw decomposition agent: *Bacillus subtilis* (wettable powder) was produced by Byvo Co., LTD. (Beijing, China), and the active ingredient content was 10 billion CTU/g.

(3) *Pseudomonas* (wettable powder) was produced by Shandong Huimin Zhonglian Biotechnology Co., LTD. (Shandong, China), and the active ingredient content was 300 billion/g.

(4) The yeast quick rot agent was produced by Huaian Dahua Biotechnology Co., LTD. (Jiangsu, China), and was mainly composed of *Bacillus subtilis*, Rhizopus oryzae, lactic acid bacteria community, and tablet pentose, with an effective viable bacteria number $\geq$ 0.5 million.

(5) The EM bacteria was produced by Jiangsu Werner Biotechnology Co., LTD. (Jiangsu, China), and the main components were lactic acid bacteria, yeast, actinomyces, and photosynthetic bacteria, totaling more than 80 varieties of functional beneficial microbial flora. The effective viable bacteria number was 200 million/mL.

5.1.2. Experiment Design

A total of 20 g wheat straw was evenly sprayed with water to reach 60% water content; then, the straw was wrapped and sealed with plastic wrap and placed at room temperature to facilitate rot decomposition (20 °C–25 °C). In total, 5 treatments were used: control group (CK), straw + Bacillus subtilis (T1), straw + enzyme quick rot agent (T2), straw + Pseudomonas (T3), and straw + fecal treasure (T4). Each treatment used 3 replicates (shown in Figure 7).

Figure 7. Treatment design.

5.1.3. Research Trends of the Straw Decomposition Agent

On the 7th, 15th, and 30th d of decomposition, the straw was put in an oven until dried (75 °C) to a constant weight, and its dry weight was determined. The straw was then ground and screened (0.1 mm) to determine the content of total N, total P, and total K.

The cumulative decomposition rate and decomposition rate of straw were calculated using Formulas (1) and (2) [99,100], and Formula (3) was used to calculate the straw nutrient release rate [101]. To further compare the decomposition dynamics of straw under different treatments, a modified Olson exponential decay model (4) was used in this study for fitting [102]:

$$D_t = \frac{(M_0 - M_t)}{M_0} \times 100\% \tag{1}$$

$$V_t = \frac{(M_0 - M_t)}{t} \tag{2}$$

$$Y_t = \frac{C_0 \times M_0 - C_t \times M_t}{C_0 \times M_0} \tag{3}$$

$$D = \frac{M_t}{M_0} = e^{-kt} \tag{4}$$

where D_t is the cumulative decomposition rate of straw on the T day (%); V_t is the straw decomposition rate (%); M_0 is the initial dry weight of straw (g); M_t is the dry weight of straw on the t day of decomposition (g); C_0 is the initial nutrient content of straw (mg/g); Y_t is the percentage of nutrient release in the process of straw decomposition in t days to the total initial straw nutrient (%); C_t is the nutrient content of residual straw after t days of decomposition (mg/g); D is the residue rate of straw decomposition (%); and K is the decomposition rate constant.

5.1.4. Data Analysis

The straw decomposition rate and nutrient release rate were analyzed using one-way analysis of variance (ANOVA), and the means were compared using a least significant difference (LSD) test at a significance level of $p < 0.05$ with SPSS.

5.2. Results and Discussion

5.2.1. Effect of the Decomposition Agent on the Wheat Straw Decomposition Rate

As shown in Figure 8, with the straw decomposition process, the decomposition rate of the treatments presented a downward trend, and the difference in the straw decomposition rate between each treatment and the CK was reduced. In the first 7 days of decomposition, the straw decomposition rates of all treatments were higher than 360 mg/d, and the straw decomposition rate of T3 (569.52 mg/d) was significantly higher than that of the other treatments, which was 26.96% higher than that of the CK. During 7–15 d, the straw decomposition rates of all treatments were less than 200 mg/d, among which the decomposition rate of T1 was the fastest (186.25 mg/d), and that of T4 was the slowest (134.17 mg/d). However, there was a significant difference between CK and other treatments. During 15–30 d, there was no significant difference between treatments. The reason for this phenomenon is that a large number of easily degradable substances and carbon sources were present in the straw during the early stages of decomposition, while the microbial activity gradually decreased in the later stages of decomposition, which gradually decelerated the straw decomposition rate [103–105]. After adding different decomposition agents, the straw decomposition rate was T3 (straw + *Pseudomonas*) > T1 (straw + *Bacillus subtilis*) > T4 (straw + fecal treasure) > T2 (straw + Enzyme quick rot agent) because bacteria have the advantages of fast reproduction and a short fermentation cycle, and a large number of bacteria can accelerate the decomposition of lignocellulosic components in straw after entering the organic material [106].

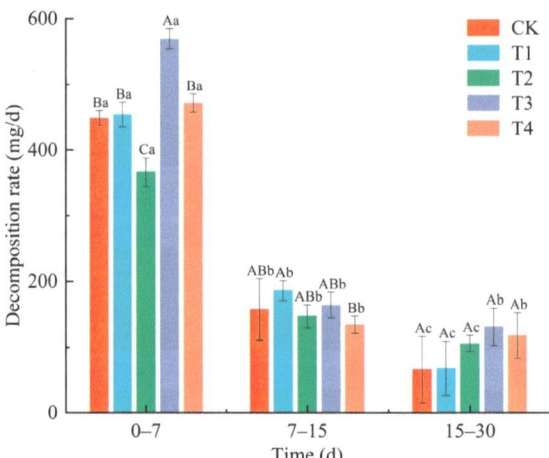

Figure 8. Decomposition rates of the treatments. The different capital letters indicate significant differences among treatments measured on the same day ($p < 0.05$), while the different lowercase letters indicate significant differences on different days in the same treatment ($p < 0.05$).

Table 1 presents the Olson exponential decay model of the straw decomposition rate. The results showed that T3 had the highest k, and the residue ratio of T3 was 63.63% after 30 d, which reduced by 13.17% compared with the CK. T2 had the lowest k, but the residue ratio of T2 was slightly higher than that of the CK after 30 days.

Table 1. Olson exponential decay equation fitting the straw decomposition residual rate with time.

Treatment	D (%)			k	R^2
	7 d	15 d	30 d		
CK	84.30	78.00	73.05	0.013 ± 0.004	0.995
T1	84.12	76.67	71.57	0.014 ± 0.004	0.996
T2	87.17	81.27	73.33	0.012 ± 0.002	0.998
T3	80.07	73.50	63.63	0.018 ± 0.002	0.994
T4	83.50	78.13	69.28	0.014 ± 0.002	0.996

The straw decomposition rate is related to the dosage of the bactericide and temperature [107,108]. Microorganisms are the main decomposers of straw, and their abundance and community composition affect the decomposition rate of straw [109]. As the dominant strain of straw decomposition, Pseudomonas increases the activity of decomposition enzymes, thereby accelerating the decomposition rate of straw [110–112].

5.2.2. Effect of the Decomposing Agent on Straw Nutrient Release

As shown in Figure 9, the straw decomposition agent promoted the nutrient release of straw, and the nutrient release rate of each treatment showed a gradually decreasing trend. The nutrient release rates of different treatments were roughly as follows: T3 (straw + *Pseudomonas*) > T4 (straw + fecal treasure) > T1 (straw + *Bacillus subtilis*) > T2 (enzyme quick rot agent). The release rates of different nutrients were K > P > N. In the first 7 days of straw decomposition, the N release rate of all treatments was higher than 13%, the P release rate was higher than 20%, and the K release rate was higher than 21%. The N release rate of T3 (18.99%) was higher than that of the other treatments and was 41.61% higher than that of the CK. The P and K release rates of T2 were higher than those of other treatments and increased, respectively, by 21.00% and 22.94% compared with the CK. The N release rate, P release rate, and K release rate of different treatments were 11–17%, 14–20%, and 16–22%, respectively, over 7–15 d of straw decomposition. The release rates of N, P, and K in the T3 treatment were the highest and increased by 39.82%, 32.99%, and 32.73%, respectively, compared with the CK. Over 15–30 d, the N release rate, P release rate, and K release rate of the different treatments were 5–11%, 8–15%, and 9–18%, respectively. The release rates of N, P, and K in the T3 treatment were the highest, showing increases of 85.09%, 66.54%, and 90.00%, respectively, compared with the CK. There was no significant differences in the nutrient release rates among the different treatments with the same decomposition period. The different nutrient release rates among the different treatments were mainly due to the morphology of the elements in straw. Among them, K mainly exists in an ionic state in plant tissues and is easily soluble in water. Thus, the release rate of K is the fastest and the final cumulative release rate is also the highest, P mainly exists in an insoluble organic state, with a smaller presence than K [40,113], and N is mainly composed of insoluble organic matter, which decomposes quickly in the early stages and has difficulty decomposing in the later stages, resulting in a decreased decomposition rate [114]. N, which is the main component of straw, has a high degree of cementation and does not easily decompose, resulting in slow release [115,116]. For different nutrients, there was a significant difference in the nutrient release rate of the CK between the early stage of decomposition (0–7 d) and at the end of decomposition (15–30 d) but no significant difference in the nutrient release rate of T1 and T2; we observed significant differences only in the P release rate of T3 and the K release rate of T4. In the early stage of straw decomposition, the decomposition agent caused a sudden increase in the microbial flora in straw. In addition, soluble organic matter and inorganic nutrients in straw provided a large amount of energy and nutrients for microorganisms, stimulated microbial activity, and accelerated straw decomposition, which promoted nutrient release [117]. At the end of decomposition, the amount of microorganisms gradually decreased, their activity became inhibited, and the effect of promoting decay decelerated, which produced significant differences [118].

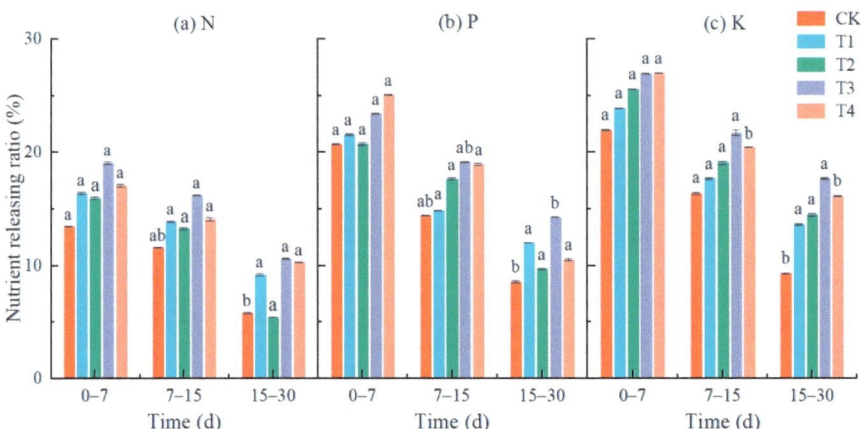

Figure 9. The nutrient (N, P, and K) release rate of straw under different treatments. The different lowercase letters indicate significant differences on different days during the same treatment ($p < 0.05$).

6. Conclusions

This paper summarized the trends in research on straw returning, analyzed the effects of straw returning decomposition on soil properties and crop growth, and verified the straw decomposition rate and nutrient release rate with a quantitative characterization experiment. The experiment showed that straw combined with *Pseudomonas* is suitable for application in soil degradation areas, such as those characterized by nutrient deficiency, low yield, and barrenness, and provides more nutrients for crop growth. Straw returning can optimize soil properties and provide nutrients needed for crop growth, and application of a straw decomposition agent will further increase the decomposition rate of straw and promote the release of nutrients therefrom. The overall analysis of straw decomposition agents showed an increasing trend, mainly focusing on straw types, soil fertility changes, and crop growth responses. The regulating effect of straw returning on soil properties is related to the amount of straw returning to the field, the time of straw returning, the length of the straw, temperature, and other factors. To exert the maximum effect of straw returning, it is necessary to conduct in-depth research and establish an optimal model for straw returning treatment suitable for different soils and regions. However, the types of straw for a field are not comprehensive enough at present. Most studies are on the application of crops such as corn and wheat. There are few studies on the utilization of crop straw such as fruit, vegetable, and tea. Therefore, it is necessary to expand the practical research of various crop straw returning to the field.

Author Contributions: Data curation, Q.S.; project administration, Y.W.; supervision, S.F. and Y.L.; writing—original draft preparation, Q.S.; writing—review and editing, Y.W., S.F., Y.L. and C.W. All authors have read and agreed to the published version of the manuscript.

Funding: This research was funded by the Jiangsu Natural Science Fund, grant numbers: BK20210824 and BK20200941; Yangzhou Talent Project of Jiangsu Province, grant number: YZLYJF2020PHD096; and College Student Academic Technological Innovation Fund Supported Project, grant number: X20220517.

Data Availability Statement: Data can be obtained upon request of the corresponding authors.

Acknowledgments: The authors of this paper express our most sincere gratitude to all the staff who help during our experiment and writing.

Conflicts of Interest: The authors declare no conflict of interest.

References

1. Cui, X.H.; Guo, L.Y.; Li, C.H.; Liu, M.Z.; Wu, G.L.; Jiang, G.M. The total biomass nitrogen reservoir and its potential of replacing chemical fertilizers in China. *Renew. Sustain. Energ. Rev.* **2021**, *135*, 110215. [CrossRef]
2. Zhang, X.Q.; Wang, Z.F.; Can, M.Y.; Bai, H.; Ta, N. Analysis of yield and current comprehensive utilization of crop straws in China. *J. China Agric. Univ.* **2021**, *26*, 30–41.
3. Zhang, Q.F. Problems and Countermeasures of Straw Returning. *Sci. Technol. West China* **2014**, *13*, 73+126.
4. Saroj, D.; Charu, G.; Shankar Lal, J.; Parmar, M.S. Crop residue recycling for economic and environmental sustainability: The case of India. *Open Agric.* **2017**, *2*, 486–494.
5. Bhuvaneshwari, S.; Hettiarachchi, H.; Meegoda, J.N. Crop residue burning in India: Policy challenges and potential solutions. *Int. J. Environ. Health* **2019**, *16*, 832. [CrossRef]
6. Mahmoud, I.; Esawy, M.; Doaa, I. Assessing the impact of water treatment residuals and rice straw compost on soil physical properties and wheat yield in saline sodic Soil. *Commun. Soil Sci. Plant Anal.* **2020**, *51*, 2388–2397.
7. Xu, X.; Pang, D.W.; Chen, J.; Luo, Y.L.; Zheng, M.J.; Yin, Y.P.; Li, Y.X.; Li, Y.; Wang, Z.L. Straw return accompany with low nitrogen moderately promoted deep root. *Field Crop Res.* **2018**, *221*, 71–80. [CrossRef]
8. Wang, Z.; Ma, L.J.; Lv, X.B.; Meng, Y.L.; Zhou, Z.G. Straw returning coupled with nitrogen fertilization increases canopy photosynthetic capacity, yield and nitrogen use efficiency in cotton. *Eur. J. Agron.* **2021**, *126*, 126–267. [CrossRef]
9. Zhang, C.F.; Lin, Z.L.; Que, Y.X.; Fallah, N.; Tayyab, M.; Li, S.Y.; Luo, J.; Zhang, Z.C.; Abubakar, A.Y. Straw retention efficiently improves fungal communities and functions in the fallow ecosystem. *BMC Microbiol.* **2021**, *21*, 13. [CrossRef]
10. Ali, U.; Shaaban, M.; Bashir, S.; Fu, Q.L.; Zhu, J.; Shoffikul Islam, M.; Hu, H.Q. Effect of rice straw, biochar and calcite on maize plant and Ni bio-availability in acidic Ni contaminated soil. *Environ. Manag.* **2020**, *259*, 109674. [CrossRef]
11. Qiu, Y.; Lv, W.C.; Wang, X.P.; Xie, Z.K.; Wang, Y.J. Long-term effects of gravel mulching and straw mulching on soil physicochemical properties and bacterial and fungal community composition in the Loess Plateau of China. *Eur. J. Soil Biol.* **2020**, *98*, 103188. [CrossRef]
12. Ma, Y.C.; Liu, D.L.; Schwenke, G.; Yang, B. The global warming potential of straw-return can be reduced by application of straw-decomposing microbial inoculants and biochar in rice-wheat production systems. *Environ. Pollut.* **2019**, *252*, 835–845. [CrossRef] [PubMed]
13. Witt, C.; Cassman, K.G.; Olk, D.C.; Biker, U.; Liboon, S.P.; Samson, M.I.; Ottow, J.C.G. Crop rotation and residue management effects on carbon sequestration, nitrogen cycling and productivity of irrigated rice systems. *Plant Soil.* **2000**, *225*, 263–278. [CrossRef]
14. Chen, J.; Yi, T.; Ye, Y.L.; Zhang, X.L. Research Progress and Development Trend of Straw Decomposing Agent. *Hunan Agric. Sci.* **2021**, 108–110.
15. Chen, C.M. Searching for intellectual turning points: Progressive knowledge domain visualization. *Proc. Natl. Acad. Sci. USA* **2004**, *101*, 5303–5310. [CrossRef] [PubMed]
16. Hou, J.H.; Hu, Z.G. Review on the Application of CiteSpace at Home and Abroad. *J. Mod. Inf.* **2013**, *33*, 99–103. (In Chinese)
17. Hu, L.N.; Zhang, J.F.; Xing, J. Research Hotspot and Evolution Trend of Green Economic Efficiency at Home and Abroad-Visual Analysis Based on CiteSpace. *J. Commer. Econ.* **2022**, *4*, 189–192.
18. Chu, W.W.; Hafiz, N.R.M.; Mohamad, U.A.; Ashamuddin, H.; Tho, S.W. A review of STEM education with the support of visualizing its structure through the CiteSpace software. *Int. J. Technol. Des. Educ. Online* **2022**, *33*, 1–23. [CrossRef]
19. Zong, X.; Wen, L.; Wang, Y.; Li, L. Research progress of glucoamylase with industrial potential. *J. Food Biochem.* **2022**, *46*, e14099. [CrossRef]
20. Zhou, Q.; Wang, L.J.; Xing, Y.; Ma, S.M.; Zhang, X.D.; Chen, J. Effects of Chinese milk vetch intercropped with rape under straw mulching on soil aggregate and organic carbon character. *Chin. J. Appl. Ecol.* **2019**, *30*, 1235–1242.
21. Lu, W.L. Effects of Tillage and Straw Return on Soil Physical and Chemical Properties and Flue-cured Tobacco Growth. Ph.D. Thesis, Chinese Academy of Agricultural Sciences, Beijing, China, 2019.
22. Li, Q.; Li, H.B.; Zhang, L.; Zhang, S.Q.; Chen, Y.L. Mulching improves yield and water-use efficiency of potato cropping in China: A meta-analysis. *Field Crop Res.* **2018**, *221*, 50–60. [CrossRef]
23. Jin, Z.Q.; Shah, T.R.; Zhang, L.; Liu, H.Y.; Peng, S.B.; Nie, L.X. Effect of straw returning on soil organic carbon in rice-wheat rotation system: A review. *Food Energy Secur.* **2020**, *9*, 13. [CrossRef]
24. Li, S.Y.; Li, Y.; Lin, H.X.; Feng, H.; Dyck, M. Effects of different mulching technologies on evapotranspiration and summer maize growth. *Agric. Water Manag.* **2018**, *201*, 309–318. [CrossRef]
25. Yang, Y.H.; Ding, J.L.; Zhang, Y.H.; Wu, J.C.; Zhang, J.M.; Pan, X.Y.; Gao, C.M.; Wang, Y.; He, F. Effects of tillage and mulching measures on soil moisture and temperature, photosynthetic characteristics and yield of winter wheat. *Agric. Water Manag.* **2018**, *201*, 299–308. [CrossRef]
26. Chang, L.; Han, F.X.; Chai, S.X.; Cheng, H.B.; Yang, D.L.; Chen, Y.Z. Straw strip mulching affects soil moisture and temperature for potato yield in semiarid regions. *Agron. J.* **2020**, *112*, 1126–1139. [CrossRef]
27. Hou, X.Q.; Li, R. Interactive effects of autumn tillage with mulching on soil temperature, productivity and water use efficiency of rainfed potato in loess plateau of China. *Agric. Water Manag.* **2019**, *224*, 105747. [CrossRef]
28. Deligios, P.A.; Farina, R.; Tiloca, M.T.; Francaviglia, R.; Ledda, L. C-sequestration and resilience to climate change of globe artichoke cropping systems depend on crop residues management. *Agron. Sustain. Dev.* **2021**, *41*, 20. [CrossRef]

29. Hu, Y.F. Study on the improvement of farmland soil organic matter by straw returning. *Agric. Ecosyst. Environ.* **2020**, *1*, 131.
30. Nie, S.A.; Lei, X.M.; Zhao, L.X.; Brookes, P.C.; Wang, F.; Chen, C.R.; Yang, W.H.; Xing, S.H. Fungal communities and functions response to long-term fertilization in paddy soils. *Appl. Soil Ecol.* **2018**, *130*, 251–258. [CrossRef]
31. Tian, L.; Shi, S.H.; Zhang, J.F.; Gao, Q.; Tian, C.J. Effects of Long-term Fertilization and Straw Return on Diversity Indices of AMF and Bacteria in Maize Rhizosphere. *Soils Crops* **2017**, *6*, 291–297.
32. Lai, R.; Lagomarsino, A.; Ledda, L.; Roggero, P.P. Variation in soil C and microbial functions across tree canopy projection and open grassland microenvironments Turk. *J. Agric. For.* **2014**, *38*, 62–69.
33. Song, Y.; Song, C.; Shi, F.; Wang, M.; Ren, J.; Wang, X.; Jiang, L. Linking plant community composition with the soil C pool, N availability and enzyme activity in boreal peatlands of Northeast China. *Appl. Soil Ecol.* **2019**, *140*, 144–154. [CrossRef]
34. Jin, Y.T.; Li, X.F.; Cai, Y.; Hu, H.X.; Liu, Y.F.; Fu, S.W.; Zhang, B.R. Effects of Straw Returning with Chemical Fertilizer on Soil Enzyme Activities and Microbial Community Structure in Rice-Rape Rotation. *Environ. Sci.* **2021**, *42*, 3985–3996.
35. Zhuang, M.H.; Zhang, J.; Kong, Z.Y.; Fleming, R.M.; Zhang, C.Y.; Zhang, Z.Y. Potential environmental benefits of substituting nitrogen and phosphorus fertilizer with useable crop straw in China during 2000–2017. *J. Clean. Prod.* **2020**, *267*, 122125. [CrossRef]
36. Yang, S.Q.; Han, R.Y.; Xing, L.; Liu, H.Y.; Wu, H.J.; Yang, Z.L. Effect of slope farmland soil and water and soil nitrogen and phosphorus loss based on different crop and straw applications and ridge patterns in the basin of the main stream of the Songhua River. *Acta Petrol. Sin.* **2018**, *38*, 42–47. [CrossRef]
37. Xu, C.; Han, X.; Zhuge, Y.P.; Xiao, G.P.; Ni, B.; Xu, X.C.; Meng, F.Q. Crop straw incorporation alleviates overall fertilizer-N losses and mitigates N_2O emissions per unit applied N from intensively farmed soils: An in situ 15N tracing study. *Sci. Total Environ.* **2020**, *764*, 142884–142894. [CrossRef]
38. Hua, K.K.; Zhu, B.; Li, C.C. Pathways of Dissolved Unreactive Phosphorus Loss under Long-Term Crop Straw and Manure Application. *Nutr. Cycl. Agroecosyst.* **2021**, *120*, 161–175. [CrossRef]
39. Wang, Y.; Liang, B.Q.; Bao, H.; Chen, Q.; Cao, Y.L.; He, Y.Q.; Li, L.Z. Potential of crop straw incorporation for replacing chemical fertilizer and reducing nutrient loss in Sichuan Province, China. *Environ. Pollut.* **2023**, *320*, 121034. [CrossRef]
40. Dai, W.C.; Gao, M.; Lan, M.L.; Huang, R.; Wang, J.Z. Nutrient release patterns and decomposition characteristics of different crop straws in drylands and paddy fields. *Chin. J. Eco-Agric.* **2017**, *25*, 188–199.
41. FAO. *Global Map of Salt Affected Soils*; Version 1.0; FAO: Rome, Italy, 2021.
42. Luo, S.S.; Wang, S.J.; Tian, L.; Shi, S.H.; Xu, S.Q.; Yang, F.; Li, X.J.; Wang, Z.C.; Tian, C.J. Aggregate-related changes in soil microbial communities under different ameliorant applications in saline-sodic soils. *Geoderma* **2018**, *329*, 108–117. [CrossRef]
43. Wang, Z.J.; Zhuang, J.J.; Zhao, A.P.; Li, X.X. Types, harms and improvement of saline soil in Songnen Plain. *IOP Conf. Ser. Mater. Sci. Eng. C.* **2018**, *322*, 052059. [CrossRef]
44. Xian, X.X.; Pang, M.Y.; Zhang, J.L.; Zhu, M.K.; Kong, F.L.; Xi, M. Assessing the effect of potential water and salt intrusion on coastal wetland soil quality: Simulation study. *J. Soils Sediments* **2019**, *19*, 2251–2264. [CrossRef]
45. Santos, J.S.; Introíni, G.O.; Veiga-Menoncello, A.C.P.; Blasco, A.; Rivera, M.; Recco-Pimentel, S.M. Comparative sperm ultrastructure of twelve leptodactylid frog species with insights into their phylogenetic relationships. *Micron* **2016**, *91*, 1–10. [CrossRef] [PubMed]
46. Song, X.L.; Sun, R.J.; Chen, W.F.; Wang, M.H. Effects of surface straw mulching and buried straw layer on soil water content and salinity dynamics in saline soils. *Can. J. Soil Sci.* **2019**, *100*, 58–68. [CrossRef]
47. Xie, W.J.; Wu, L.F.; Zhang, Y.P.; Wu, T.; Li, X.P.; Ouyang, Z. Effects of straw application on coastal saline topsoil salinity and wheat yield trend. *Soil Tillage Res.* **2017**, *169*, 1–6. [CrossRef]
48. Zhang, H.Y.; Pang, H.C.; Zhao, Y.G.; Lu, C.; Zhang, X.L.; Li, Y.Y. Water and salt exchange flux and mechanism in a dry saline soil amended with buried straw of varying thicknesses. *Geoderma* **2020**, *365*, 114213. [CrossRef]
49. Liao, X.; Yang, F.; Wang, Z.C.; Guan, Q.J.; He, M.L.; An, F.H.; Yang, H.T.; Zhao, D.D.; Zhu, W.D. Effects of decomposed straw and desulfurized gypsum on salt leaching in saline-sodic soils. *Soils Crops* **2020**, *9*, 74–82.
50. Yang, H.J.; Xia, J.B.; Xie, W.J.; Wei, S.C.; Cui, Q.; Shao, P.S.; Sun, J.K.; Dong, K.K.; Qi, X.C. Effects of straw returning and nitrogen addition on soil quality of a coastal saline soil: A field study of four consecutive wheat-maize cycles. *Land Degrad. Dev.* **2022**, *34*, 2061–2072. [CrossRef]
51. Lenka, N.K.; Lal, R. Soil aggregation and greenhouse gas flux after 15 years of wheat straw and fertilizer management in a no-till system. *Soil Tillage Res.* **2013**, *126*, 78–89. [CrossRef]
52. Zhao, H.; Shar, A.G.; Li, S.; Chen, Y.L.; Shi, J.L.; Zhang, X.Y.; Tian, X.H. Effect of straw return mode on soil aggregation and aggregate carbon content in an annual maize-wheat double cropping system. *Soil Tillage Res.* **2018**, *175*, 178–186. [CrossRef]
53. Akhtar, K.; Wang, W.Y.; Ren, G.X.; Khan, A.; Feng, W.Z.; Yang, G.H.; Wang, H.Y. Integrated use of straw mulch with nitrogen fertilizer improves soil functionality and soybean production. *Environ. Int.* **2019**, *132*, 105092. [CrossRef] [PubMed]
54. Su, Y.; Yu, M.; Xi, H.; Lv, J.L.; Ma, Z.H.; Kou, C.L.; Shen, A. Soil microbial community shifts with long-term of different straw return in wheat-corn rotation system. *Sci. Rep.* **2020**, *10*, 6350. [CrossRef] [PubMed]
55. Ahmed, W.; Qaswar, M.; Jing, H.; Wenjun, D.; Geng, S.; Kailou, L.; Ying, M.; Ao, T.; Mei, S.; Chao, L.; et al. Tillage practices improve rice yield and soil phosphorus fractions in two typical paddy soils. *Soil Sediments* **2020**, *20*, 850–861. [CrossRef]
56. Bai, Y.L.; Wang, L.; Lu, Y.L.; Yang, L.P.; Zhou, L.P.; Ni, L.; Cheng, M.F. Effects of long-term full straw return on yield and potassium response in wheat-maize rotation. *J. Integr. Agric.* **2015**, *14*, 2467–2476. [CrossRef]

57. Mao, L.L.; Guo, W.J.; Yuan, Y.C.; Qin, D.L.; Wang, S.L.; Nie, J.J.; Zhao, N.; Song, X.L.; Sun, X.Z. Cotton stubble effects on yield and nutrient assimilation in coastal saline soil. *Field Crop. Res.* **2019**, *239*, 71–81. [CrossRef]
58. Huo, L.L.; Yao, Z.L.; Zhao, L.X.; Luo, J.; Zhang, P.Z. Contribution and Potential of Comprehensive Utilization of Straw in GHG Emission Reduction and Carbon Sequestration. *Trans. Chin. Soc. Agric. Mach.* **2022**, *53*, 349–359.
59. Piao, S.L.; He, Y.; Wang, X.H.; Chen, F.H. Estimation of China's terrestrial ecosystem carbon sink: Methods, progress and prospects. *Sci. China (Earth Sci.)* **2022**, *65*, 641–651. [CrossRef]
60. Liu, P.; He, J.; Li, H.W.; Wang, Q.J.; Lu, C.Y.; Zheng, K.; Liu, W.Z.; Lou, S.Y. Effect of straw retention on crop yield, soil properties, water use efficiency and greenhouse gas emission in China: A meta-analysis. *Int. J. Plant Sci.* **2019**, *13*, 347–367. [CrossRef]
61. Jin, L.; Li, Y.E.; Gao, Q.Z.; Liu, Y.T.; Wan, Y.F.; Qin, X.B.; Shi, F. Estimate of carbon sequestration under cropland management in China. *Sci. Agric. Sin.* **2008**, *41*, 734–743.
62. Sun, J.F.; Zheng, J.F.; Cheng, K.; Pan, G.X. Estimate of the quantity of collectable straw resources and competitive utilization potential. *J. Plant Nutr. Fertil.* **2018**, *24*, 404–413.
63. Chen, H.X.; Liu, J.J.; Zhang, A.F.; Chen, J.; Cheng, G.; Sun, B.H.; Pi, X.M.; Dyck, M.; Si, B.C.; Zhao, Y.; et al. Effects of straw and plastic film mulching on greenhouse gas emissions in Loess Plateau, China: A field study of 2 consecutive wheat-maize rotation cycles. *Sci. Total Environ.* **2017**, *579*, 814–824. [CrossRef] [PubMed]
64. Gao, H.H.; Yan, C.G.; Liu, Q.; Ding, W.L.; Chen, B.Q.; Li, Z. Effects of plastic mulching and plastic residue on agricultural production: A meta-analysis. *Sci. Total Environ.* **2019**, *651*, 484–492. [CrossRef] [PubMed]
65. Wang, L.; Qin, T.; Liu, T.Q.; Guo, L.J.; Li, C.F.; Zhai, Z.B. Inclusion of microbial inoculants with straw mulch enhances grain yields from rice fields in central China. *Food Energy Secur.* **2020**, *9*, e230. [CrossRef]
66. Czyrnek-Deletre, M.M.; Smyth, B.M.; Murphy, J.D. Beyond carbon and energy: The challenge in setting guidelines for life cycle assessment of biofuel systems. *Renew. Energy* **2017**, *105*, 436–448. [CrossRef]
67. Poeschl, M.; Ward, S.; Owende, P. Evaluation of energy efficiency of various biogas production and utilization pathways. *Appl. Energy* **2010**, *87*, 3305–3321. [CrossRef]
68. Wall, D.M.; O'Kiely, P.; Murphy, J.D. The potential for biomethane from grass and slurry to satisfy renewable energy targets. *Bioresour. Tech.* **2014**, *149*, 425–431. [CrossRef] [PubMed]
69. Arthurson, V. Closing the global energy and nutrient cycles through application of biogas residue to agricultural land-Potential benefits and drawbacks. *Energies* **2009**, *2*, 226–242. [CrossRef]
70. Mancini, M.; Volpe, M.L.; Badaracco, P.; Cravero, V.P. Lignocellulosic materials characterization of wild and cultivated cardoon. *Acta Hortic.* **2016**, *1147*, 183–187. [CrossRef]
71. De Menna, F.; Malagnino, R.A.; Vittuari, M.; Molari, G.; Seddaiu, G.; Deligios, P.A.; Solinas, S.; Ledda, L. Potential Biogas Production from Artichoke Byproducts in Sardinia, Italy. *Energies* **2016**, *9*, 92. [CrossRef]
72. Niu, W.J.; Huang, G.Q.; Liu, X.; Chen, L.J.; Han, L.J. Chemical Composition and Calorific Value Prediction of Wheat Straw at Different Maturity Stages Using Near-Infrared Reflectance Spectroscopy. *Energy Fuels* **2014**, *28*, 7474–7482. [CrossRef]
73. Li, M.H.; Tang, C.G.; Chen, X.; Huang, S.W.; Zhao, W.W.; Cai, D.Q.; Wu, Z.Y.; Wu, L.F. High Performance Bacteria Anchored by Nanoclay to Boost Straw Degradation. *Materials* **2019**, *12*, 1148. [CrossRef] [PubMed]
74. Zheng, H.B.; Fu, R.G.; Jia, W.; Tang, Q.Y. Rapid decomposition of straw: Effect on rice yield and nitrogen use efficiency in the rice region of southern China. *J. Agric.* **2019**, *9*, 1–6.
75. Wu, C.L.; Tang, C.S.; Yu, L.; Zhang, J.Y.; Wang, P. Application of straw decomposing agent in straw returning to field in northeast China. *Mod. Agric.* **2022**, *12*, 29–31.
76. Su, Y.; Jia, S.Q.; He, Z.C.; Yang, Y.H.; Yu, M.; Chen, X.J.; Shen, A.L. Optimization of straw decomposition with inoculants by using response surface method. *Acta Agric. Zhejiangensis* **2019**, *31*, 798–805.
77. Zhang, Z.Y.; He, J.; Fan, X.P.; Xia, Y.; Zhang, F.L.; Liu, D.B.; Wu, M.Q. Characteristics of straw decomposition and nutrient release in rice and wheat rotation system. *Soil Fertil. Sci. China* **2022**, *8*, 221–230.
78. Malhi, S.S.; Nyborg, M.; Goddard, T.; Puurveen, D. Long-term tillage, straw and N rate effects on quantity and quality of organic C and N in a Gray Luvisol soil. *Nutr. Cycl. Agroecosyst.* **2011**, *90*, 1–20. [CrossRef]
79. Marschner, P.; Umar, S.; Baumann, K. The microbial community composition changes rapidly in the early stages of decomposition of wheat residue. *Soil Biol. Biochem.* **2011**, *43*, 445–451. [CrossRef]
80. Li, G.Y.; Yan, Z.L.; Li, Q.; Wei, L.; Guan, X.K.; Wang, T.C. Effects of Straw Returning with Fertilizer and Decomposition Inoculants on Soil Enzyme Activity and Yield of Winter Wheat. *J. Agric. Sci.* **2016**, *45*, 59–63.
81. Hu, C.; Chen, Y.F.; Qiao, Y.; Liu, D.H.; Zhang, S.T.; Li, S.L. Effect of returning straw added with straw-decomposing inoculants on soil melioration in low-yielding yellow clayey soil. *J. Plant Nutr.* **2016**, *22*, 59–66.
82. Wang, J.; Xiao, G.J.; Zhang, F.J.; Wang, J.; Xu, X. Effect of returning straw with straw-decomposing inoculants on saline-alkali soil in North Yinchuan of China. *Agric. Res. Arid. Areas* **2017**, *35*, 209–215+283.
83. He, Z.F.; Yang, X.R.; Xiang, J.; Wu, Z.L.; Shi, X.Y.; Gui, Y.; Liu, M.Q.; Kalkhajeh, Y.K.; Gao, H.J.; Ma, C. Does Straw Returning Amended with Straw Decomposing Microorganism Inoculants Increase the Soil Major Nutrients in China's Farmlands? *Agronomy* **2022**, *12*, 890. [CrossRef]
84. Yang, G.H.; Zhang, J.J.; Yang, G.L. Preliminary study on effect of straw decomposition additive in rice/rape pattern. *Soil Tillage Res.* **2013**, *4*, 20–22.

85. Zhao, S.C.; Qiu, S.J.; Xu, X.P.; Ciampitti, I.A.; Zhang, S.Q.; He, P. Change in straw decomposition rate and soil microbial community composition after straw addition in different long-term fertilization soils. *Appl. Soil Ecol.* **2019**, *138*, 123–133. [CrossRef]
86. Li, D.D.; Li, Z.Q.; Zhao, B.Z.; Zhang, J.B. Relationship between the chemical structure of straw and composition of main microbial groups during the decomposition of wheat and maize straws as affected by soil texture. *Biol. Fert. Soils Coop. J. Int. Soc. Soil Sci.* **2020**, *56*, 11–24. [CrossRef]
87. Wang, X.; Wang, X.X.; Geng, P.; Yang, Q.; Chen, K.; Liu, N.; Fan, Y.L.; Zhan, X.M.; Han, X.R. Effects of different returning method combined with decomposer on decomposition of organic components of straw and soil fertility. *Sci. Rep.* **2021**, *11*, 15495. [CrossRef]
88. Liu, S.R.; Hu, R.G.; Cai, G.C.; Lin, S.; Zhao, J.S.; Li, Y.Y. The role of UV-B radiation and precipitation on straw decomposition and topsoil C turnover. *Soil Boil. Biochem.* **2014**, *77*, 197–202. [CrossRef]
89. Naruo, M.; Wanida, N.; Nongluck, P.; Tomohide, S.; Praison, R.; Suphakarn, L.; Kensuke, K. Soil carbon sequestration on a maize-mung bean field with rice straw mulch, no-tillage, and chemical fertilizer application in Thailand from 2011 to 2015. *Soil Sci. Plant Nutr.* **2020**, *67*, 190–196.
90. Qin, S.J.; Jiao, K.B.; Lyu, D.G.; Shi, L.; Liu, L.Z. Effects of maize residue and cellulose-decomposing bacteria inocula on soil microbial community, functional diversity, organic fractions, and growth of *Malus hupehensis* Rehd. *Arch. Agron. Soil Sci.* **2015**, *61*, 173–184. [CrossRef]
91. Zhao, J.; Ni, T.; Xun, W.B.; Huang, X.L.; Huang, Q.W.; Ran, W.; Shen, B.; Zhang, R.F.; Shen, Q.R. Influence of straw incorporation with and without straw decomposer on soil bacterial community structure and function in a rice-wheat cropping system. *Appl. Microbiol. Biot.* **2017**, *101*, 4761–4773. [CrossRef]
92. Zhang, P.; Chen, X.L.; Wei, T.; Yang, Z.; Jia, Z.K.; Yang, B.P.; Han, Q.F.; Ren, X.L. Effects of straw incorporation on the soil nutrient contents, enzyme activities, and crop yield in a semiarid region of China. *Soil Tillage Res.* **2016**, *160*, 65–72. [CrossRef]
93. Du, X.Z.; Hao, M.; Guo, L.J.; Li, S.H.; Hu, W.L.; Sheng, F.; Li, C.F. Integrated assessment of carbon footprint and economic profit from paddy fields under microbial decaying agents with diverse water regimes in central China. *Agric Water Manag.* **2022**, *262*, 107403. [CrossRef]
94. Liu, G.; Yu, H.Y.; Ma, J.; Xu, H.; Wu, Q.Y.; Yang, J.H.; Zhuang, Y.Q. Effects of straw incorporation along with microbial inoculant on methane and nitrous oxide emissions from rice fields. *Sci. Total Environ.* **2015**, *518–519*, 209–216. [CrossRef] [PubMed]
95. Ma, L.J.; Kong, F.X.; Wang, Z.; Luo, Y.; Lv, X.B.; Zhou, Z.G.; Meng, Y.L. Growth and yield of cotton as affected by different straw returning modes with an equivalent carbon input. *Field Crop Res.* **2019**, *243*, 107616. [CrossRef]
96. Zhang, Z.Y.; Liao, G.X.; Chen, Y.C.; Li, S.Q.; Yu, C.Q.; Chen, L. Application effect of rape straw returning rotten technology on rice. *Mod. Agric. Sci. Technol.* **2020**, *14*, 18+20.
97. Yang, X.R.; Xu, B.; He, Z.F.; Wu, J.; Zhuang, R.H.; Ma, C.; Chai, R.S.; Yusef, K.K.; Ye, X.X.; Zhu, L. lmpacts of Decomposing Microorganism Inoculum on Straw Decomposition and Crop Yield in China: A Meta-Analysis. *Sci. Agric. Sin.* **2020**, *53*, 1359–1367.
98. Xiong, W.; Guo, S.; Jousset, A.; Zhao, Q.Y.; Wu, H.S.; Li, R.; Kowalchuk, G.A.; Shen, Q.R. Bio-fertilizer application induces soil suppressiveness against Fusarium wilt disease by reshaping the soil microbiome. *Soil Boil. Biochem.* **2017**, *114*, 238–247. [CrossRef]
99. Kalkhajeh, Y.K.; He, Z.F.; Yang, X.R.; Lu, Y.; Zhou, J.; Gao, H.J.; Ma, C. Co-application of nitrogen and straw-decomposing microbial inoculant enhanced wheat straw decomposition and rice yield in a paddy soil. *J. Agric. Food Res.* **2021**, *4*, 100134. [CrossRef]
100. He, H.; Li, J.H.; Wei, C.Z.; Zhuang, Y.T. Decomposition characteristics and nutrient release rules of maize straw under different returning amounts. *Appl. Ecol. Environ. Res.* **2019**, *17*, 3695–3707. [CrossRef]
101. Wang, K.K.; Hu, W.S.; Xu, Z.Y.; Xue, Y.H.; Zhang, Z.; Liao, S.P.; Zhang, Y.Y.; Li, X.K.; Ren, T.; Cong, R.H.; et al. Seasonal Temporal Characteristics of In Situ Straw Decomposition in Different Types and Returning Methods. *J. Soil Sci. Plant Nut.* **2022**, *22*, 4228–4240. [CrossRef]
102. Zhang, L.P.; Liu, Z.W.; Gao, X.B.; Du, H.X.; Gao, W.J. Study on Decomposition of Different Mixed Leaf Litter. *J. Northwest For. Univ.* **2006**, *21*, 57–60.
103. Heitkamp, F.; Wendland, M.; Offenberger, K.; Gerold, G. Implications of input estimation, residue quality and carbon saturation on the predictive power of the Rothamsted Carbon Model. *Geoderma* **2012**, *170*, 168–175. [CrossRef]
104. Li, F.Y.; Sun, X.F.; Feng, W.Q.; Qin, Y.S.; Wang, C.Q.; Tu, S.H. Nutrient release patterns and decomposing rates of wheat and rapeseed straw. *J. Plant Nutr. Fertil.* **2009**, *15*, 374.
105. Wang, Y.Q.; Guo, X.S. Decomposition characteristics of crop-stalk under different incorporation methods. *Chin. J. Eco-Agric.* **2008**, *16*, 607.
106. Li, M.H. *Study on the Degradation of Wheat Straw by Nano-Carrier Bacteria*; University of Science and Technology of China: Anhui, China, 2019.
107. Chen, S.; Liu, Z.R.; Zeng, K. Effect of straw-decomposing inoculant on decomposition of rice straw. *Chin. J. Environ. Eng.* **2016**, *10*, 840–844.
108. Wang, J.Z.; Lu, C.A.; Zhang, W.J.; Feng, G.; Wang, X.J.; Xu, M.G. Decomposition of organic materials in cropland soils across China: A meta analysis. *Acta Pedol. Sin.* **2016**, *53*, 16–27.
109. Kamble, P.N.; Baath, E. Comparison of fungal and bacterial growth after alleviating induced N-limitation in soil. *Soil Boil. Biochem.* **2016**, *103*, 97–105. [CrossRef]

110. Baldrian, P. Increase of laccase activity during interspecific interactions of white-rot fungi. *Fems. Microbiol. Ecol.* **2004**, *50*, 245–253. [CrossRef]
111. Liang, J.J.; Fang, X.X.; Lin, Y.Q.; Wang, D.H. A new screened microbial consortium OEM2 for lignocellulosic biomass deconstruction and chlorophenols detoxification. *J. Haz. Mat.* **2018**, *347*, 341–348. [CrossRef]
112. Guo, T.F. The Mechanism of Carbon and Nitrogen Interaction During Rice Straw Decomposition. Ph.D. Thesis, Chinese Academy of Agricultural Sciences, Beijing China, 2019.
113. Zhao, S.C.; Ignacio, A.C.; Qiu, S.J.; Xu, X.P.; He, P. Characteristics of maize residue decomposition and succession in the bacterial community during decomposition in Northeast China. *J. Integr. Agric.* **2021**, *20*, 2–11. [CrossRef]
114. Murayama, S. Decomposition kinetics of straw saccharides and synthesis of microbial saccharides under field conditions. *J. Soil Sci.* **2010**, *35*, 231–242. [CrossRef]
115. Wu, J.; Guo, X.S.; Lu, J.W.; Wan, S.X.; Wang, Y.Q.; Xu, Z.Y.; Zhang, X.L. Decomposition characteristics of wheat straw and effects on soil biological properties and nutrient status under different rice cultivation. *Acta. Ecol. Sin.* **2013**, *33*, 565–575.
116. Devêvre, O.C.; Horwath, W.R. Decomposition of rice straw and microbial carbon use efficiency under different soil temperatures and moistures. *Soil Boil. Biochem.* **2000**, *32*, 1773–1785. [CrossRef]
117. Borjigin, Q.; Yu, X.F.; Gao, J.L.; Wang, Z.G.; Borjigin, N.; Wang, Z.; Hu, S.P.; Gao, L.; Hu, H.H. Study on degradation of corn stalk by decomposing microbial inoculants. *J. Northwest A F Univ. Nat. Sci. Ed.* **2016**, *44*, 107–116.
118. Li, H.D.; Liu, Y.; Cong, R.H. The characteristics of returning straw to the field with different carbon-nitrogen ratios. *Soil Fertil. Sci. China* **2022**, *9*, 102–106.

Disclaimer/Publisher's Note: The statements, opinions and data contained in all publications are solely those of the individual author(s) and contributor(s) and not of MDPI and/or the editor(s). MDPI and/or the editor(s) disclaim responsibility for any injury to people or property resulting from any ideas, methods, instructions or products referred to in the content.

MDPI AG
Grosspeteranlage 5
4052 Basel
Switzerland
Tel.: +41 61 683 77 34

Water Editorial Office
E-mail: water@mdpi.com
www.mdpi.com/journal/water

Disclaimer/Publisher's Note: The title and front matter of this reprint are at the discretion of the Guest Editors. The publisher is not responsible for their content or any associated concerns. The statements, opinions and data contained in all individual articles are solely those of the individual Editors and contributors and not of MDPI. MDPI disclaims responsibility for any injury to people or property resulting from any ideas, methods, instructions or products referred to in the content.

www.ingramcontent.com/pod-product-compliance
Lightning Source LLC
LaVergne TN
LVHW072333090526
838202LV00019B/2415